# THE CHEMICAL FORMULARY

# The
# Chemical Formulary

*A Collection of Valuable, Timely, Practical,*
*Commercial Formulae and Recipes for*
*Making Thousands of Products in*
*Many Fields of Industry*

## VOLUME XI

*Editor-in-Chief*

## H. BENNETT
*Technical Director, Cheminform Institute*

1961
CHEMICAL PUBLISHING CO., INC.
212 FIFTH AVENUE      NEW YORK, N. Y.

11,976

© 1961 BY H. BENNETT

Editor-in-Chief

# H. BENNETT

## Board of Editors

5

# PREFACE

Chemistry, as taught in our schools and colleges, concerns chiefly synthesis, analysis, and engineering — and properly so. It is part of the right foundation for the education of the chemist.

Many a chemist entering an industry soon finds that most of the products manufactured by his concern are not synthetic or definite chemical compounds, but are mixtures, blends, or highly complex compounds of which he knows little or nothing. The literature in this field, if any, may be meager, scattered, or obsolete.

Even chemists with years of experience in one or more industries spend considerable time and effort in acquainting themselves with any new field which they may enter. Consulting chemists similarly have to solve problems brought to them from industries foreign to them. There was a definite need for an up-to-date compilation of formulae for chemical compounding and treatment. Since the fields to be covered are many and varied, an editorial board of chemists and engineers engaged in many industries was formed.

Many publications, laboratories, manufacturing firms, and individuals have been consulted to obtain the latest and best information. It is felt that the formulae given in this volume will save chemists and allied workers much time and effort.

Manufacturers and sellers of chemicals will find, in these formulae, new uses for their products. Nonchemical executives, professional men, and interested laymen will make through this volume a "speaking acquaintance" with products which they may be using, trying, or selling.

It often happens that two individuals using the same ingredients in the same formula get different results. This may be due to slight deviations in the raw materials or unfamiliarity with the intricacies of a new technique. Accordingly, repeated experiments may be

7

necessary to get the best results.  Although many of the formulae given are being used commercially, many have been taken from the literature and may be subject to various errors and omissions.  This should be taken into consideration.  Wherever possible, it is advisable to consult with other chemists or technical workers regarding commercial production.  This will save time and money and help avoid trouble.

A formula will seldom give exactly the results which one requires. Formulae are useful as starting points from which to work out one's own ideas.  Also, formulae very often give us ideas which may help us in our specific problems.  In a compilation of this kind, errors of omission, commission, and printing may occur.  I shall be glad to receive any constructive criticism.

H. BENNETT

# PREFACE TO VOLUME XI

This new volume of the CHEMICAL FORMULARY series is a collection of new, up-to-date formulae. The only repetitious material is the introduction (Chapter I) which is used in every volume for the benefit of those who may have bought only one volume and who have no educational background or experience in chemical compounding. The simple basic formulae and compounding methods given in the introduction will serve as a guide for beginners and students. It is suggested that they read the introduction carefully and even make a few preparations described there before compounding the more intricate formulae included in the later chapters.

The list of chemicals and their suppliers has been enlarged with new trade-mark chemicals, so that buying the required ingredients will present no problem.

Grateful acknowledgement is made to the Board of Editors for their valuable suggestions and contributions.

H. BENNETT

---

NOTE: All the formulae in volumes I, II, III, IV, V, VI, VII, VIII, IX, X, and XI (except in the introduction) are different. Thus, if you do not find what you are looking for in this volume, you may find it in one of the others.

---

NOTE: This book is the result of cooperation of many chemists and engineers who have given freely of their time and knowledge. It is their business to act as consultants and to give advice on technical matters for a fee. As publishers, we do not maintain a laboratory or consulting service to compete with them. Therefore, please do not ask *us* for advice or opinions, but confer with a chemist.

Formulae for which patent numbers are listed can be manufactured only after obtaining a license from the patentees.

# CONTENTS

# ABBREVIATIONS

| | |
|---|---|
| amp. | ampere |
| amp./dm$_2$ | amperes per square decimeter |
| amp./sq. ft. | amperes per square foot |
| anhydr. | anhydrous |
| avoir. | avoirdupois |
| bbl. | barrel |
| Bé. | Baumé |
| B.P. | boiling point |
| °C | degrees Centigrade |
| cc. | cubic centimeter |
| c.d. | current density |
| cm. | centimeter |
| cm$_3$ | cubic centimeter |
| conc. | concentrated |
| c.p. | chemically pure |
| cp. | centipoise |
| cu. ft. | cubic foot |
| cu. in. | cubic inch |
| cwt. | hundredweight |
| d. | density |
| dil. | dilute |
| dm. | decimeter |
| dm$^2$ | square decimeter |
| dr. | dram |
| E. | Engler |
| °F. | degrees Fahrenheit |
| f.f.c. | free from chlorine |
| f.f.p.a. | free from prussic acid |
| fl. dr. | fluid dram |
| fl. oz. | fluid ounce |
| fl. pt. | flash point |
| F.P. | freezing point |
| ft. | foot |
| ft$^2$ | square foot |
| g. | gram |

11

gal..................gallon
gr..................grain
hl..................hectoliter
hr..................hour
in..................inch
kg..................kilogram
l..................liter
lb..................pound
liq..................liquid
m..................meter
min..................minim, minute
ml..................milliliter (cubic centimeter)
mm..................millimeter
M.P..................melting point
N..................Normal
N.F..................National Formulary
oz..................ounce
pH..................hydrogen-ion concentration
p.p.m..................parts per million
pt..................pint
pwt..................pennyweight
q.s..................a quantity sufficient to make
qt..................quart
r.p.m..................revolutions per minute
sec..................second
sp..................spirits
Sp. Gr..................specific gravity
sq. dm..................square decimeter
tech..................technical
tinc..................tincture
tr..................tincture
Tw..................Twaddell
U.S.P..................United States Pharmacopeia
v..................volt
visc..................viscosity
vol..................volume
wt..................weight

## Chapter I

# INTRODUCTION

The following introductory matter has been included at the suggestion of teachers of chemistry and home economics.

This section will enable anyone, with or without technical education or experience, to start making simple products without any complicated or expensive machinery. For commercial production, however, suitable equipment is necessary.

Chemical specialties are composed of pigments, gums, resins, solvents, oils, greases, fats, waxes, emulsifying agents, dyestuffs, perfumes, water, and chemicals of great diversity. To compound certain of these with some of the others requires definite and well-studied procedures, any departure from which will inevitably result in failure. The steps for successful compounding are given with the formulae. Follow them rigorously. If the directions require that (a) is added to (b), carry this out literally, and do not reverse the order. The preparation of an emulsion is often quite as tricky as the making of mayonnaise. In making mayonnaise, you add the oil to the egg, slowly, with constant and even stirring. If you do it correctly, you get mayonnaise. If you depart from any of these details: If you add the egg to the oil, or pour the oil in too quickly, or fail to stir regularly, the result is a complete disappointment. The same disappointment may be expected if the prescribed procedure of any other formulation is violated.

The point next in importance is the scrupulous use of the proper ingredients. Substitutions are sure to result in inferior quality, if not in complete failure. Use what the formula calls for. If a cheaper

product is desired, do not prepare it by substituting a cheaper ingredient for the one prescribed: use a different formula.  Not infrequently, a formula will call for an ingredient which is difficult to
obtain.  In such cases, either reject the formula or substitute a similar
substance only after a preliminary experiment demonstrates its
usability.  There is a limit to which this rule may reasonably be
extended.  In some cases, substitution of an equivalent ingredient
may be made legitimately.  For example, when the formula calls
for white wax (beeswax), yellow wax can be used, if the color of the
finished product is a matter of secondary importance.  Yellow beeswax can often replace white beeswax, making due allowance for color,
but paraffin wax will not replace beeswax, even though its light color
seems to place it above yellow beeswax.

And this leads to the third point: the use of good-quality ingredients, and ingredients of the correct quality.  Ordinary lanolin
is not the same thing as anhydrous lanolin.  The replacement of one
with the other, weight for weight, will give discouragingly different
results.  Use exactly what the formula calls for: if you are not ac-
quainted with the substance and you are in doubt as to just what
is meant, discard the formula and use one you understand.  Buy
your chemicals from reliable sources.  Many ingredients are obtain-
able in a number of different grades: if the formula does not desig-
nate the grade, it is understood that the best grade is to be used.
Remember that a formula and the directions can tell you only part
of the story.  Some skill is often required to attain success.  Practice
with a small batch in such cases until you are sure of your technique.
Many examples can be cited.  If the formula calls for steeping quince
seed for 30 minutes in cold water, steeping for 1 hour may yield a
mucilage of too thin a consistency.  The originator of the formula
may have used a fresher grade of seed, or his conception of what
"cold" water means may be different from yours.  You should have
a feeling for the right degree of mucilaginousness, and if steeping the
seed for 30 minutes fails to produce it, steep them longer until you
get the right kind of mucilage.  If you do not know what the right
kind is, you will have to experiment until you find out.  This is the
reason for the recommendation to make small experimental batches
until successful results are obtained.  Another case is the use of

dyestuffs for coloring lotions and the like. Dyes vary in strength; they are all very powerful in tinting value; it is not always easy to state in quantitative terms how much to use. You must establish the quantity by carefully adding minute quantities until you have the desired tint. Gum tragacanth is one of those products which can give much trouble. It varies widely in solubility and bodying power; the quantity listed in the formula may be entirely unsuitable for your grade of tragacanth. Therefore, correction is necessary, which can be made only after experiments with the available gum.

In short, if you are completely inexperienced, you can profit greatly by experimenting. Such products as mouth washes, hair tonics, and astringent lotions need little or no experience, because they are, as a rule, merely mixtures of simple liquid and solid ingredients, which dissolve without difficulty and the end product is a clear solution that is ready for use when mixed. However, face creams, tooth pastes, lubricating greases, wax polishes, etc., whose formulation requires relatively elaborate procedure and which must have a definite final viscosity, need some skill and not infrequently some experience.

## *Figuring*

Some prefer proportions expressed by weight or volume, others use percentages. In different industries and foreign countries different systems of weights and measures are used. For this reason, no one set of units could be satisfactory for everyone. Thus divers formulae appear with different units, in accordance with their sources of origin. In some cases, parts are given instead of percentage or weight or volume. On the pages preceding the index, conversion tables of weights and measures are listed. These are used for changing from one system to another. The following examples illustrate typical units:

EXAMPLE No. 1

### Ink for Marking Glass

| | | | |
|---|---|---|---|
| Glycerin | 40 | Ammonium Sulfate | 10 |
| Barium Sulfate | 15 | Oxalic Acid | 8 |
| Ammonium Bifluoride | 15 | Water | 12 |

Here no units are mentioned. In this case, it is standard practice to use parts by weight throughout. Thus here we may use ounces, grams, pounds, or kilograms as desired. But if ounces are used for one item, the ounce must be the unit for all the other items in the formula.

EXAMPLE No. 2

### Flexible Glue

| | | | |
|---|---|---|---|
| Powdered Glue | 30.90% | Glycerin | 5.15% |
| Sorbitol (85%) | 15.45% | Water | 48.50% |

Where no units of weight or volume, but percentages are given, forget the percentages and use the same method as given in Example No. 1.

EXAMPLE No. 3

### Antiseptic Ointment

| | | | |
|---|---|---|---|
| Petrolatum | 16 parts | Benzoic Acid | 1 part |
| Coconut Oil | 12 parts | Chlorothymol | 1 part |
| Salicylic Acid | 1 part | | |

The instructions given for Example No. 1 also apply to Example No. 3. In many cases, it is not wise to make up too large a quantity of a product before making a number of small batches to first master the necessary technique and also to see whether the product is suitable for the particular purpose for which it is intended. Since, in many cases, a formula may be given in proportions as made up on a factory scale, it is advisable to reduce the quantities proportionately.

EXAMPLE No. 4

### Neutral Cleansing Cream

| | | | |
|---|---|---|---|
| Mineral Oil | 80 lb. | Water | 90 lb. |
| Spermaceti | 30 lb. | Glycerin | 10 lb. |
| Glyceryl Monostearate | 24 lb. | Perfume | To suit |

Here, instead of pounds, ounces or even grams may be used. This formula would then read:

| | | | |
|---|---|---|---|
| Mineral Oil | 80 g. | Water | 90 g. |
| Spermaceti | 30 g. | Glycerin | 10 g. |
| Glyceryl Monostearate | 24 g. | Perfume | To suit |

Reduction in bulk may also be obtained by taking the same fractional part or portion of each ingredient in a formula. Thus in the following formula:

EXAMPLE No. 5

### Vinegar Face Lotion

| Acetic Acid (80%) | 20 | Alcohol | 440 |
| Glycerin | 20 | Water | 500 |
| Perfume | 20 | | |

We can divide each amount by ten and then the finished bulk will be only one tenth of the original formula. Thus it becomes:

| Acetic Acid (80%) | 2 | Alcohol | 44 |
| Glycerin | 2 | Water | 50 |
| Perfume | 2 | | |

## Apparatus

For most preparations, pots, pans, china, and glassware, which are used in every household, will be satisfactory. For making fine mixtures and emulsions, a malted- iilk mixer or egg beater is necessary. For weighing, a small, low-priced scale should be purchased from a laboratory-supply house. For measuring fluids, glass graduates or measuring glasses may be purchased from your local druggist. Where a thermometer is necessary, a chemical thermometer should be obtained from a druggist or chemical-supply firm.

## Methods

To understand better the products which you intend to make, it is advisable that you read the complete section covering such products. You may learn different methods that may be used and also to avoid errors which many beginners are prone to make.

## Containers for Compounding

Where discoloration or contamination is to be avoided, as in light-colored, or food and drug products, it is best to use enameled or earthenware vessels. Aluminum is also highly desirable in such cases, but it should not be used with alkalis as these dissolve and corrode aluminum.

## Heating

To avoid overheating, it is advisable to use a double boiler when temperatures below 212°F. (temperature of boiling water) will suffice. If a double boiler is not at hand, any pot may be filled with water and the vessel containing the ingredients to be heated placed in the water. The pot may then be heated by any flame without fear of overheating. The water in the pot, however, should be replenished from time to time; it must not be allowed to "go dry." To get uniform higher temperatures, oil, grease, or wax is used in the outer container in place of water. Here, of course, care must be taken to stop heating when thick fumes are given off as these are inflammable. When higher uniform temperatures are necessary, molten lead may be used as a heating medium. Of course, with chemicals which melt uniformly and are nonexplosive, direct heating over an open flame is permissible, with stirring, if necessary.

Where instructions indicate working at a certain temperature, it is important to attain the proper temperature not by guesswork, but by the use of a thermometer. Deviations from indicated temperatures will usually result in spoiled preparations.

*Temperature Measurement*

In the United States and in Great Britain, the Fahrenheit scale of temperature is used. The temperature of boiling water is 212° Fahrenheit (212°F.); the temperature of melting ice is 32° Fahrenheit (32°F.).

In scientific work, and in most foreign countries, the Centigrade scale is used, on which the temperature of boiling water is 100 °Centigrade (100°C.) and the temperature of melting ice is 0° Centigrade (0°C.).

The temperature of liquids is measured by a glass thermometer. This is inserted as deeply as possible in the liquid and is moved about until the temperature reading remains steady. It takes a short time for the glass of the thermometer to reach the temperature of the liquid. The thermometer should not be placed against the bottom or side of the container, but near the center of the liquid in the vessel. Since the glass of the thermometer bulb is very thin, it breaks easily when striking it against any hard surface. A cold thermometer should be warmed gradually (by holding it over the surface of a hot liquid) before immersion. Similarly the hot thermometer when taken out of the liquid should not be put into cold water suddenly. A sharp change in temperature will often crack the glass.

## Mixing and Dissolving

Ordinary dissolution (e.g., that of sugar in water) is hastened by stirring and warming. Where the ingredients are not corrosive, a clean stick, a fork, or spoon may be used as a stirring rod. These may also be used for mixing thick creams or pastes. In cases where very thorough stirring is necessary (e.g., in making mayonnaise, milky polishes, etc.), an egg beater or a malted-milk mixer is necessary.

## Filtering and Clarification

When dirt or undissolved particles are present in a liquid, they are removed by settling or filtering. In the first procedure, the solution is allowed to stand and if the particles are heavier than the liquid they will gradually sink to the bottom. The liquid may be poured or siphoned off carefully and, in some cases, it is then sufficiently clear for use. If, however, the particles do not settle out, then they must be filtered off. If the particles are coarse they may be filtered or strained through muslin or other cloth. If they are very small, filter paper is used. Filter papers may be obtained in various degrees of fineness. Coarse filter paper filters rapidly but will not retain extremely fine particles. For fine particles, a very fine grade of filter paper should be used. In extreme cases, even this paper may not be fine enough. Then, it will be necessary to add to the liquid 1 to 3% infusorial earth or magnesium carbonate. These are filter aids that clog up the pores of the filter paper and thus reduce their size and hold back undissolved material of extreme fineness. In all such filtering, it is advisable to take the first portions of the filtered liquid and pour them through the filter again as they may develop cloudiness on standing.

## Decolorizing

The most commonly used decolorizer is decolorizing carbon. This is added to the liquid to the extent of 1 to 5% and the liquid is heated, with stirring, for ½ hour to as high a temperature as is feasible. The mixture is then allowed to stand for a while and filtered. In some cases, bleaching must be resorted to.

## Pulverizing and Grinding

Large masses or lumps are first broken up by wrapping in a clean cloth, placing between two boards, and pounding with a hammer. The smaller pieces are then pounded again to reduce their size. Finer grinding is done in a mortar with a pestle.

## Spoilage and Loss

All containers should be closed when not in use to prevent evaporation or contamination by dust; also because, in some cases, air affects the product adversely. Many chemicals attack or corrode the metal containers in which they are kept. This is particularly true of liquids. Therefore, liquids should be transferred into glass bottles which should be as full as possible. Corks should be covered with aluminum foil (or dipped in melted paraffin wax when alkalis are present).

Glue, gums, olive oil, or other vegetable or animal products may ferment or become rancid. This produces discoloration or unpleasant odors. To avoid this, suitable antiseptics or preservatives must be used. Cleanliness is of utmost importance. All containers must be cleaned thoroughly before use to avoid various complications.

## Weighing and Measuring

Since, in most cases, small quantities are to be weighed, it is necessary to get a light scale. Heavy scales should not be used for weighing small amounts as they are not accurate enough for this type of weighing.

For measuring volumes of liquids, measuring glasses or cylinders (graduates) should be used. Since this glassware cracks when heated or cooled suddenly, it should not be subjected to sudden changes of temperature.

## Caution

Some chemicals are corrosive and poisonous. In many cases, they are labeled as such. As a precautionary measure, it is advised not to inhale them and, if smelling is absolutely necessary, only to sniff a few inches from the cork or stopper. Always work in a well-ventilated room when handling poisonous or unknown chemicals. If anything is spilled, it should be wiped off and washed away at once.

## Where to Buy Chemicals and Apparatus

Many chemicals and most glassware can be purchased from your druggist. A list of suppliers of all products is at the end of this book.

## Advice

This book is the result of cooperation of many chemists and engineers who have given freely of their time and knowledge. It is their business to act as consultants and to give advice on technical matters for a fee. As publishers, we do not maintain a laboratory or consulting service to compete with them.

Please, therefore, do not ask us for advice or opinions, but confer with a chemist in your vicinity.

## Extra Reading

Keep up with new developments of materials and methods by reading technical magazines. Many technical publications are listed under references in the back of this book.

## Calculating Costs

Raw materials, purchased in small quantities, are naturally higher in price than when bought in large quantities. Commercial prices, as given in the trade papers and catalogs of manufacturers, are for large quantities such as barrels, drums, or sacks. For example, 1 lb. epsom salts, bought at retail, may cost 10 or 15 cents. In barrel lots its price is about 2 or 3 cents per pound.

### Typical Costing Calculation
#### Formula for Beer- or Milk-Pipe Cleaner

| | | |
|---|---|---|
| Soda Ash | 25 lb. @ $0.02½ per lb. = | $ 0.63 |
| Sodium Perborate | 75 lb. @ 0.16 per lb. = | 12.00 |
| Total 100 lb. | Total | $12.63 |

If 100 lb. cost $12.63, 1 lb. will cost $12.63 divided by 100 or about $0.126, assuming no loss.

Always weigh the amount of finished product and use this weight in calculating costs. Most compounding results in some loss of material because of spillage, sticking to apparatus, evaporation, etc. Costs of making experimental lots are always high and should not be used for figuring costs. To meet competition, it is necessary to buy in large quantities and manufacturing costs should be based on these.

## Elementary Preparations

The simple recipes that follow have been selected because of their importance and because they are easy to make.

The succeeding chapters go into greater detail and give many different types and modifications of these and other recipes for home and commercial use.

### Cleansing Creams

Cleansing creams, as the name implies, serve as skin cleaners. Their basic ingredients are oils and waxes which are rubbed into the skin. When wiped off, they carry off dirt and dead skin. The liquefying type cleansing cream contains no matter and melts or liquefies when rubbed on the skin. To suit different climates and likes and dislikes harder or softer products can be made.

Cleansing Cream (Liquefying)

| | |
|---|---|
| Liquid Petrolatum | 5.5 |
| Paraffin Wax | 2.5 |
| Petrolatum | 2.0 |

Melt the ingredients together, with stirrings, in an aluminum or enamelled dish and allow to cool. Then stir in a perfume oil. Allow to stand until it becomes hazy and then pour into jars, which should be allowed to stand undisturbed overnight.

### Cold Creams

The most important facial cream is the cold cream. This type of cream contains mineral oil and wax which are emulsified in water with a small amount of borax or glycosterin. The function of a cold cream is to form a film that takes up dirt and waste tissue, which are removed when the skin is wiped thoroughly. Many modifications of this basic cream are encountered in stores. They vary in color, odor, and in claims, but, essentially, they are not more useful than this simple cream. The latest type of cold cream is the nongreasy cold cream which is of particular interest because it is nonalkaline and, therefore, nonirritating for sensitive skins.

### Cold Cream

| | |
|---|---|
| Liquid Petrolatum | 52 g. |
| White Beeswax | 14 g. |

Heat this in an aluminum or enamelled double boiler. (The water in the outer pot should be brought to a boil.) In a separate aluminum or enamelled pot dissolve:

| | |
|---|---|
| Borax | 1 g. |
| Water | 33 cc. |

and bring this to a boil. Add this in a thin stream to the melted wax, while stirring vigorously in one direction only. When the temperature drops to 140°F., add 0.5 cc. perfume oil and continue stirring until the temperature drops to 120°F. At this point, pour into jars, where the cream will set after a while. If a harder cream is desired, reduce the amount of liquid petrolatum. If a softer cream is wanted, increase it.

### Nongreasy Cold Cream

| | |
|---|---|
| White Paraffin Wax | 1.25 |
| Petrolatum | 1.50 |
| Glycosterin or Glyceryl Monostearate | 2.25 |
| Liquid Petrolatum | 3.00 |

Heat this mixture in an aluminum or enamelled double boiler. (The water in the outer pot should be boiling.) Stir until clear. To this slowly add, while stirring vigorously:

| | |
|---|---|
| Boiling Water | 10 |

Continue stirring until smooth and then add, with stirring, perfume oil. Pour into jars at 110 to 130°F. and cover the jars as soon as possible.

---

### Vanishing Creams

Vanishing creams are nongreasy soapy creams which have a cleansing effect. They are also used as a powder base.

### Vanishing Cream

| | |
|---|---|
| Stearic Acid | 18 oz. |

Melt this in an aluminum or enamelled double boiler. (The water in the outer pot must be boiling.) Add, in a thin stream, while stirring vigorously, the following boiling solution made in an aluminum or enamelled pot:

| | |
|---|---|
| Potassium Carbonate | ¼ oz. |
| Glycerin | 6½ oz. |
| Water | 5 lb. |

Continue stirring until the temperature falls to 135°F., then mix in a perfume oil and stir from time to time until cold. Allow to stand overnight and stir again the next day. Pack into jars and close these tightly.

---

### Hand Lotions

Hand lotions are usually clear or milky liquids or salves which are useful in protecting the skin from roughness and redness because of exposure to cold, hot water, soap, and other agents. Chapped hands are common. The use of a good hand lotion keeps the skin smooth, soft, and in a healthy condition. The lotion is best applied at night, rather freely, and cotton gloves may be worn to prevent soiling. During the day, it should be put on sparingly and the excess wiped off.

## Hand Lotion
### (Salve)

Boric Acid                1
Glycerin                  6

Warm these in an aluminum or enamelled dish and stir until dissolved (clear). Then allow to cool and work this liquid into the following mixture, adding only a little at a time.

Lanolin                   6
Petrolatum                8

To impart a pleasant odor a little perfume may be added and worked in.

## Hand Lotion
### (Milky liquid)

Lanolin               ¼ tsp.
Glycosterin or
  Glyceryl
  Monostearate     1   oz.
Tincture of Ben-
  zoin             2   oz.
Witch Hazel       25   oz.

Melt the first two items together in an aluminum or enamelled double boiler. If no double boiler is at hand, improvise one by placing a dish in a small pot containing boiling water. When the mixture becomes clear, remove from the double boiler and add slowly, while stirring vigorously, the tincture of benzoin and then the witch hazel. Continue stirring until cool and then put into one or two large bottles and shake vigorously. The finished lotion is a milky liquid comparable to the best hand lotions on the market sold at high prices.

---

## Brushless Shaving Creams

Brushless or latherless shaving creams are soapy in nature and do not require lathering or water. The formula given here is of the latest type being free from alkali and nonirritating. It should be borne in mind, however, that certain beards are not softened by this type of cream and require the old-fashioned lathering shaving cream.

## Brushless Shaving Cream

White Mineral Oil        10
Glycosterin or Glyceryl
  Monostearate           10
Water                    50

Heat the first two ingredients together in a Pyrex or enamelled dish to 150°F. and run in slowly, while stirring, the water which has been heated to boiling. Allow to cool to 150°F. and, while stirring, add a few drops of perfume oil. Continue stirring until cold.

---

## Mouth Washes

Mouth washes and oral antiseptics are of practically negligible value. However, they are used because of their refreshing taste and slight deodorizing effect.

### Mouth Wash

| | |
|---|---|
| Benzoic Acid | ⅝ |
| Tincture of Rhatany | 3 |
| Alcohol | 20 |
| Peppermint Oil | ⅛ |

Mix together in a dry bottle until the benzoic acid is dissolved. One teaspoonful is used to a small-wine-glassful of water.

### Tooth Powders

The cleansing action of tooth powders depends on their contents of soap and mild abrasives, such as precipitated chalk and magnesium carbonate. The antiseptic present is practically of no value. The flavoring ingredients mask the taste of the soap and give the mouth a pleasant aftertaste.

### Tooth Powder

| | |
|---|---|
| Magnesium Carbonate | 420 g. |
| Precipitated Chalk | 565 g. |
| Sodium Perborate | 55 g. |
| Sodium Bicarbonate | 45 g. |
| Powdered White Soap | 50 g. |
| Powdered Sugar | 90 g. |
| Wintergreen Oil | 8 cc. |
| Cinnamon Oil | 2 cc. |
| Menthol | 1 g. |

Dissolve the last three ingredients together and then rub well into the sugar. Add the soap and perborate, mixing well. Add the chalk, with good mixing, and then the sodium bicarbonate and magnesium carbonate. Mix thoroughly and sift through a fine wire screen. Keep dry.

### Foot Powders

Foot powders consist of talc or starch with or without an antiseptic or deodorizer. In the following formula the perborates liberate oxygen, when in contact with perspiration, which tends to destroy unpleasant odors. The talc acts as a lubricant and prevents friction and chafing.

### Foot Powder

| | |
|---|---|
| Sodium Perborate | 3 |
| Zinc Peroxide | 2 |
| Talc | 15 |

Mix thoroughly in a dry container until uniform. This powder must be kept dry or it will spoil.

### Liniments

Liniments usually consist of an oil and an irritant, such as methyl salicylate or turpentine. The oil acts as a solvent and tempering agent for the irritant. The irritant produces a rush of blood and warmth which is often slightly helpful.

### Sore-Muscle Liniment

| | |
|---|---|
| Olive Oil | 6 fl. oz. |
| Methyl Salicylate | 3 fl. oz. |

Mix together and keep in a well-stoppered bottle. Apply externally, but do not use on chafed or cut skin.

---

### Chest Rubs

In spite of the fact that chest rubs are practically useless, countless sufferers use them. Their action is similar to that of liniments and they differ only in that they are in the form of a salve.

### Chest-Rub Salve

| | | |
|---|---|---|
| Yellow Petro- | | |
| latum | 1 | lb. |
| Paraffin Wax | 1 | oz. |
| Eucalyptus Oil | 2 | fl. oz. |
| Menthol | ½ | oz. |
| Cassia Oil | ⅛ | fl. oz. |
| Turpentine | ½ | fl. oz. |

Melt the petrolatum and paraffin wax together in a double boiler and then add the menthol. Remove from the heat, stir, and cool a little; then mix in the oils, and turpentine. When it begins to thicken, pour into tins and cover.

---

### Insect Repellents

Preparations of this type may irritate sensitive skins and they will not always work.

### Mosquito-Repelling Oil

| | |
|---|---|
| Cedar Oil | 2 fl. oz. |
| Citronella Oil | 4 fl. oz. |
| Spirits of Camphor | 8 fl. oz. |

Mix in a dry bottle and the oil is ready for use. This preparation may be smeared on the skin as often as is necessary.

---

### Fly Sprays

Fly sprays usually consist of deodorized kerosene, perfume, and an active insecticide. In some cases, they merely stun the flies who may later recover and begin buzzing again.

### Fly Spray

| | |
|---|---|
| Deodorized | |
| Kerosene | 80 fl. oz. |
| Methyl Salicylate | 1 fl. oz. |
| Pyrethrum Powder | 10 oz. |

Mix thoroughly by stirring from time to time; allow to stand covered overnight and then filter through muslin.

This spray is inflammable and should not be used near open flames.

---

### Deodorant Spray

(For public buildings, sick rooms, lavatories, etc.)

| | |
|---|---|
| Pine-Needle Oil | 2 |
| Formaldehyde | 2 |
| * Acetone | 6 |

* Isopropyl Alcohol 20

One ounce of this mixture is diluted with 1 pt. water for spraying.

---

### Cresol Disinfectant

† Caustic Soda 25.5 g.

Water 140.0 cc.

Dissolve in a Pyrex or enamelled dish and warm. To this, add slowly the following warmed mixture:

‡ Cresylic Acid 500.0 cc.

Rosin 170.0 g.

Stir until dissolved and add water to make 1,000 cc.

---

### Ant Poison

Sugar 1 lb.

Water 1 qt.

‡ Arsenate of Soda 125 g.

Boil and stir until uniform; strain through muslin and add 1 spoonful honey.

---

### Bedbug Exterminator

* Kerosene 90 fl. oz.

Clove Oil 5 fl. oz.

‡ Cresol 1 fl. oz.

Pine Oil 4 fl. oz.

Simply mix and bottle.

---

### Nonstaining Mothproofing Fluid

Sodium Aluminum
  Silicofluoride 0.50

Water 98.00

Glycerin 0.50

"Sulfatate" (Wetting
  Agent) 0.25

Stir until dissolved.

---

### Fly Paper

Rosin 32

Rosin Oil 20

Castor Oil 8

Heat this mixture in an aluminum or enamelled pot on a gas stove, with stirring, until all the rosin has melted and dissolved. While hot, pour on firm paper sheets of suitable size which have been brushed with soap water just before coating. Smooth out the coating with a long knife or piece of thin flat wood and allow to cool. If a heavier coating is desired, increase the amount of rosin. Similarly, a thinner coating results by reducing the amount of rosin. The finished paper should be laid flat and not exposed to undue heat.

---

### Baking Powder

Bicarbonate of Soda 28

Monocalcium Phosphate 35

Corn Starch 27

Mix these powders thoroughly

---

* Inflammable.

† Do not get this on the skin as it is corrosive.

‡ Poison.

in a dry can by shaking and rolling for ½ hour. Pack into dry
airtight tins as moisture will
cause lumping.

---

### Malted-Milk Powder

| | |
|---|---|
| Powdered Malt Extract | 5 |
| Powdered Skim Milk | 2 |
| Powdered Sugar | 3 |

Mix thoroughly by shaking
and rolling in a dry can. Pack
in an airtight container.

---

### Cocoa-Malt Powder

| | |
|---|---|
| Corn Sugar | 55 |
| Powdered Malt | 19 |
| Powdered Skim Milk | 12½ |
| Cocoa | 13 |
| Vanillin | ⅛ |
| Powdered Salt | ⅜ |

Mix thoroughly and then run
through a fine wire sieve.

---

### Sweet Cocoa Powder

| | |
|---|---|
| Cocoa | 17½ oz. |
| Powdered Sugar | 32½ oz. |
| Vanillin | ¾ g. |

Mix thoroughly and sift.

---

### Pure Lemon Extract

| | |
|---|---|
| Lemon Oil | |
| U.S.P. | 6½ fl. oz. |
| Alcohol | 121½ fl. oz. |

Shake together in 1 gal. jug
until dissolved.

---

### Artificial Vanilla Flavor

| | |
|---|---|
| Vanillin | ¾ oz. |
| Coumarin | ¼ oz. |
| Alcohol | 2 pt. |

Stir the ingredients in a glass
or china pitcher until dissolved.
Then mix into the following
solution:

| | |
|---|---|
| Sugar | 12 oz. |
| Water | 5¼ pt. |
| Glycerin | 1 pt. |

Color brown by adding sufficient burnt-sugar coloring.

---

### Canary Food

| | |
|---|---|
| Dried and Chopped | |
| Egg Yolk | 2 |
| Poppy Heads (Coarse | |
| Powder) | 1 |
| Cuttlefish Bone | |
| (Coarse Powder) | 1 |
| Granulated Sugar | 2 |
| Powdered Soda | |
| Crackers | 8 |

Mix well together.

---

### Blue-Black Writing Ink

| | |
|---|---|
| Naphthol Blue | |
| Black | 1 oz. |
| Powdered Gum | |
| Arabic | ½ oz. |
| Carbolic Acid | ¼ oz. |
| Water | 1 gal. |

Stir together in a glass or
enamelled vessel until dissolved.

Indelible Laundry-Marking Ink

a. Soda Ash      1 oz.
   Powdered Gum
      Arabic      1 oz.
   Water      10 fl. oz.

Stir until dissolved.

b. Silver Nitrate      4 oz.
   Powdered Gum
      Arabic      4 oz.
   Lampblack      2 oz.
   Water      40 fl. oz.

Stir this in a glass or porcelain dish until dissolved. Do not expose the mixture to strong light or it will spoil. Then pour into a brown glass bottle. In using these solutions, wet the cloth with solution a and allow to dry. Then write on it with solution b using a quill pen.

---

Green Marking Crayon

| | |
|---|---|
| Ceresin | 8 |
| Carnauba Wax | 7 |
| Paraffin Wax | 4 |
| Beeswax | 1 |
| Talc | 10 |
| Chrome Green | 3 |

Melt the first four ingredients in a container and then add the last two slowly, while stirring. Remove from the heat and continue stirring until thickening begins. Then pour into molds. If other-color crayons are desired, other pigments may be used. For example, for black, use carbon black or bone black; for blue, Prussian blue; for red, orange chrome yellow.

---

Antique Coloring for Copper

| | |
|---|---|
| Copper Nitrate | 4 oz. |
| Acetic Acid | 1 oz. |
| Water | 2 oz. |

Dissolve by stirring together in a glass or porcelain vessel. Pack into glass bottles.

Wet the copper to be colored and apply the coloring solution hot.

---

Blue-Black Finish on Steel

a. Place the object in molten sodium nitrate at 700 to 800°F. for 2 to 3 minutes. Remove and allow to cool somewhat, wash in hot water, dry, and oil with mineral or linseed oil.

b. Then put the object in the following solution for 15 minutes:

| | | |
|---|---|---|
| Copper Sulfate | ½ | oz. |
| Iron Chloride | 1 | lb. |
| Hydrochloric Acid | 4 | oz. |
| Nitric Acid | ½ | oz. |
| Water | 1 | gal. |

Allow to dry for several hours. Place in a solution again for 15 minutes, remove and dry for 10 hours. Place in boiling water for ½ hour, dry, and scratch-brush very lightly. Oil with mineral or linseed oil and wipe dry.

Rust-Prevention Compound

| Lanolin | 1 |
|---|---|
| * Naphtha | 2 |

Mix until dissolved.

The metal to be protected is cleaned with a dry cloth and then coated with the composition.

---

### Metal Polish

| Naphtha | 62 | oz. |
|---|---|---|
| Oleic Acid | ⅓ | oz. |
| Abrasive | 7 | oz. |
| Triethanolamine Oleate | ⅓ | oz. |
| Ammonia (26°) | 1 | oz. |
| Water | 1 | gal. |

In one container mix together the naphtha and oleic acid to a clear solution. Dissolve the triethanolamine oleate in the water separately, stir in the abrasive, and then add the naphtha solution. Stir the resulting mixture at a high speed until a uniform creamy emulsion results. Then add the ammonia and mix well, but do not agitate so vigorously as before.

---

### Glass-Etching Fluid

| Hot Water | 12 |
|---|---|
| † Ammonium Bifluoride | 15 |
| Oxalic Acid | 8 |
| Ammonium Sulfate | 10 |

| Glycerin | 40 |
|---|---|
| Barium Sulfate | 15 |

Warm the washed glass slightly before writing on it with this fluid. Allow the fluid to act on the glass for about 2 minutes.

---

### Leather Preservative

| Cold-Pressed Neatsfoot Oil | 10 |
|---|---|
| Castor Oil | 10 |

Mix.

This is an excellent preservative for leather book bindings, luggage, and other leather goods.

---

### White-Shoe Dressing

| Lithopone | 19 | oz. |
|---|---|---|
| Titanium Dioxide | 1 | oz. |
| Bleached Shellac | 3 | oz. |
| Ammonium Hydroxide | ¼ | fl. oz. |
| Water | 25 | fl. oz. |
| Alcohol | 25 | fl. oz. |
| Glycerin | 1 | oz. |

Dissolve the last four ingredients by mixing in a porcelain vessel. When dissolved, stir in the first two pigments. Keep in stoppered bottles and shake before using.

---

### Waterproofing for Shoes

| Wool Grease | 8 |
|---|---|
| Dark Petrolatum | 4 |
| Paraffin Wax | 4 |

Melt together in any container.

---

* Inflammable — keep away from flames.

† Corrosive.

## Polishes

Polishes are generally used to restore the original luster and finish of a smooth surface. They are also expected to clean the surface and to prevent corrosion or deterioration. There is no one polish which will give good results on all surfaces.

Most polishes contain oil or wax for their lustering or polishing properties. Oil polishes are easy to apply, but the surfaces on which they are used attract dust and show finger marks. Wax polishes are more difficult to apply, but are more lasting.

Oil or wax polishes are of two types: waterless and aqueous. The former are clear or translucent, the latter are milky in appearance.

For use on metals, abrasives of various kinds, such as tripoli, silica dust, or infusorial earth, are incorporated to grind away oxide films or corrosion products.

### Black Shoe Polish

| | |
|---|---|
| Carnauba Wax | 5½ oz. |
| Crude Montan Wax | 5½ oz. |

Melt together in a double boiler. (The water in the outer container should be boiling.) Then stir in the following melted and dissolved mixture:

| | |
|---|---|
| Stearic Acid | 2 oz. |
| Nigrosine Base | 1 oz. |

Then stir in

| | |
|---|---|
| Ceresin | 15 oz. |

Remove all flames and run in slowly, while stirring,

| | |
|---|---|
| Turpentine | 90 fl. oz. |

Allow the mixture to cool to 105°F. and pour into airtight tins which should stand undisturbed overnight.

### Clear Oil-Type Auto Polish

| | | |
|---|---|---|
| Paraffin Oil | 5 | pt. |
| Raw Linseed Oil | 2 | pt. |
| China-Wood Oil | ½ | pt. |
| * Benzol | ¼ | pt. |
| * Kerosene | ¼ | pt. |
| Amyl Acetate | 1 | tbsp. |

Mix together in a glass jar and keep it stoppered.

### Paste-Type Auto and Floor Wax

| | | |
|---|---|---|
| Yellow Beeswax | 1 | oz. |
| Ceresin | 2½ | oz. |
| Carnauba Wax | 4½ | oz. |
| Montan Wax | 1¼ | oz. |
| * Naphtha or | | |
| Mineral Spirits | 1 | pt. |
| * Turpentine | 2 | oz. |
| Pine Oil | ½ | oz. |

Melt the waxes together in a double boiler. Turn off the heat and run in the last three ingredients in a thin stream, with stirring. Pour into cans, cover, and allow to stand undisturbed overnight.

---

* Inflammable — keep away from flames.

Oil-and-Wax Type Furniture
Polish

Paraffin Oil          1    pt.
Powdered Carnauba
  Wax               ¼ oz.
Ceresin Wax         ⅛ oz.

Heat together until all of the wax is melted. Allow to cool and pour into bottles before the mixture turns cloudy.

Liquid Polishing Wax

Yellow Beeswax       1 oz.
Ceresin Wax          4 oz.

Melt together and then cool to 130°F.; turn off all flames and stir in slowly:

* Turpentine         17   fl. oz.
Pine Oil             ½ fl. oz.

Pour into cans or bottles which are closed tightly to prevent evaporation.

Floor Oil

Mineral Oil          46   fl. oz.
Beeswax             ½ oz.
Carnauba Wax         1   oz.

Heat together in double boiler until dissolved (clear). Turn off the flame and stir in

* Turpentine         3 fl. oz.

---

Lubricants

Lubricants, in the form of oils or greases, are used to prevent friction and wearing of parts which are rubbed together. Lu-

---

* Inflammable.

---

bricants must be chosen to fit specific uses. They consist of oils and fats often compounded with soaps and other unctuous substances. For heavy duty, heavy oils or greases are used and light oils are suitable for light duty.

Gum Lubricant

White Petrolatum     15 oz.
Acid-Free Bone Oil    5 oz.

Warm gently and mix together.

Graphite Grease

Ceresin              7 oz.
Tallow               7 oz.

Warm together and gradually work in with a stick:

Graphite             3 oz.

Stir until uniform and pack in tins when thickening begins.

Penetrating Oil

(For loosening rusted bolts,
screws, etc.)

Kerosene             2 oz.
Thin Mineral Oil     7 oz.
Secondary Butyl
  Alcohol            1 oz.

Mix and keep in a stoppered bottle.

---

Molding Compound

White Glue           13   lb.
Rosin                13   lb.
Raw Linseed Oil      ⅓ qt.
Glycerin             1    qt.
Whiting              19   lb.

Heat the white glue until it

melts. Then cook separately the rosin and raw linseed oil until the rosin is dissolved. Add the rosin, oil, and glycerin to the glue, stirring in the whiting until the mass reaches the consistency of a putty. Keep the mixture hot.

Press this mass into the die firmly and allow it to cool slightly before removing. The finished product is ready to use within a few hours after removal. Suitable pigments may be added to secure brown, red, black, or other color.

In applying ornaments made of this composition to a wood surface, they are first steamed to make them flexible; in this condition, they will adhere to the wood easily and securely. They can be bent to any shape, and no nails are required for applying them.

### Grafting Wax

| | |
|---|---|
| Wool Grease | 11 |
| Rosin | 22 |
| Paraffin Wax | 6 |
| Beeswax | 4 |
| Japan Wax | 1 |
| Rosin Oil | 9 |
| Pine Oil | 1 |

Melt together until clear and pour into tins. This composition can be made thinner by increasing the amount of rosin oil and thicker by decreasing it.

### Candles

| | |
|---|---|
| Paraffin Wax | 30.0 |
| Stearic Acid | 17.5 |
| Beeswax | 2.5 |

Melt together and stir until clear. If colored candles are desired, add a very small amount of any oil-soluble dye. Pour into vertical molds in which wicks are hung.

---

### *Adhesives*

Adhesives are sticky substances used to unite two surfaces. Adhesives are specifically called glues, pastes, cements, mucilages, lutes, etc. For different uses different types are required.

### Wall-Patching Plaster

| | |
|---|---|
| Plaster of Paris | 32 |
| Dextrin | 4 |
| Pumice Powder | 4 |

Mix thoroughly by shaking and rolling in a dry container. Keep away from moisture.

### Cement-Floor Hardener

| | |
|---|---|
| Magnesium Fluosilicate | 1 lb. |
| Water | 15 pt. |

Mix until dissolved.

The cement should first be washed with clean water and then drenched with this solution.

### Paperhanger's Paste

| | |
|---|---|
| White or Fish Glue | 4 oz. |
| Cold Water | 8 oz. |

Venice Turpentine    2 fl. oz.
Rye Flour            1 lb.
Cold Water           16 fl. oz.
Boiling Water        64 fl. oz.

Soak the glue in the first amount of cold water for 4 hours. Dissolve on a water-bath (glue-pot) and while hot stir in the Venice turpentine. Use a cheap grade of rye or wheat flour, mix thoroughly with the second amount of cold water to about the consistency of dough or a little thinner, being careful to remove all lumps. Stir in 1 tbsp. of powdered alum to 1 qt. flour, then pour in the boiling water, stirring rapidly until the flour is thoroughly cooked. Let this cool and finally add the glue solution. This makes a very strong paste which will also adhere to a painted surface, owing to the Venice turpentine content.

### Aquarium Cement

| | |
|---|---|
| Litharge | 10 |
| Plaster of Paris | 10 |
| Powdered Rosin | 1 |
| Dry White Sand | 10 |
| Boiled Linseed Oil | |
| | Sufficient |

Mix all together in the dry state, and make a stiff putty with the oil just before use.

Do not fill the aquarium for 3 days after cementing. This cement hardens under water, and will stick to wood, stone, metal, or glass and as it resists the action of sea water, it is useful for marine *aquaria*.

### Wood-Dough Plastic

| | |
|---|---|
| * Collodion | 86 |
| Powdered Ester Gum | 9 |
| Wood Flour | 30 |

Allow the first two ingredients to stand until dissolved, stirring from time to time. Then, while stirring, add the wood flour, a little at a time, until uniform. This product can be made softer by adding more collodion.

### Putty

| | |
|---|---|
| Whiting | 80 |
| Raw Linseed Oil | 16 |

Rub together until smooth. Keep in a closed container.

---

### Wood-Flour Bleach

| | |
|---|---|
| Sodium Metasilicate | 90 |
| Sodium Perborate | 10 |

Mix thoroughly and keep dry in a closed can. Use 1 lb. to 1 gal. boiling water. Mop or brush on the floor, allow to stand ½ hour, then rub off and rinse well with water.

---

### * Paint Remover

| | | |
|---|---|---|
| Benzol | 5 | pt. |
| Ethyl Acetate | 3 | pt. |
| Butyl Acetate | 2 | pt. |
| Paraffin Wax | ½ | lb. |

* Inflammable.

Stir together until dissolved.

---

*Soaps and Cleaners*

Soaps are made from a fat or fatty acid and an alkali. They lather and produce a foam which entraps dirt and grease. There are many kinds of soaps.

Cleaners contain a solvent, such as naphtha, with or without a soap. Abrasive cleaners are soap pastes containing powdered pumice, stone, silica, etc.

Concentrated Liquid Soap

| | |
|---|---|
| Water | 11 |
| * Solid Caustic Potash | 1 |
| Glycerin | 4 |
| Red Oil (Oleic Acid) | 4 |

Dissolve the caustic soda in water, add the glycerin, and bring to a boil in an enamelled pot. Remove from the heat, add the red oil slowly, while stirring. If a more neutral soap is wanted, use more red oil.

Saddle Soap

| | |
|---|---|
| Beeswax | 5.0 |
| * Caustic Potash | 0.8 |
| Water | 8.0 |

Boil for 5 minutes, while stirring. In another vessel heat:

| | |
|---|---|
| Castile Soap | 1.6 |
| Water | 8.0 |

Mix the two solutions with

---

* Do not get on the skin as it is corrosive.

good stirring; remove from the heat and add, while stirring:

| | |
|---|---|
| Turpentine | 12 |

Mechanics' Hand-Soap Paste

| | |
|---|---|
| Water | 1.8 qt. |
| White Soap Chips | 1.5 lb. |
| Glycerin | 2.4 oz. |
| Borax | 6.0 oz. |
| Dry Sodium Carbonate | 3.0 oz. |
| Coarse Pumice Powder | 2.2 lb. |
| Safrol | To suit |

Dissolve the soap in two thirds of the water by heat. Dissolve the last three ingredients in the rest of the water. Pour the two solutions together and stir well. When it begins to thicken, sift in the pumice, stirring constantly till thick, then pour into cans. Vary the amount of water, for heavier or softer paste. Water cannot be added to the finished soap.

Dry-Cleaning Fluid

| | |
|---|---|
| Glycol Oleate | 2 fl. oz. |
| Carbon Tetrachloride | 60 fl. oz. |
| Naphtha | 20 fl. oz. |
| Benzine | 18 fl. oz. |

This is an excellent cleaner that will not injure the finest fabrics.

Wall-Paper Cleaner

| | |
|---|---|
| Whiting | 10 lb. |
| Calcined Magnesia | 2 lb. |

Fuller's Earth          2 lb.
Powdered Pumice          12 oz.
Lemenone or
    Citronella Oil          4 oz.
Mix well together.

### Household Cleaner

Soap Powder          2
Soda Ash          3
Trisodium Phosphate          40
Finely Ground Silica          55
Mix well and pack in the usual containers.

### Window Cleanser

Castile Soap          2
Water          5
Chalk          4
French Chalk          3
Tripoli Powder          2
Petroleum Spirits          5
Mix well and pack in tight containers.

### Straw-Hat Cleaner

Sponge the hat with a solution of:

Sodium Hyposulfite          10 oz.
Glycerin          5 oz.
Alcohol          10 oz.
Water          75 oz.

Lay the hat aside in a damp place for 24 hours and then apply a mixture of:

Citric Acid          2 oz.
Alcohol          10 oz.
Water          90 oz.

Press with a moderately hot iorn after stiffening with gum water, if necessary.

### Grease, Oil, Paint, and Lacquer Spot Remover

Alcohol          1
Ethyl Acetate          2
Butyl Acetate          2
Toluol          2
Carbon Tetrachloride          3

Place the garment with the spot over a piece of clean paper or cloth and wet the spot with this fluid; rub with a clean cloth toward the center of the spot. Use a clean section of cloth for rubbing and clean paper or cloth for each application of the fluid. This cleaner is inflammable and should be kept away from flames. Cleaners of this type should be used out of doors or in well-ventilated rooms as the fumes are toxic.

### Paint-Brush Cleaner

a. Kerosene          2.00
   Oleic Acid          1.00
b. Strong Liquid
   Ammonia (28%)          0.25
   Denatured Alcohol          0.25

Slowly stir b into a until a smooth mixture results. To clean brushes, pour into a can and leave the brushes in it overnight. In the morning, wash out with warm water.

### Rust and Ink Remover

Immerse the part of the fabric with the rust or ink spot alternately in solutions a and b, rinsing with water after each immersion.

a. Ammonium Sulfide
   Solution ............ 1
   Water ............... 19
b. * Oxalic Acid ........ 1
   Water ............... 19

---

### Javelle Water
### (Laundry bleach)

Bleaching Powder      2 oz.
Soda Ash              2 oz.
Water                 5 gal.

Mix well until the reaction is completed. Allow to settle overnight and siphon off the clear liquid.

---

### Liquid Laundry Blue

Prussian Blue           1
Distilled Water         32
* Oxalic Acid           ¼

Dissolve by mixing in a crock or wooden tub.

---

### Glassine Paper

Paper is coated with or dipped in the following solution and then hung up to dry.

Copal Gum            10 oz.
Alcohol              30 fl. oz.
Castor Oil           1 fl. oz.

Dissolve by letting stand overnight in a covered jar and stirring the next day.

---

* Poisonous.

### Waterproofing Paper and Fiberboard

The following composition and method of application will make uncalendered paper, fiberboard, and similar porous material waterproof.

Paraffin (M.P.
   about 130°F.)        22.5
Trihydroxyethylamine
   Stearate             3.0
Water                   74.5

The paraffin wax is melted and the stearate added to it. The water is then heated to nearly the boiling point and vigorously agitated with a suitable mechanical stirring device while the mixture of melted wax and emulsifier is being slowly added. This mixture is cooled while it is stirred.

The paper or fiberboard is coated on the side which is to be in contact with water. This method works most effectively on paper-pulp molded containers and has the advantage of being much cheaper than dipping in melted paraffin as only about one tenth as much paraffin is needed. In addition, the outside of the container is not greasy and can be printed on after treatment which is not the case when treating with melted wax.

## * Waterproofing Liquid

| | |
|---|---|
| Paraffin Wax | ⅖ oz. |
| Gum Dammar | 1⅕ oz. |
| Pure Rubber | ⅛ oz. |
| Benzol | 13   oz. |
| Carbon Tetrachloride | |
| | To make 1 gal. |

Dissolve the rubber in the benzol, add the other ingredients, and allow to dissolve.

This liquid is suitable for wearing apparel and wood. It is applied by brushing on two or more coats, allowing each to dry before applying another coat. Apply outdoors as vapors are inflammable and toxic.

---

## Waterproofing Heavy Canvas

| | |
|---|---|
| Raw Linseed Oil | 1 gal. |
| Crude Beeswax | 13 oz. |
| White Lead | 1 lb. |
| Rosin | 12 oz. |

Heat, while stirring, until all lumps are removed and apply warm to the upper side of the canvas, wetting it with a sponge on the underside before application.

---

## Waterproofing Cement

| | | |
|---|---|---|
| China-Wood Oil | | |
|   Fatty Acids | 10   | oz. |
| Paraffin Wax | 10   | oz. |
| Kerosene | 2½ | gal. |

* Inflammable.

Stir until dissolved. Paint or spray on cement walls, which must be dry.

---

## Oil- and Greaseproofing Paper and Fiberboard

This solution, applied by brush, spray, or dipping, will leave a thin film which is impervious to oil and grease. Applied to paper or fiber containers, it will enable them to retain oils and greases.

| | |
|---|---|
| Starch | 6.6 |
| Caustic Soda | 0.1 |
| Glycerin | 2.0 |
| Sugar | 0.6 |
| Water | 90.5 |
| Sodium Salicylate | 0.2 |

The caustic soda is dissolved in the water. Then the starch is made into a thick paste by adding a portion of this solution. The paste is then added to the water. The resulting mixture is placed on a water bath and heated to about 85°C. until all the starch granules have broken. The temperature is maintained about ½ hour longer at 85°C. The other substances are then added and thoroughly mixed. The composition is now ready for application. Less water may be used if applied hot and then a thicker coating will result.

## Fireproof Paper

| Ammonium Sulfate | 8.00 |
|---|---|
| Boric Acid | 3.00 |
| Borax | 1.75 |
| Water | 100.00 |

The ingredients are mixed together in a gallon jug by shaking until dissolved.

The paper to be treated is dipped into this solution in a pan, until uniformly saturated. It is then taken out and hung up to dry. Wrinkles can be prevented by drying between cloths in a press.

## Fireproofing Canvas

| Ammonium Phosphate | 1 | lb. |
|---|---|---|
| Ammonium Chloride | 2 | lb. |
| Water | ½ | gal. |

Impregnate with the solution; squeeze out the excess, and dry. Washing or exposure to rain will remove fireproofing salts.

## Fireproofing Light Fabrics

| Borax | 10 oz. |
|---|---|
| Boric Acid | 8 oz. |
| Water | 1 gal. |

Impregnate, squeeze, and dry. Fabrics so impregnated must be treated again after washing or exposure to rain as the fireproofing salts wash out easily.

## Dry Fire Extinguisher

| Ammonium Sulfate | 15 |
|---|---|
| Sodium Bicarbonate | 9 |
| Ammonium Phosphate | 1 |
| Red Ochre | 2 |
| "Silex" | 23 |

Use powdered substances only. Mix well and pass through a fine sieve. Pack in tight containers to prevent lumping.

## Fire-Extinguishing Liquid

| Carbon Tetrachloride | 95 |
|---|---|
| Solvent Naphtha | 5 |

The naphtha minimizes the development of toxic fumes when extinguishing fires.

## Fire Kindler

| Rosin or Pitch | 10 |
|---|---|
| Sawdust | 10 or more |

Melt, mix, and cast in forms.

## Solidified Gasoline

| * Gasoline | ½ | gal. |
|---|---|---|
| Fine-Shaved White Soap | 12 | oz. |
| Water | 1 | pt. |
| Ammonia | 5 | oz. |

Heat the water, add the soap, mix and, when cool, add the ammonia. Then slowly work in the gasoline to form a semisolid mass.

* Inflammable.

## Boiler Compound

| Soda Ash | 87 |
|---|---|
| Trisodium Phosphate | 10 |
| Starch | 1 |
| Tannic Acid | 2 |

Use powders, mix well, and then pass through a fine sieve.

---

## Noncorrosive Soldering Flux

| Powdered Rosin | 1 |
|---|---|
| Denatured Alcohol | 4 |

Soak overnight and mix well.

---

### *Photographic Solutions*

## Developing Solution

Stock Solution *a*

| Pyro | 4 oz. |
|---|---|
| Pure Sodium Bisulfite | 280 gr. |
| Potassium Bromide | 32 gr. |
| Distilled Water | 64 oz. |

Dissolve in a glass or enamel dish.

Stock Solution *b*

| Pure Sodium Sulfite | 7 oz. |
|---|---|

Pure Sodium Carbonate 5 oz.
Distilled Water 64 oz.

Dissolve separately in a glass or enamel dish.

Use the following proportions:

| Stock Solution *a* | 2 |
|---|---|
| Stock Solution *b* | 2 |
| Distilled Water | 16 |

At 65°F., this developer requires about 8 minutes.

## Acid-Hardening Fixing Bath

*a.* Sodium Hyposulfite 32
    Distilled Water 8

Stir until dissolved and then add the following chemicals in the order given, stirring each until dissolved:

*b.* Warm Distilled Water 2½
    Pure Sodium Sulfite ½
    Pure Acetic Acid (28%) 1½
    Potassium Alum Powder ½

Add *b* to *a* and store in dark bottles away from light.

## CHAPTER II

# ADHESIVES

Leather or Metal to Cloth,
Paper Adhesive

(Hot Melt)

| | |
|---|---|
| "Paradene" No. 35 | 82 |
| Ethyl Cellulose | 9 |
| "Nevinol" | 9 |

The ethyl cellulose is sifted into 30 parts of melted "Paradene" No. 35 and mixed until uniform. The remaining "Paradene" No. 35 and "Nevinol" are then added.

---

Tile Adhesives

Formula No. 1
(Solvent)

| | |
|---|---|
| "Vinylite" VYHH | 40 |
| "Nevillac" Hard | 40 |
| Butyl Acetate | 220 |

A uniform mixture is obtained by slowly adding the resins to the solvent under agitation. If more flexibility is desired, then either "Nevillac" 10° or a blend of "Nevillac" 10° with dioctylphthalate should be added.

**Applications.** Floor-tile adhesive (ceramic, cork, vinyl).

No. 2
(Emulsion)

| | |
|---|---|
| R-7 "Neville" Resin, 108-112 | 162 |
| "Ethomeen" T-15 | 10 |
| Dibutylphthalate | 162 |
| "Elvacet" 81-900 | 578 |
| "Dixie Clay" | 50 |
| "Calcene" TM | 50 |

The R-7 "Neville" resin, "Ethomeen" and dibutylphthalate are melted and mixed at 100 °C. until uniform. This mixture is allowed to cool to 80 to 90 °C. and the "Elvacet" 81-900 slowly added. The mastic base is cooled to room temperature and the fillers are then added in the order shown.

**Applications.** Acoustical-tile adhesive. Floor-tile adhesive (asphalt, ceramic, cork, rubber, vinyl). Vinyl to brick, concrete. Wall-tile adhesive (ceramic, polystyrene).

### No. 3
### (Solvent)

| | |
|---|---|
| "Nevillac" Hard | 80 |
| Methyl Ethyl Ketone | 20 |
| Plaster of Paris | 6 |
| Asbestos Floats | 3 |
| "Dixie Clay" | 70 |

The "Nevillac" Hard is added to the methyl ethyl ketone under agitation. The other ingredients are then added in the order shown.

**Applications.** Floor-tile adhesive (asphalt, ceramic, cork). Wall-tile adhesive (ceramic, polystyrene).

### No. 4
### (Solvent)

| | |
|---|---|
| Smoked Sheets | 45.0 |
| R-16-A "Neville" Resin | 75.0 |
| "Calcene" TM | 25.0 |
| "McNamee Clay" | 150.0 |
| Mineral Spirits | 550.0 |
| "Nevastain" B | 0.5 |

The elastomer and fillers are masticated on a cool tight mill for 10 minutes. This is added to the solvent with agitation. The R-16-A "Neville" resin and "Ne-vastain" B are added and dissolved.

**Applications.** Floor-tile adhesive (asphalt, ceramic, cork, rubber, vinyl). Wall-tile adhesive (ceramic).

### No. 5
### (Solvent)

| | |
|---|---|
| R-14 "Neville" Resin | 40 |
| Refined Tall Oil | 35 |
| Blown Soya Oil (1000 cps. visc.) | 10 |
| VM&P Naphtha | 15 |

The resin and oils are melted together before adding the VM&P naphtha. The usual fillers such as whiting, china clay or asbestine can be finally added.

**Applications.** Floor-tile adhesive (asphalt).

### No. 6
### (Solvent)

| | |
|---|---|
| R-16-A "Neville" Resin | 288 |
| Refined Tall Oil | 2010 |
| Magnesite | 75 |
| Blown Soya Oil | 288 |
| 6% Manganese Tallate | 42 |
| VM&P Naphtha | 317 |

The resin and oils are melted together before adding the VM&P naphtha. The usual fillers such as whiting, china clay or asbestine can be finally added.

**Applications.** Floor-tile adhesive (asphalt).

## No. 7
### (Solvent)

| | |
|---|---|
| R-29 "Neville" Resin, 35° | 45 |
| Refined Tall Oil | 28 |
| Blown Soya Oil | 12 |
| VM&P Naphtha | 15 |

The resin and oils are melted together before adding the VM&P naphtha. The usual fillers such as whiting, china clay or asbestine can be finally added.

**Applications.** Floor-tile adhesive (asphalt, vinyl).

## No. 8
### (Solvent)

| | |
|---|---|
| "Neville" LX-685, 135 | 30 |
| Refined Tall Oil | 45 |
| Blown Soya Oil | 10 |
| VM&P Naphtha | 15 |

The resin and oils are melted together before adding the VM&P naphtha. The usual fillers such as whiting, china clay or asbestine can be finally added.

**Applications.** Floor-tile adhesive (asphalt, cork, vinyl).

## No. 9
### (Solvent)

| | |
|---|---|
| Smoked Sheets | 25.0 |
| "Nevastain" A | 0.3 |
| "Calcene" TM | 66.0 |
| "Titanox" RCHT | 50.0 |
| "Neville" LX-685, 125 | 25.0 |
| R-29 "Neville" Resin, 35° | 22.0 |
| Mineral Spirits | 75.0 |
| Asbestos Floats (Johns-Manville 7R) | 10.0 |

The smoked sheets are masticated on cool, tight mill rolls for ten minutes. The masticated rubber is added to the mineral spirits under agitation. The rubber solution is then heated to 65 to 70°C. in a Sigma Blade Mixer and the "Neville" LX-685, 125 is added under agitation. The "Calcene," "Titanox" RCHT, asbestos, and R-29 "Neville" resin, 35° are added in that order.

**Applications.** Floor-tile adhesive (asphalt, ceramic, cork, rubber, vinyl). Wall-tile adhesive (ceramic, polystyrene).

## No. 10
### (Solvent)

| | |
|---|---|
| Smoked Sheets | 25.0 |
| "Nevastain" A | 0.3 |
| "Calcene" TM | 50.0 |
| "Titanox" RCHT | 50.0 |
| "Neville" LX-685, 125 | 25.0 |
| R-29 "Neville" Resin, 35° | 22.0 |
| Mineral Spirits | 75.0 |
| Soap | 4.0 |
| Water | 30.0 |

The smoked sheets are masticated on cool, tight mill rolls for ten minutes. The masticated rubber is added to the mineral spirits under agitation. The rubber solution is then heated to 65 to 70°C. in a Sigma-Blade mixer and the "Neville" LX-685, 125 is added under agitation. The "Calcene," "Titanox" RCHT, and R-29 "Neville" resin, 35° are added in that order. The premixed water and soap are added to give the mixture more viscosity.

**Applications.** Acoustical-tile adhesive. Floor-tile adhesive (asphalt, ceramic, cork, rubber, vinyl). Wall-tile adhesive (ceramic, polystyrene).

### No. 11
### (Solvent)

| | |
|---|---|
| Gelled Soybean Oil | 13.0 |
| "Abitol" | 13.0 |
| "Neville" LX-685 (70% Solution) | 20.0 |
| Titanium Dioxide | 4.5 |
| "SnoBrite" Clay | 43.0 |
| Denatured Alcohol | 6.5 |

The pigment and filler are mixed with the gelled soybean oil. The "Abitol" and "Neville" LX-685 solution are then added. The alcohol should be added last to minimize evaporation.

**Applications.** Acoustical-tile adhesive. Wall-tile adhesive (polystyrene).

### No. 12
### (Solvent)

| | |
|---|---|
| R-21 "Neville" Resin | 43 |
| Pine Oil | 2 |
| Mineral Spirits | 10 |
| Fillers (Talc, Whiting, Silica Gel) | 45 |

The resin, pine oil and mineral spirits are mixed until uniform with the fillers then being added. The fillers should be proportioned to give the desired properties. Talc causes easier spreading, while whiting imparts bulk and lighter color. Silica gel imparts body and strength to the adhesive when freshly applied.

**Applications.** Floor-tile adhesive (asphalt). Wall-tile adhesive (polystyrene).

### No. 13
### (Solvent)

| | |
|---|---|
| R-17 "Neville" Resin | 35.0 |
| Oleic Acid | 3.5 |
| Deodorized Kerosene | 9.0 |
| Methyl Alcohol | 1.6 |

Inerts such as clay, slate flour, etc., should be added to this mixture until the desired spreading consistency is obtained.

**Applications.** Floor-tile adhesive (asphalt, cork, vinyl).

Transmission-Belt Adhesive
("Neoprene" Solvent Cement)

| "Neoprene" GN-M2 | 100 |
| Benzothiazyl Disulfide | 1 |
| Magnesium Oxide | 10 |
| "Thermoflex" A | 2 |
| Medium Thermal Black | 15 |
| Fine Thermal Black | 15 |
| Stearic Acid | 1 |
| Zinc Oxide | 10 |
| R-16 "Neville" Resin | 10 |
| "X-1 Resinous Oil" | 28 |
| Toluene | 154 |

The "Neoprene" is added to cool, tight rolls and masticated for 20 minutes. The benzothiazyl disulfide and magnesium oxide are added together and dispersed. The "Thermoflex A" is added with both thermal blacks and mixed well. The stearic acid and zinc oxide are added last and together. The batch is then slabbed off and dissolved in toluene. The R-16 "Neville" resin and "X-1 Resinous Oil" are added under agitation to the rubber solution.

---

Vinyl Copolymers to Glass
Adhesive
(Solvent)

| Polyvinyl Acetate ("Vinylite" AYAF) | 75 |
| "Nevillac" Hard | 28 |
| Butyl Acetate | 432 |

The polyvinyl acetate is sifted into the butyl acetate under agitation. The "Nevillac" Hard is then added along with a heat or light stabilizer if needed.

---

Leather or Metal to Cloth and
Paper Adhesive
(Hot Melt or Solvent)

| R-28 "Neville" Resin, 35° | 45 |
| Coating-Grade Polystyrene ("KTPL-3") | 5 |

A hot-melt mixture is made of the polystyrene and 10 parts of the R-28 "Neville" resin, 35°. When uniform, the remainder of the R-28 "Neville" resin, 35° is added. This mixture can be dissolved in a suitable aromatic solvent for solution application.

**Applications.** Pressure-sensitive adhesive.

---

Vinyl to Vinyl (Noncuring)
(Solvent)

| "Neoprene" GRT | 100 |
| Zinc Oxide | 5 |
| Magnesium Oxide | 4 |
| "Neozone" A | 2 |
| "Silene" EF | 10 |
| R-12 "Neville" Resin | 20 |
| Toluene | 423 |

*Milling:*

Mill Roll

   Temperature  90-100°F

"Neoprene" Mastication

   Time        6 minutes

"Neozone" A   3 minutes

Magnesium Oxide and

   "Silene"     3 minutes

Zinc Oxide    3 minutes

*Mixing:*

The milled rubber with fillers is added to the toluene under agitation. The R-12 "Neville" resin is added after the rubber compound is in solution.

------

### Sprayable—Fast Drying Adhesives

An interesting possibility, from the method of application standpoint, is the solvent type adhesive that is of a suitable spraying viscosity. When correctly applied, a smooth uniform adhesive film is obtained from which the solvents will evaporate quite evenly and rapidly. Adhesives of this type are widely used in those applications where a high rate of production and efficiency is required.

The following formulations illustrate adhesives with improved initial tack and tack retention properties of the ad-hesive film. These adhesives are strongly recommended for use in those applications where the bonding of glass, leather, metal, plastics, rubber, and wood is involved.

|  | Formula | |
|---|---|---|
|  | No. 1 | No. 2 |
| "Neoprene"AC | 100 | 100 |
| Magnesium Oxide | 4 | 4 |
| Zinc Oxide | 5 | 5 |
| "Nevastain" A | 1 | 1 |
| "Durez" 12603 | 55 | — |
| "Arofene" 700 | — | 45 |
| "Neville" LX-685, 125 | — | 45 |
| R-13 "Neville" Resin | 30 | — |
| Methyl Ethyl Ketone | 130 | 130 |
| Mineral Spirits | As required | |

The "Neoprene" AC is broken down on a cool, tight mill for 10 minutes. The magnesium oxide, "Nevastain" A, and zinc oxide are then added. The batch is removed from the mill, cut into small sections and dissolved in the methyl ethyl ketone–mineral spirits mixture along with the modifying resins. Generally a ratio of 2:1 mineral spirits–methyl ethyl ketone blend is needed to bring the finished adhesive solution into a suitable spraying viscosity range.

Other previously shown formulations can be made suitable for spray-type applications with adjustments in solids content and solvent balance.

---

### Emulsion-Type Adhesive

In the adhesive field, as in the case of many other industries, the latex or emulsion adhesive that can be thinned with water is rapidly gaining wide-spread recognition. The majority of these compounds are based on high molecular weight resins that have been polymerized in a water medium. Here, too, "Neville" Resins can be used to great advantage. Their inherent tackiness improves the initial adhesion and also aids in prolonging the tack life of the adhesive film. Their trial is strongly recommended to those who are actively engaged in or planning on producing latex adhesives.

It is necessary that the highly hydrophobic "Neville" resins be emulsified or made water miscible so that they need only be stirred into the latex adhesive. The following emulsification information is given to this end.

If a hydrocarbon solvent, such as toluene, mineral spirits, etc., can be tolerated in the adhesive

formulation, then the emulsification of the high melting "Neville" resins can be accomplished quite readily. Dispersion, without a solvent, is somewhat more difficult. When a solvent is used, the "Neville" resin is first put into solution and then the solution emulsified at room temperature by any of a number of methods with perhaps the most well known being the *in-situ* type. The following example of an anionic system illustrates this technique:

### Formula No. 1
(40% Resin Solids)

| | |
|---|---|
| Hard "Neville" Resin | 42 |
| Solvent | 18 |
| Oleic Acid | 4 |
| Triethanolamine | 2 |
| Water | 40 |

Initially, the resin is dissolved in the solvent. The oleic acid and triethanolamine are then added to the resin solution and mixed until uniform using an "Eppenbach Homomixer," "Cowles Dissolver," or some other high-speed mixer. The water is then added in small portions under high speed mixing. Initially, a water-in-oil emulsion is formed with a noticeable increase in viscosity. However, with the continued addition of water, the emulsion will reach a maximum in viscosity and invert to an oil-in-water type with some decrease in viscosity. After inversion, the remainder of the water can be added more rapidly. This emulsion is water dispersible and can be further reduced with water to any desired solids concentration.

In general, the majority of "Neville" resins can be emulsified in this fashion. Of course, the "Neville" resin being emulsified should be completely soluble in the organic solvent used in the formulation.

Another method of emulsifying a high melting "Neville" resin (in solution) is shown in the following anionic-nonionic system:

## No. 2
### (48% Resin Solids)

| | |
|---|---|
| "Nevillac" Hard | 60.0 |
| 1:1 Vol. Mixture of | |
| Mineral Spirits- | |
| Toluol | 15.0 |
| "Aerosol" OT (75%) | 2.4 |
| "Tergitol" NPX | 2.0 |
| Water | 44.0 |

Initially, the resin is dissolved in the solvent mixture. The surfactants are then added to the resin solution and mixed until uniform. The water is added

with continued high speed mixing. The resultant oil-in-water emulsion can be thinned as desired.

The lower melting "Neville" resins such as R-27 "Neville" resin, 10°, R-29 "Neville" resin, 10° and "Nevillac" 10° can also be emulsified in a similar manner without the use of a solvent. Gentle heating of the resin phase is advisable to keep the viscosity increase of the emulsion before inversion at a minimum.

The dispersing in water of the high-melting, hard "Neville" resins (without solvent) is not quite so easily done. The most important factors to be desired are very fine particle size of the resin being dispersed and suitable dispersing agents and thickeners or stabilizers to keep the resin particles from settling.

The following formulas illustrate the dispersing of the high-melting "Neville" resins in both anionic and nonionic systems:

## No. 3
### (47% Resin Solids)

| | |
|---|---|
| Hard "Neville" Resin | 45 |
| "Methocel" (400 cps) | |
| (3% Solution) | 30 |
| "Tamol" 731 (25%) | 2 |
| Water | 20 |

Charge resin, "Tamol" 731 and water to a ball mill and mill until a desired particle size is obtained. The "Methocel" solution can then be added.

## No. 4
### (37% Resin Solids)

| | |
|---|---|
| Hard "Neville" Resin | 300.0 |
| Water | 500.0 |
| "Triton" NE | 1.5 |
| "Tween" 20 | 1.5 |

The resin is initially powdered by ball milling or any other conventional means. A slurry is made of the above and passed through a colloid mill at 7200 r.p.m. (Premier Mill, 3½ in. pulley). Multiple passes are made reducing the mill setting from 0.010 in. to 0.002 in.

---

## Shoe Adhesive
### (Vulcanizing Type)

| | Wet | Dry |
|---|---|---|
| Natural Rubber Latex | 167.0 | 100.0 |
| Potassium Hydroxide (10% Solution) | 10.0 | 0.5 |
| Ammonium Caseinate (10% Solution) | 10.0 | 0.5 |
| Zinc Oxide | — | 2.0 |
| Sulfur | — | 1.0 |
| "Nevastain" B | — | 2.0 |
| "Neville" LX-685, 125, Resin Solution Emulsion | 187.0 | 75.0 |
| "Setsit"—5 | 3.0 | 1.5 |

The above ingredients are slowly stirred into the natural rubber latex in the order shown. The zinc oxide, sulfur, and "Nevastain" B are combined and ball-milled at about 50% solids using a suitable dispersing agent, following the general formula and technique shown for Emulsion Type Adhesive, Formula No. 4. The resultant dispersion of these ingredients is slowly stirred into the batch along with the "Neville" LX-685, 125, Resin Solution emulsion, prepared according to Emulsion-Type Adhesive, Formula No. 1. Finally the accelerator is added after being diluted with an equal volume of water.

A suggested cure for this adhesive would be 20 minutes at 300°F.

---

## Corrugating Starch
### Formula No. 1
### (Single Facer)

| | | |
|---|---|---|
| a | Water | 200 gal. |
| | "ASP" 400 | 200 lb. |
| | Corrugating Starch | 200 lb. |

*b* Caustic Soda
(in 10 gal.
Water)                29 lb.

*c* Water               330 gal.

*d* Borax               30 lb.
Corrugating
Starch              900 lb.

Heat *a* to 140°F. Add *b*, heat to 160°F. and hold for 15 minutes. Add *c* and cool to 100°F. Add *d* and mix until smooth.

(Double Backer)

*a* Water               200 gal.
"ASP" 400             200 lb.
Corrugating
Starch              220 lb.

*b* Caustic Soda
(in 10 gal.
Water)                32 lb.

*c* Water               330 gal.

*d* Borax               30 lb.
Corrugating
Starch              880 lb.

Heat *a* to 140°F. Add *b*, heat to 160°F. and hold for 10 minutes. Add *c* and cool to 100°F. Add *d* and mix until smooth.

No. 2

(Upper Mixer—140 gal.)

Add water to 23 in. from top; heat to 130°F. Add 115 lb. corrugating starch; mix 5 minutes and add 16 lb. caustic soda dissolved in water (5 gal.). Mix 5 minutes. Heat to 160°F., hold for 5 minutes and add water to 6 in. from top. Mix 5 minutes and add 3 pt. formaldehyde.

(Secondary Mixer, lower—
445 gal.)

Add water to 18 in. from top, heat to 90°F., add 80 to 100 lb. "ASP" 400, and mix 5 minutes. Add 485 lb. corrugating starch, mix 5 minutes, add 11 lb. borax #5 mol, and mix 5 minutes. Slowly drop contents of upper mixer into lower mixer (Should require 20 to 30 minutes). Add water, if necessary, to 8 in. from top. Check viscosity. If viscosity is above 40 seconds at 100 to 110°F., add water to 4 in. from top, if necessary, to adjust viscosity to 35 to 40 seconds. Should viscosity be below 35 or above 40 seconds at the 6 in. level, increase or decrease the starch in upper mixer by 5 or 10 lb. Pump to storage.

Temperature of the starch in the upper mixer should be less than 130°F. before dropping into lower mixer.

Viscosity of finished paste should be 31 to 34 seconds. Temperature should be 100 to 110°F.

## No. 3
### (Upper Mixer)

Add cold water to 23¾ in. from top. Heat to 130°F. and add 180 lb. corrugating starch. Mix 10 minutes, then add 34 lb. caustic soda in 4 gal. water and mix 15 minutes. Cook with live steam to 165°F. and then add water to 6 in. from top.

### (Secondary Mixer)

Add cold water to 26 in. from top and heat to 90°F. Add 175 to 200 lb. "ASP" 400, mix 5 minutes and add 1020 lb. corrugating starch. Mix 10 minutes, add 34 lb. borax and 8 lb. formaldehyde. Add contents of upper mixer, allowing 20 to 25 minutes. Mix 10 minutes, then check viscosity which should be 40 to 45 seconds. Check level which should be 8½ to 9 in. from top. Add 6 lb. sulfonated castor oil then pump to storage.

Operating viscosity should be about 33 to 35 seconds. Quantity of corrugating starch in carrier (upper mixer) may be adjusted to obtain correct viscosity.

---

### Tube and Core Winding
### Formula No. 1

| | |
|---|---|
| Hide Glue | 110 |
| Dextrin | 55 |

| | |
|---|---|
| Water | 200 |
| "ASP" 400 or "ASP" 900 | 40 |

### No. 2

| | |
|---|---|
| Tube Winding Paste | 75.00 |
| Water | 187.50 |
| "ASP" 400 | 18.75 |

No cooking.

### No. 3

| | |
|---|---|
| Tube Winding Paste | 258 |
| "ASP" 400 | 19⅛ |

No cooking.

---

### Palletizing Adhesive

| | |
|---|---|
| Palletizing Dextrin | 1.0 |
| Water | 2.0 |
| "ASP" 600 or "ASP" 900 | 1.2 |

---

### Multipurpose Adhesives
### Formula No. 1
### (Hot Melt)

| | |
|---|---|
| Ethyl Cellulose | 5 |
| "Paradene" No. 35 | 45 |
| "Nevinol" or 1-D Heavy Oil | 9 |

A hot mix is made of the ethyl cellulose and a small amount of melted "Paradene" No. 35. When this mixture is uniform, the remaining "Paradene" No. 35 is added. "Nevinol" or 1-D heavy oil can be added in varying amounts to give the desired degree of tackiness.

**Applications.** Aluminum foil to: aluminum foil, chipboard, fabric, paper, wood. Paper to glass, paper. Pressure-sensitive adhesive.

No. 2
(Hot Melt)

| | |
|---|---|
| "Nevillac" Hard | 40 |
| "Nevillac" 10° | 70 |
| Polyvinyl Acetate ("Vinylite" AYAT) | 16 |

The "Nevillac" Hard is heated to a temperature of 180 to 190°C. The polyvinyl acetate is slowly sifted into the melted "Nevillac" with agitation. This mixture is agitated until a uniform and clear mix is obtained. The "Nevillac" 10° is then added slowly with continued agitation.

**Applications.** Aluminum foil to: aluminum foil, chipboard, fabric, paper, wood. Asphalt to vinyl coated objects. Box-top adhesive, polyethylene coated. Cellulose acetate to paper. Formica to wood. Glass to glass. Label adhesive. Leather to: cloth, paper. Metal to: cloth, paper. Mylar to mylar. Paper to: glass, paper. Tape, plastisol coated to cloth backing.

No. 3
(Hot Melt)

| | |
|---|---|
| Ethyl Cellulose | 9 |
| "465" Resin | 81 |

| | |
|---|---|
| "X-743" Plasticizing Oil | 10-15 |

A small portion of the "465" resin is melted at 180°C. and the ethyl cellulose is slowly added with vigorous agitation. Agitation is continued until a uniform mixture is obtained. The remainder of the "465" resin and "X-743" plasticizing oil is then added.

**Applications.** Cellulose acetate to paper. Cloth-coated adhesive.

No. 4
(Emulsion)

| | |
|---|---|
| R-7 "Neville" Resin, 108-112 | 20.0 |
| Dibutylphthalate | 15.0 |
| "Tergitol" NPX | 0.8 |
| "Aerosol" OT, 75% Active | 1.0 |
| Polyvinyl Acetate Emulsion ("Gelva" TS-22) | 68.0 |

The R-7 "Neville" resin, dibutylphthalate, "Tergitol" NPX, and "Aerosol" OT are mixed at 100°C. until homogeneous. This mixture is then cooled to 55°C. and the polyvinyl acetate emulsion is very slowly added with vigorous agitation. Additional water can be added if desired to adjust the solids content and the viscosity.

**Applications.** Cellulose acetate to paper. Clothcoated adhesive. Label adhesive.

### No. 5
### (Emulsion)

| | |
|---|---|
| 70% (wt.) R-7 "Neville" Resin, 108-112 in Toluene | 29.0 |
| Dibutylphthalate | 10.0 |
| "Tergitol" NPX | 0.8 |
| "Aerosol" OT, 75% Active | 1.0 |
| Polyvinyl Acetate Emulsion ("Gelva" TS-22) | 68.0 |

This formula is a modification of Formula No. 4 and is somewhat easier to prepare. The polyvinyl acetate emulsion is added to the mixture of the other components with agitation. Additional water can be added if desired to adjust the solids content and viscosity.

**Applications.** Cellulose acetate to paper. Cloth-coated adhesive.

### No. 6
### (Hot Melt)

| | |
|---|---|
| R-16-A "Neville" Resin | 50 |
| "Indopol" No. H35 | 20 |
| "Nevinol" | 12 |

The R-16-A "Neville" resin is melted and the "Indopol" added to it with agitation. The "Nevinol" is then added.

**Applications.** Label adhesive. Paper to glass, paper. Pressure-sensitive adhesive.

### No. 7
### (Solvent)

| | |
|---|---|
| GR-S 1023 | 100 |
| 1-D Heavy Oil | 10 |
| "Nevastain" B | 1 |

The GR-S is broken down on a cold mill and the 1-D heavy oil and "Nevastain" B are added. This mix can then be dissolved in a paraffinic solvent (mineral spirits, n-hexane) to a suitable consistency for application. For additional tackiness, the addition of R-28 "Neville" resins, 35° is suggested.

**Applications.** Glass to glass. Leather to cloth, paper. Metal to cloth, paper. Paper to paper.

### No. 8
### (Hot Melt)

| | |
|---|---|
| "901 Compound" | 52 |
| "Nuba" No. 1 | 38 |
| "X-743" Plasticizer | 40 |

The "901 Compound" and "Nuba" No. 1 are melted and mixed until uniform. The "X-743" plasticizer is then added under agitation.

**Applications.** Leather to cloth, paper. Metal to cloth, paper. Paper to paper. Paper to tin.

## No. 9
### (Solvent)

| | |
|---|---|
| "Neoprene" AC Soft | 100 |
| R-12 "Neville" Resin | 15 |
| Litharge | 20 |
| Magnesium Oxide | 4 |
| Zinc Oxide | 5 |
| "Silene" EF | 15 |
| Sulfur (Blackbird) | 4 |
| "Neozone" A | 2 |
| Stearic Acid | 1 |
| "Accelerator 833" | 3 |

| Milling Addition | Milling Time, Minutes |
|---|---|
| "Neoprene" AC breakdown | 10 |
| "Neozone" A | 2 |
| Magnesium Oxide | 3 |
| "Silene" EF | 3 |
| Litharge and Stearic Acid | 2 |
| Sulfur and Zinc Oxide | 2 |

The R-12 "Neville" resin is added directly to the ball mill when putting this compound into solution. The above formulation (with the exception of the accelerator) is dissolved in toluene at 40% solids by ball milling for approximately 5 hours. After complete solution is effected, the accelerator is added.

The best application procedure is to apply the adhesive directly to the road surface and allow it to air-dry for about 15 minutes. The vinyl strip is then pressed firmly onto the partially dried adhesive.

**Applications.** Sponge adhesive. Vinyl traffic marking strip adhesive.

---

### Urethane Caulking Compounds

| Can A | Two-Can System | | One-Can System |
|---|---|---|---|
| "Polycin" U-56 Prepolymer | 100 | 100 | — | 100 |
| "Polycin" U-63 Prepolymer | — | — | 100 | — |
| Tributyl Phosphate | 5 | 5 | 5 | 5 |
| VM&P Naphtha | 30 | 30 | 30 | 15 |
| "Atomite" | 155 | 155 | 155 | 100 |
| Colloidal Silica ("Syloid" 244) | 10 | 10 | 10 | 5 |

### Can B

| | | | | |
|---|---|---|---|---|
| DB Castor Oil | 150 | — | — | — |
| "Polycin" 52 | — | — | 70 | — |
| "Polycin" 53 | — | 90 | — | — |
| Tributyl Phosphate | 5 | 2 | 3 | — |
| VM&P Naphtha | 6 | 2 | 5 | — |
| "Atomite" | 400 | 160 | 195 | — |
| Colloidal Silica | 20 | 8 | 10 | — |

In each instance, the "Atomite" is first incorporated with the "Polycin" U prepolymer (Can A) and with the DB Oil or "Polycin" curing polyol (Can B), using an efficient mixer. Addition of solvent (VM&P naphtha), plasticizer (tributyl phosphate), and antislump agent (colloidal silica) is then made in that order.

For best shelf life, the "Atomite" and other materials should be thoroughly dried before use, since the reactive—NCO groups of the prepolymer component react with any contaminating moisture.

Recent work indicates that improved physical properties can be obtained by completely eliminating the solvent and plasticizer from the calk formulation. The following are typical sealants formulated at a 100% solids content. Note that a reduced filler content is necessary to maintain a satisfactory consistency which in turn makes these upgraded calks somewhat more expensive on a weight basis.

### Two-Can System

#### Can A

| | | |
|---|---|---|
| "Polycin" U-63 | | |
| Prepolymer | 100 | 100 |
| "Atomite" | 100 | 100 |

#### Can B

| | | |
|---|---|---|
| DB Castor Oil | 130 | — |
| "Polycin" 52 | — | 65 |
| "Atomite" | 130 | 65 |

Certain formulating ideas have gradually evolved from the experimental work to date. An excess of curing polyol in the final mixture appears to aid calk adhesion and lowers the calk raw material cost. A 20% excess is suggested as a starting value although a 75% excess still gives acceptable calks without detracting too much from physical properties. In general, "Polycin" curing polyols impart better adhesion, tensile strength and elongation than DB castor oil. However, the presence of some DB castor oil is of value in giving a significant amount of resilience. Plasticizers tend to reduce calk adhesion to substrates. Satisfactory bonding to glass and aluminum generally means that acceptable bonding to alternate substrates (wood, steel, etc.) will also be achieved.

---

## Solid Fibre Laminating

### Formula No. 1
#### (Protein-Vegetable)

1 part protein to 1 part "ASP" 900

### No. 2
#### (Starch)

| | |
|---|---|
| Laminating Starch Derivative | 2000 |
| "ASP" 400 | 100 |
| Borax | 6 |
| Water | 4111 |

Cook to 195°F., hold 10 minutes, cool to 100°F. for use.

---

## Bag Seam Pastes

### Formula No. 1
#### (Bottom Paste)

| | |
|---|---|
| Water | 750 lb. |
| Seam Gum | 200 lb. |
| Sodium Silicate | 5 gal. |
| "ASP" 400 | 45 lb. |

#### (Multiwall Seam Paste)

| | |
|---|---|
| Water | 300 lb. |
| Seam Gum | 166 lb. |
| "ASP" 400 | 35 lb. |

Cook both pastes to 200° F. and cool to room temperature for use.

### No. 2
#### (Bottom Paste)

| | |
|---|---|
| Water | 108 gal. |
| Seam Gum | 250 lb. |
| Tapioca | 20 lb. |
| Caustic Soda (in 50 gal. Water) | 28 lb. |
| Sulfonated Castor Oil | 4¼ lb. |
| "ASP" 400 | 50 lb. |

Cook to 195° F., hold 5 minutes, and cool to room temperature for use.

#### (Multiwall Seam Paste)

| | |
|---|---|
| Water | 300 lb. |
| Prepared Seam Mix | 120 lb. |
| "ASP" 400 | 25 lb. |

---

## Film Metal Adhesive

| | |
|---|---|
| Vinyl Acetate Solution in Methyl Acetate (28% Solids) | 1 |
| 6-Second Nitrocellulose in Acetone (12% Solids) | 1 |

## Adhesive for Dust Filters
### U. S. Patent 2,538,187

| | |
|---|---|
| Ethylene Glycol | 12.4 |
| Glycerin | 37.2 |
| Sorbitol | 27.4 |
| Phosphoric Acid | 23.0 |

## Caulking Compound

### Formula No. 1

| | |
|---|---|
| Heavy-Bodied Linseed Oil | 34½ |
| Silica | 65 |
| Diglycol Oleate S | ½ |

### No. 2
### U. S. Patent 2,618,569

| | |
|---|---|
| Blown Soybean Oil | 32 |
| Kerosene | 10 |
| Ground Asbestos | 10 |
| Titanium Calcium Sulfate | 9 |
| Calcium Magnesium Carbonate | 36 |
| Calcium Linoleate | ½ |
| Cobalt Drier | ½ |
| Acetic Acid (55%) | 1½ |
| Chlorinated Paraffin Wax | ½ |

## Glass-Fiber Binder

### Formula No. 1

| | |
|---|---|
| GR–S Latex Type 4 (75% Solids) | 187.5 |
| "Indulin" A | 30.0 |
| Water | 50.0 |
| 25% Ammonium Hydroxide | 12.0 |

### No. 2

| | |
|---|---|
| GR–S–Latex Type 3 (76% Solids) | 200 |
| "Indulin" A | 50 |
| 25% Ammonium Hydroxide | 17 |

Add the "Indulin" directly to the latex. The ammonium hydroxide dissolves the "Indulin" by forming an ammonium salt. On drying, heat causes decomposition of the ammonium salt and precipitation of the "Indulin."

## Polyethylene Adhesive
### U. S. Patent 2,622,056

| | |
|---|---|
| Polyethylene | 1 |
| Chlorinated Diphenyl | 3 |

Heat at 180°F. for 3 hours. Cool, grind, heat to 175°F., and spread on a metal plate at 190°F. To obtain a firm bond, press a polyethylene or metal sheet against this.

## Polystyrene Adhesive
### U. S. Patent 2,628,180

| | |
|---|---|
| Styrene | 95.0000 |
| Butyl Stearate | 5.0000 |
| Quinol | 0.0005 |

This is used for uniting polystyrene with other solids.

### Glass, Wood, Metal, Putty
*Belgian Patent 487,162*

| | |
|---|---|
| Litharge | 2 |
| Aluminum Powder | 2 |
| Washed Chalk | 81 |
| Boiled Linseed Oil | 15 |

This dries quickly and retains elasticity.

---

### Quick-Setting Wood Adhesive
*U. S. Patent 2,680,078*

| | |
|---|---|
| Brazilian Tapioca Flour | 320.00 |
| Water | 500.00 |
| 3% Hydrogen Peroxide | 10.00 |
| Sodium Bicarbonate | 0.65 |
| Caustic Soda ⎫ Solution | 12.00 |
| Water ⎭ | 112.00 |

Stir for 8 hours at 60 to 65°F.

---

### Wallpaper Paste
*U. S. Patent 2,689,184*

| | |
|---|---|
| Methylcellulose | 10 - 11 |
| Sodium Chloride | 7 - 8 |

Add water and mix before use.

---

### Quick-Setting Adhesive

| | |
|---|---|
| Dextrin | 47.5 |
| Water | 47.5 |
| Sorbitol | 4.8 |
| "Tergitol" Anionic 7 or P-28 | 0.2 |

### Nonstaining Veneer Adhesive
*German Patent 925,787*

| | |
|---|---|
| Casein | 50 |
| Urea | 30 |
| Calcium Carbonate | 70 |
| Water | 100 |

---

### Stabilized Casein for Adhesives
*U. S. Patent 2,757,171*

| | |
|---|---|
| *a* Casein | 100.0 |
| Ammonium Chloride | 5.6 |
| *b* 30% Hydrogen Peroxide | 5.0 |
| Water | 30.0 |

Mix *a* well and add *b*. Put in a sealed jar and heat for 18 hours at 65°C. This product exhibits no microbial growth for 6 months.

---

### Dielectric Adhesive

| | |
|---|---|
| Polyphenylpoly-siloxane | 20 |
| Barium Oxide $3H_2O$ | 8 |
| Silica | 60 |
| Lead Peroxide | ½ |
| Chromium Oxide | 1 |
| Phosphate Cement | 4½ |

---

### Tire-Cord Adhesive

| | |
|---|---|
| Styrene-Butadiene Rubber | 67 |
| Smoked Sheets (Rubber) | 33 |

Phenyl Beta
Naphthylamine          1
SRF Black             20
HAF Black             20
Zinc Oxide             5
"Coray" 230           10
Sulfur                 3
"Altax"                1

All cords are dried at 250°F. for 5 minutes and cured as follows; Butyl at 320°F. for 25 minutes and SBR at 300°F. for 20 minutes.

---

Adhesive for Plastics

"Vistanex" L-100        100
"Vistanex" LM
type MH              25-50
Synthetic Resin*        125
Solvent                 312

---

*The most commonly used resins are those derived from alkylated phenols, terpenes and hydrogenated ester gum.

---

Paper to Polyethylene Adhesive
Swiss Patent 295,557

Polyisobutylene
(M.W. 50,000)         20
Heptane                100

---

Pressure-Sensitive
Adhesive for "Teflon"
U. S. Patent 2,770,842

Polyisobutylene
(M.W. >10,000)        50
Toluene                50

Adhesive for Silicone Surfaces

Formula No. 1
U. S. Patent 2,782,949

a Water                347
Phenol                  3
Dextrin               650
b Hydrofluoric
Acid (60%)           500

Mix a and heat at 90 to 100°C until uniform. Then add b.

No. 2
German Patent 957,155

a Yellow Dextrin       150
Water                 100
b Ammonium
Bifluoride            7½

Mix a to dissolve and add b.

---

"Paracril" Adhesive

"Paracril" BJ          100.0
Zinc Oxide              5.0
Stearic Acid            1.0
SRF Carbon Black       75.0
Dibutyl Phthalate       7.5
Coumarone Indene
Resin (M.P. 25°C.)    7.5
"Monex"                 0.6
Sulfur                  1.0

Cure 20 minutes at 310°F. in ASTM mold.

---

Pressure-Sensitive Adhesive

This adhesive is especially

useful as a gum coating for plastic, paper, fabric tapes and labels. Phenolic resins or hydrogenated rosin and its derivatives may be used in place of the coumarone indene resin.

| | |
|---|---|
| "Paracril" CV | 100 |
| Coumarone Indene Resin (M.P. 45 to 55°C.) | 100 |
| Solvent: Methyl Ethyl Ketone and Toluene | |
| Solids Concentration: 20% by weight | |

Blend "Paracril" CV and resin in a "Precision" Equipoise Shaker until a smooth, free-flowing cement is obtained.

---

### Noncuring Adhesive for Vinyl Chloride

| | |
|---|---|
| "Paracril" C | 100 |
| Calcium Silicate | 40 |
| Titanium Dioxide | 10 |
| Refined Coal Tar Oil | 25 |
| Coumarone Indene Resin (M.P. 45-55°C.) | 25 |
| "Parlon", 128 cps. | 100 |
| Solvents: Methyl Ethyl Ketone and Toluene | |
| Solids Content: 20% by weight | |

Breakdown "Paracril" C by passing at least ten times through cool, tight rolls of an open mill.

Mix the compound, except "Parlon," and solvate in a "Precision" Equipoise Shaker until a smooth, free-flowing cement is obtained. Make a 20% solids solution of the "Parlon" in toluene. Blend 100 parts of the "Parlon" solution with 200 parts of the "Paracril" compound solution.

---

### Self-Curing Adhesive for Vinyl Chloride

| | |
|---|---|
| "Paracril" C | 100 |
| Zinc Oxide | 10 |
| Extralight Calcined Magnesium Oxide | 5 |
| Calcium Silicate | 24 |
| Silicon Dioxide | 36 |
| D-B-A | 4 |
| "Ethazate" | 4 |
| Z-B-X | 4 |
| Sulfur | 4 |
| "Parlon", 20 cp. | 40 |
| Solvents: Methyl Ethyl Ketone and Toluene | |
| Solids Content: 20% by weight | |

Breakdown "Paracril" C by passing at least ten times through cool, tight rolls of an open mill. Mix the compound, omitting the D-B-A, "Ethazate," Z-B-X, sulfur and "Parlon." Split the batch in equal parts and add the sulfur, D-B-A in one batch and "Ethazate",Z-B-X in the other. Prepare separately 20%-solids solutions of each compounded batch in methyl ethyl ketone and a 20% solution of "Parlon" (20 cps.) in toluene in a "Precision" Equipoise Shaker until smooth, free-flowing solutions are obtained. Just prior to use, blend 100 parts of each solution and add 40 parts of the "Parlon" solution.

---

Adhesive for Plastics (Shoes)

The following cement is an excellent adhesive for bonding a variety of materials either to themselves or to each other. Such adhesives have been used extensively by the shoe industry and are particularly satisfactory for applications involving the bonding of vinyls. Because the bond is flexible, it has good impact resistance. For most applications only room temperature cures are necessary which become even stronger on aging. For some applications where greater heat resistance is required, curing at temperatures of 250°F., or lower, gives marked improvement.

| | |
|---|---|
| "Paracril" CV | 100 |
| "Durez" 12687 | |
| (Phenolic Resin) | 100 |

Solvent: Methyl Ethyl Ketone

Solids Content: 25% by weight

Blend "Paracril" CV, phenolic resin and solvent by agitation in a "Precision" Equipoise Shaker until a smooth, free-flowing cement is obtained.

---

Optical Cement for Crystals

| | |
|---|---|
| Polystyrene | 23-45% |
| Benzol | to make 100% |

---

Tin-Can Label Adhesive

| | |
|---|---|
| Polyvinyl Acetate | |
| Emulsion | 60 cc. |
| Polvinyl Methyl | |
| Ether | 14 cc. |
| Water | 26 cc. |

---

Glass-Joint Mastic
U. S. Patent 2,870,036

| | |
|---|---|
| Ammonium Chloride | 2-6 |
| Aluminum Stearate | 1-3 |
| Asphalt | To make 100 |

Melt and mix until uniform.

---

Silicate Adhesive
U. S. Patent 2,772,177

| | |
|---|---|
| Sodium Silicate Solution (35-45%) | 75-85 |
| Urea | 2-10 |
| Sugar | 0.5-3 |
| Sodium Bichromate | 0.1-1 |

| | |
|---|---|
| Magnesium Sulfate | 1-4 |
| Barden Clay | 0-8 |
| Water | To make 100 |

---

### Sealing Cement for Spark Plugs
*U. S. Patent 2,829,063*

| | |
|---|---|
| Calcium Aluminate | 60 |
| Phosphoric Acid | 19 |
| Water | 21 |

After application, air-dry for 2 hours and at 150°C. for 2 hours.

---

### Increasing Adhesion of Asphalt to Aggregate

| | |
|---|---|
| Slaked Lime | 1 |
| Low-Temperature Tar or Petroleum | 1 |

Mix 5 to 10% of the above into asphalt to improve swelling and compressibility. It also improves freezing and thawing cycles.

---

### Splice Adhesive

A "Paracril" BJ adhesive designed for splicing cured or uncured "Paracril" compounds to themselves with heat and slight pressure.

| | |
|---|---|
| "Paracril" BJ | 100 |
| Zinc Oxide | 15 |
| "Aminox" | 2 |
| EPC Carbon Black | 50 |
| Coumarone Indene | |

| | |
|---|---|
| Resin (M.P. 25°C.) | 25 |
| "Durez" 12687 (Phenolic Resin) | 20 |
| M-B-T | 2 |
| Sulfur | 2 |
| Solvent: | Methyl Ethyl Ketone |
| Solids Content: 20% by weight | |

Breakdown "Paracril" BJ by passing at least ten times through cool, tight rolls of an open mill. Mix the compound, including the phenolic resin, and solvate in a "Precision" Equipoise Shaker until a smooth, free-flowing cement is obtained.

---

### Self-Curing Splice Adhesive

This self-curing splicing cement can be used for a variety of purposes, including repairing cured "Paracril" products.

| | |
|---|---|
| "Paracril" C | 100 |
| Zinc Oxide | 5 |
| SRF Carbon Black | 50 |
| Coumarone Indene Resin (M.P. 25°C.) | 25 |
| "Aminox" | 2 |
| M-B-T | 4 |
| "Beutene" | 2 |
| Sulfur | 4 |
| "Parlon", 20 cps. | 200 |
| Solvents: | Methyl Ethyl Ketone and Toluene |

Solids

Content: 20% by weight

Breakdown "Paracril" C by passing at least ten times through cool, tight rolls of an open mill. Mix the compound omitting "Beutene" and "Parlon". Prepare a 20% solution of the "Parlon". Solvate the "Paracril" C compound until a smooth, free-flowing cement is obtained. Blend equal volumes of "Paracril" C cement and "Parlon" solution. Add "Beutene" just prior to use.

---

"Paracril" to Metal Adhesives

The excellent oil and fuel resistance of "Paracril" can be used advantageously for tank and pipe linings, industrial roll covering and miscellaneous molded goods by bonding "Paracril" compounds to metals.

The bonding of both cured and uncured "Paracril" compounds to aluminum, steel, brass and copper has been greatly simplified by the use of phenolic resins and chlorinated rubbers. In adhering unvulcanized "Paracril" compounds to metals and then curing with heat and pressure, unmodified solutions of these materials in recommended solvents have been found to be excellent general purpose bonding agents. The impact resistance and flexibility of the phenolic resin bonds can be improved by blending with "Paracril" CV; however, adhesion values are reduced. The heat resistance of phenolic resin adhesions can be improved by prebaking the resin coated metal at 225 to 250°F. for 15 to 20 minutes.

To adhere a vulcanized "Paracril" compound to metal at a low curing temperature, moderate adhesions are obtained with chlorinated rubber.

The following are formulas for solutions of unmodified phenolic resins, chlorinated rubber, and "Paracril" CV-phenolic resin blends.

Formula No. 1

"Paracril" CV        100
"Durez" 12987        100
Solvent:        Methyl Ethyl
                        Ketone
Solids
Content: 30% by weight

No. 2

"Durez" 12687 or 12987  100
Solvent: Methyl Ethyl Ketone or Methyl Ethyl Ketone/-Methyl Alcohol in 50/50 Blend

Solids
Content: 30% by weight

### No. 3

"Parlon", 20 cps.          100
Solvent:          Toluene
Solids
Content: 30% by weight

Blend "Paracril" CV, phenolic resin and solvents in a "Precision" Equipoise Shaker until a smooth, free-flowing cement is obtained. For straight phenolic resin or chlorinated rubber the same method is suggested. Commercial adhesives are generally used as received unless other recommendations are made by the manufacturer.

---

### Self-Curing Cements

Cements which will cure at room temperature or slightly above are particularly useful for fabrication and repair. These require ultra accelerator combinations. In making such cements the so-called split batch technique is used. This consists of making complimentary batches, each containing one element of the accelerator combination, solvating separately and blending just prior to application. By this method precure and loss is minimized. Similarly, the cement may be made up to contain everything but the accelerator and this may be added later with a small amount of solvent.

---

### Self-Curing Cements by Split-Batch Technique

**Base Masterbatch**

"Paracril" BJ or C   100
Zinc Oxide          5
MT Carbon Black   100

#### Cement Component A
Formula

| | No. 1 | No. 2 | No. 3 | No. 4 |
|---|---|---|---|---|
| Base Masterbatch | 205 | 205 | 205 | 205 |
| D-B-A | 4 | – | – | 4 |
| M-B-T | – | – | 4 | – |
| Sulfur | 4 | 4 | 4 | 4 |

#### Cement Component B
Formula

| | No. 1 | No. 2 | No. 3 | No. 4 |
|---|---|---|---|---|
| Base Masterbatch | 205 | 205 | 205 | 205 |
| Z-B-X | 4 | 4 | – | – |
| "Ethazate" | 4 | 4 | – | – |
| "Beutene" | – | – | 4 | – |
| C-P-B | – | – | – | 4 |

Solvent:          Methyl Ethyl Ketone
Solids
Content: 20% by weight

## "Dacron" to Rubber Adhesive

| | Formula | |
|---|---|---|
| | No. 1 | No. 2 |
| "Hylene" MP | 100 | 100 |
| "Gen-Tac" Latex (4%) | 286 | 500 |
| "Aerosol" OT | 8 | 1 |
| Sodium Alginate | 11 | — |
| Water | 743 | 776 |
| "Marasperse" CB | — | 3 |
| "Pliolite 2104" (58-60%) | — | 340 |

## Adhesive for Nylon Fabric

Adhesion of a "Paracril" B compound to a 3 oz. nylon fabric is obtained using the following "Paracril" C adhesive as a primer for the fabric. The cement should be used within 24 hours of preparation, otherwise it will become unstable.

| "Paracril" C | 100 |
|---|---|
| Zinc Oxide | 5 |
| Stearic Acid | 1 |
| Dibutyl Sebacate | 20 |
| Sulfur | 1½ |
| "Beutene" | 1 |
| "Durez" 12687 (Phenolic Resin) | 50 |
| "Penacolyte" G-1131-A (Liquid Resorcinol-Formaldehyde Resin) | 10 |
| "Penacolyte" G-1131-B (Paraformaldehyde) | 10 |
| Solvents: Methyl Ethyl Ketone and Acetone | |
| Solids Content: 30% by weight | |

Breakdown "Paracril" C by passing at least ten times through cool, tight rolls of an open mill. Mix the compound, omitting the resins and paraformaldehyde. Blend the compound, phenolic resin, resorcinol resin, paraformaldehyde and solvents in a "Precision" Equipoise Shaker until a smooth free-flowing cement is obtained.

## Adhesive for Interior Car Trim

This cement is especially useful for adhering mohair fabric to asphalt board or fabrics to steel or to themselves if extreme flexibility is not required.

| "Paracril" CV | 100 |
|---|---|
| "Durez" 12987 | 100 |
| Solvent: Methyl Ethyl Ketone | |
| Solids Content: 20% by weight | |

Blend "Paracril" CV, phenolic resin and solvent in a "Precision" Equipoise Shaker until smooth, free-flowing cement is obtained.

## Carpet-Scrim Adhesive

| "Pliolite" | |
|---|---|
| Latex 140 | 100.00 |
| Water* | To desired solids |
| Potassium Hydroxide | 0.25 |
| Whiting No. 10 | 20.00 |
| "McNamee Clay" | 30.00 |
| Thickener | To desired viscosity |

* Solids in this formulation should be adjusted to the 45 to 50% level.

## Ammonia-Resistant Seal
### U. S. Patent 2,804,393

| Litharge | 4 |
|---|---|
| Anhydrous Glycerol | 1 |
| Portland Cement | ½ |

## Coating Cement

The cement below is an excellent coating for fabrics primed with a general-purpose adhesive. This coating cement also has excellent low temperature flexibility. In addition substitution of calcium silicate for the carbon black and addition of about 25 parts of titanium dioxide makes a good white coating which may be colored to any desired shade by the incorporation of color pigments.

| "Paracril" BJ | 100.0 |
|---|---|
| Zinc Oxide | 5.0 |
| SRF Black | 100.0 |
| Diisooctyl Adipate | 30.0 |
| "Aminox" | 2.0 |
| Sulfur | 1.0 |
| "Monex" | 0.6 |
| Solvent: | Methyl Ethyl Ketone |
| Solids Content: | 20% by weight |

## Oil-Well Cement

### Formula No. 1

| Calcium Lignosulfonate | 0.25 - 2.0 |
|---|---|
| Cement | 84.75-83.0 |
| Trass Gel | 15.0 - 25.0 |

### No. 2
#### U. S. Patent 2,846,327

%

| Anhydrous Calcium Sulfate | 33-67 (vol.) |
|---|---|
| Portland Cement | 67-33 (vol.) |
| Wyoming Bentonite | 2.4 (wt.) |
| Water | 45-75 (of solids) |

### No. 3
#### U. S. Patent 2,776,713

| Portland Cement | 250 |
|---|---|
| "Bentone" | 10 |
| Methanol | 3 |
| Lead Naphthenate | ½ |

Make a slurry of formula in diesel oil—1 barrel. Pump into well, forcing into formation, under pressure and set with water.

----

Illuminating Carbon
Electrode Cement
*U. S. Patent 2,710,812*

| | |
|---|---|
| Carbon Flour | 46 |
| Glucose Solution (75%) | 43 |
| Ethyleneglycol | 11 |

This cement carbonizes, in arc, to form a consumable joint.

----

Binder for Carbon and
Graphite Electrodes
*Japanese Patent 2328 (1956)*

| | |
|---|---|
| Coal-Tar Pitch | 15 tons |
| Neutralized Tar Acids | 50 kg. |

----

Tooth-Root Canal Sealer

| | | |
|---|---|---|
| *a* | Zinc Oxide | 41.20 |
| | Powdered Silver | 30.00 |
| | White Rosin | 16.00 |
| | Thymol Iodide | 12.79 |
| *b* | Clove Oil | 78.00 |
| | Canada Balsam | 22.00 |

Mix *a* and *b* before use.

----

Cork Binder

| | |
|---|---|
| Zein G200 | 100 |
| Diethyleneglycol | 180 |
| Triethyleneglycol | 120 |

In making cork binder, the zein is stirred into the glycols, with or without added glycerin to replace a portion of the triglycol. Heating at 90 to 100°C. helps to make the zein dissolve faster. Glycol solutions containing 20 to 30% zein are the most effective for cork binders. Use of glycols results in finished composition cork having maximum dimensional stability.

Plasticized with the proper glycols, zein produces a binder solution that is water free and low enough in viscosity so that it can be readily mixed with granulated cork at moderate temperature in a conventional mixer. Generally, about four times as much cork as zein binder is used. However, this depends on the particle size of the cork, and can be varied to achieve the optimum dry tensile strength in the finished composition cork.

A curing agent and frequently a catalyst to speed curing are added to the cork and binder in the mixer, as is molten wax for mold lubricant. The entire mixing operation takes only about 45 minutes.

After mixing, the composition should be slightly tacky and free of lumps. It is then ready for molding or extrusion and curing.

Composition cork bonded with zein can be compression molded into block or rod, and it can also be extruded as rod. Since zein binders, unlike those made with glue, require no water for their formulation, the risk of blowing in the mold due to the presence of water is eliminated. By varying the charge to the mold, cork block can be produced with densities ranging from 17 to 24 lb. per cu. ft.

Compressed block, still in the mold, is cured by heating in an oven at close to 150°C. for two hours or less, depending on the size of the block. Rod is also cured in a closed mold, or sometimes extruded hot and cured as it is extruded.

During curing, heat must penetrate throughout the molded or extruded composition cork so that the curing agent, usually paraformaldehyde, will react with the zein to produce the tough cross linkages that greatly improve the water resistance of zein-bonded cork. Enough curing agent must be used to insure adequate wet tensile strength in the finished composition cork. Often a catalyst, such as ammonium chloride, is added to accelerate curing.

After curing, finished composition cork is cooled. Cork block is removed from the mold and cut into the sheet. Cured rod is sliced into discs.

More recently, adapting a technique employed in the veneer industry, manufacturers have been turning out logs of composition cork that can be peeled continuously into large sheets for making gaskets or similar products.

---

## Caulking Compound

### Formula No. 1

| "Vistanex" | |
|---|---|
| LM–Type MH | 80 |
| Blanc Fixe | 105 |
| Titanium Dioxide | 40 |
| Short-Fiber Asbestos | 15 |
| Graphite | 15 |
| Mineral Spirits | 85 |

### No. 2

| "Vistanex" | |
|---|---|
| LM–Type MH | 100 |
| Lithopone | 106 |
| Carbon Black | 50 |
| Short-Fiber Asbestos | 19 |
| Graphite | 19 |
| Kerosene | 106 |

Compounds of this type have been tested in an air oven for 96 hours at 180°F., in a weatherometer for twenty cycles, in 45°F. southern exposure for 20 months, and in service for over a year. The results show excellent adhesion and retention of bond with no surface hardening or tendency to crack. Shrinkage was only slight with no tendency to crack or pull away from the surface.

### No. 3
### U. S. Patent 2,847,315

| | |
|---|---|
| Asbestos Fibers | 2-5 |
| Diatomaceous Earth | 1-3 |
| Tall Oil | 1.5-4 |
| Petroleum Residue | 1.5-4 |
| Quaternary | |
| Ammonium | |
| Compound | 0.1-0.3 |
| Lead Oxide | 0.1-0.3 |
| Balata Resin | 0.1-0.3 |

### Adhesive for Silicon-Treated Surfaces
### U. S. Patent 2,716,612

| | |
|---|---|
| Water | 347 |
| Yellow Tapioca Dextrin | 650 |
| Phenol | 3 |
| Hydrogen Fluoride | |
| (60% Solution) | 500 |

### Cement for Arc Carbons
### U. S. Patent 2,692,205

| | |
|---|---|
| 75% Glucose Solution | 45-55 |
| Powdered Carbon | 45-15 |
| Ethanolamine | 5-15 |

### Electrically Conducting Adhesive
### British Patent 716,253

| | |
|---|---|
| Flaked Silver | 20 |
| Precipitated Silver | 35 |
| Glass Enamel Powder | 27 |
| "Carbitol" | 17½ |
| Nitrocellulose | ½ |

### Film-Splicing Cement
### U. S. Patent 2,697,044

| | |
|---|---|
| Acetone | 20.5 |
| Dioxane | 19.0 |
| Dichloromethane | 55.0 |
| Methanol | 3.7 |
| Nitrocellulose | 1.5 |
| Ethanol | 0.3 |

### Photographic-Film Cement
### U. S. Patent 2,697,044

| | |
|---|---|
| Acetone | 10 |
| Methylene Chloride | 35 |
| Dioxane | 19 |
| Methanol | 3 |
| Cellulose Nitrate | 1 |

### Inorganic Wood Adhesive
### Japanese Patent 5886 (1953)

| | |
|---|---|
| Magnesium Oxide | 1.0 kg. |
| Magnesium Chlorate Solution 28-30° Bé. | 1.3 liter |
| Magnesium Fluoride | 50.0 g. |

## CHAPTER III

# CERAMICS AND GLASS

Antifogging Pads for Glass

When moistened with water, antifogging pads made from the following solution deposit a thin even film that is almost invisible yet sufficient to prevent fogging.

| | |
|---|---|
| "Carbowax" Polyethylene Glycol 4000 | 22 |
| "Tergitol" Anionic 7 | 10 |
| Water | 68 |

Dissolve "Carbowax" polyethylene glycol 4000 in the water and add Anionic 7. Soak flannelette, or other fabric to be used for pads, in the solution and extract lightly so that the cloth retains about twice its weight of solution. Dry and cut into pieces 8 to 10 in. square. Before use, moisten part of the pad with water and then wipe over the glass surface.

Antidim for Glass
*Japanese Patent 1588 (1955)*

| | |
|---|---|
| Polymethylsiloxane | 8 |
| "Tween" 80 | 20 |
| Water | 72 |

---

Ferromagnetic Ceramic
*British Patent 795,269*

| | Mole % |
|---|---|
| Ferric Oxide | 47.1 |
| Alumina | 3.5 |
| Ferrous Oxide | 3.2 |
| Manganese Oxide | 21.6 |
| Zinc Oxide | To make 100.0 |

---

Refractory Coating for Graphite Molds

Formula No. 1
*U. S. Patent 2,840,480*

| | |
|---|---|
| Ball Clay | 12.0 |
| Kaolin | 18.0 |
| Alumina Cement | 60.0 |

Fused Alumina 10.0

Sodium Silicate 0.2

Sodium Carbonate 0.1

Make into a slurry with water (25 to 40%). This coating reduces carbon pickup and increases mold life. It is useful for casting uranium, nickel, and cobalt.

### No. 2
### U. S. Patent 2,857,285

Fused Alumina 95-99

Bentonite 0.5-2

Ball Clay 0.5-3.5

Soluble Silicate 0.1-0.5

All above 3.3 to 44 microns in size. Make into a slurry with water for use.

### Brick Glaze
### U. S. Patent 2,871,132

|  | % |
|---|---|
| Flint | 25.0 |
| Kaolinic Clay | 3.6 |
| Talc | 3.6 |
| Barium Carbonate | 22.6 |
| Nephelite Syenite | 5.0-25.0 |
| Colemanite | 5.0-30.0 |
| Lepidolite | 5.0-20.0 |

This fuses below 1950°F. to give a smooth glossy, continuous, film on structural clay.

### Porcelain Repair Coating
### U.S.S.R. Patent 102,544

|  | % |
|---|---|
| Alcohol-Soluble Copal | 40-50 |
| Lithopone | 5-8 |
| Alcohol | 50-60 |

### Low-Loss Ceramic Insulator
### U. S. Patent 2,839,414

| Wollastonite | 60.2 |
|---|---|
| Kentucky Old Mine No. 4 Clay | 12.7 |
| Calcium Carbonate | 3.1 |
| Titanium Dioxide | 2.2 |
| Potter's Flint | 5.3 |
| Lead Bisilicate | 16.5 |

Ball mill to pass a 325-mesh sieve, mix with water, form in the shape of a disk, dry at 230°F., and fire at 2100°F. in air, then kiln cool.

### Graphite Crucible Protective Glaze
### U.S.S.R. Patent 104,247

| Sodium Silicate | 45 |
|---|---|
| Sand | 45 |
| Fireclay | 10 |

### Zero-Shrinkage Ceramics

Two types of nonshrinking bodies have been developed which are dependent upon lead to attain this unique property. This zero shrinkage provides for easier fabrication of parts where extremely close tolerances are prerequisite.

A typical formula for the first type of body is

| Spodumene | 60 |
| Lead Bisilicate | 40 |

This body fired to 1970°F. (1077°C.) has zero absorption. The shrinkage is also zero if the optimum pressure is used in fabrication (22 tons per sq. in. using 200-mesh materials). Additions of clay up to 15% decrease the pressure required to 10 tons per sq. in. at which zero shrinkage can be obtained. This type ceramic has a dielectric constant of 6.4, a loss factor of 0.029 and a coefficient of linear thermal expansion of $2.02 \times 10^{-6}$ (77 to 842° F., 25 to 450°C.). With the optimum forming conditions the fired part has the same dimensions as the die in which it is formed.

An example of the other type of nonshrinking ceramic is

| Lead Bisilicate | 68.6 |
| Kaolin | 26.6 |
| Alumina | 4.8 |

This ceramic depends on the formation of the crystalline phase of lead alumina silicate. It is dense and has zero absorption at 1700°F. (927°C.). At 1950°F. (1066°C.) this ceramic is still dense and not overfired, thus having an unusually long firing range. There is some self-glazing at 1950°F. (1066°C.) and this self-glazing mechanism becomes pronounced at 1975°F. (1079°C.).

This ceramic has zero shrinkage with the optimum forming conditions (33 tons per sq. in. using 200-mesh materials). The dielectric constant is 5.8 and the loss factor 0.006, thus affording a Grade L-5 ceramic with zero shrinkage.

---

### Grout for Ceramic Tile
#### U. S. Patent 2,769,720
##### Formula No. 1

| White Cement | 100 |
| Pumice | 10 |
| Kaolin | 10 |
| Salt | 2½ |
| Water | 60 |

It is put in spaces between tiles. It is slow setting, has good workability and does not sag.

##### No. 2
#### U. S. Patent 2,836,502

| Limestone | 30-50 |
| Magnesium Oxide | 1-5 |
| Portland Cement | 40-60 |
| Barium Chloride | 1-5 |

Mix with water before use.

## Tile Pointing Compound
### U. S. Patent 2,838,411

| | Formula No. 1 | No. 2 |
|---|---|---|
| Portland Cement | >90.00 | 93.75 |
| Methyl Cellulose | 0.25-2.2 | 1.25 |
| Calcium Chloride | 1.50-2.5 | 2.00 |
| Aluminum Powder | — | 0.01 |
| Titanium Dioxide | — | 3.00 |
| Water | 30-40 | 35.00 |

# Chapter IV

# COSMETICS AND DRUGS

Odorless Permanent Hair
Wave Paste
*German Patent 958,501*

| | |
|---|---|
| Sodium Cysteinate | 20 |
| Ammonium α-Thiolactate | 5 |
| Ammonium Sulfite | 5 |
| Ammonium Borate | 10 |
| Tyrosine | 5 |
| Water | 55 |

Cold-Wave Neutralizer

Formula No. 1

| | |
|---|---|
| "Texapon" Q | 15.0 |
| Citric Acid | 0.5 |
| Water | 84.5 |

No. 2

| | |
|---|---|
| *a* "Kessco" X-211 | 4 |
| Polyethyleneglycol 6000 Monostearate | 6 |
| *b* Sodium Bromate | 10 |
| Water | 80 |

Heat *a* to 75°C. and mix well. Heat *b* to 75°C. and stir until the bromate is dissolved. Add *b* to *a*. Allow the mixture to cool with continuous stirring until a clear solution is formed. The solution may be bottled as soon as it becomes clear. It will develop increased viscosity on standing overnight.

No. 3

| | |
|---|---|
| Sodium Bromate | 4.0 oz. |
| Sodium Perborate, Monohydrate | 2.0 oz. |
| Sodium Tripolyphosphate | 1.0 oz. |
| Sulfonated Castor Oil | 1.0 oz. |
| Perfume Oil | 0.5 oz. |
| Distilled Water | 119.5 oz. |
| | 1 gal. |

## No. 4

a Oxone Monoper-
  sulfate (Sealed
  in Packaged Alu-
  minum Foil En-
  velope) 10 gr.
b Sodium Bromate 5 oz.
  Distilled Water 122 oz.
  Perfume 1 oz.
  _____
  1 gal.

For use on one permanent, 4 oz. of hot water are mixed together with the oxone from the envelope. Dissolve all the powder, add 2 oz. of liquid to this mixture, then add enough water to make 1 pt. Leave the mixture on the rods about 10 minutes, then complete the wave by rinsing with tepid water. This is especially successful for difficult or fine, thin hair; it may be colored if desired. Unless a bleach-fast color is used, however, the oxone will bleach out the color upon addition to the liquid part of the neutralizer.

## No. 5
(One-Minute, One-Step, Sudsy)

Sodium Bromate 5.00 oz.
Sodium Lauryl
  Sulfate 2.00 oz.
Perfume 1.00 oz.
Ferric Ammonium
  Citrate (Green) 0.50 oz.
Cetyl Trimethyl
  Ammonium
  Bromide 0.25 oz.
Distilled Water 119.00 oz.
  _____
  1 gal.

For this formula, the bromate is dissolved in the warmed distilled water and to this the other ingredients are added, one at a time. The cetyl trimethyl ammonium bromide does not aid in the neutralization process, but is an aid to the patron's sensitive scalp. It also helps recondition the hair.

Formula 3 is far ahead of the others on the market. The ferric ammonium citrate works as a catalyst in conjunction with the bromate and digs deep down, sudsing and foaming around the rods, yet it is thin enough in solution for good adsorption. This neutralizer assures permanence of the wave. Moreover, it saves much time for the operator, because of its quick astringent action on the rod. Use about 3 oz. of this solution with enough water to make 1 pt. of finished neutralizer. This is poured over the curls on rods at the shampoo bowl, and should remain on the rods 1 to 2 minutes, a procedure familiar to the average hairdresser.

## Clean Liquid Hair Shampoo

| | |
|---|---|
| "Duponol" EP | 40 |
| "Kessco" X-209 | 6 |
| Sodium Chloride | 1 |
| Water | 50 |
| Polyethyleneglycol 400 | |
| Distearate | 3 |

Dissolve the sodium chloride in the water, then add all of the other ingredients. Heat to 70°C. and mix gently until a clear solution is formed. The formula may be bottled while hot.

---

## Liquid Cream Shampoo

### Formula No. 1

| | |
|---|---|
| Polyethyleneglycol 400 | |
| Distearate | 3 |
| Ethyleneglycol | |
| Monostearate 70 | 1 |
| Propyleneglycol | 2 |
| Water | 49 |
| Sodium Lauryl Sulfate | 45 |

Heat all of the ingredients to 65 to 70°C. Stir slowly until the mixture cools to 35°C.

### No. 2

| | |
|---|---|
| a "Amerchol" L-101 | 3.0 |
| Cetyl Alcohol | 3.0 |
| Propyleneglycol | 2.0 |
| Beeswax | 0.4 |
| b Sodium Lauryl Sulfate | 0.4 |
| Water | 14.2 |
| c Sodium Lauryl | |
| Sulfate | 10.0 |

| | |
|---|---|
| Water | 65.5 |
| "Veegum" | 1.5 |
| Preservative | To suit |

Add b to a at 75°C. and mix slowly until emulsified. Add c. Mix slowly until cool.

---

## Nonfoaming Shampoo

| | |
|---|---|
| "Nonic" 218 | 20 |
| Lanolin | 5 |
| "Arlacel" C | 1 |
| Sodium Carboxymethyl- | |
| cellulose (High | |
| Viscosity) | 2 |
| Water | 72 |
| Preservative, | |
| Perfume, etc. | To suit |

Mix the oils together and then add to the carboxymethylcellulose-water gel with vigorous stirring. The viscosity may be varied with the type and quantity of carboxymethylcellulose used.

---

## Solid Cream Shampoo

### Formula No. 1

| | |
|---|---|
| a Polyethyleneglycol | |
| 400 Distearate | 4.1 |
| Lanolin | 0.3 |
| "Stenol" | 0.5 |
| Stearic Acid | 5.1 |
| Pure Glyceryl | |
| Monostearate | 2.5 |
| b Potassium Hydroxide | 0.5 |
| Water | 19.0 |

*c* Sodium Lauryl
    Sulfate        68.0

Heat parts *a* and *b* at 90 to 95°C. in separate containers. Add *b* to *a* and agitate until the mixture cools to 60°C. Heat part *c* to 60°C. and add it to the mixture of *a* and *b*. Allow the final mixture to cool to 45 to 47°C. and then package.

### No. 2

| | |
|---|---|
| "Amerchol" L-101 | 4.0 |
| "Modulan" | 2.0 |
| Stearic Acid | 8.0 |
| Neutral Glyceryl Monostearate | 2.0 |
| "Veegum" | 1.5 |
| "Duponol" WA Paste | 60.0 |
| Water | 20.0 |
| Triethanolamine | 1.5 |
| Sodium Hydroxide | 0.5 |
| Potassium Hydroxide | 0.5 |
| Preservative | To suit |

Mix all ingredients together slowly while heating to 85°C. Continue mixing until cool and remix the following day.

---

### Cream Rinse

#### Formula No. 1

| | |
|---|---|
| "Pendit" CA | 7.00 |
| Polyethyleneglycol 400 Distearate | 1.00 |
| Sodium Chloride | 0.25 |
| Perfume | As desired |
| Color | As desired |
| Water | To make 100.00 |

Heat and mix the "Pendit" CA and the polyethyleneglycol 400 distearate until a thick jellylike material having a clear or translucent appearance is formed. Dissolve the salt in a small portion of the water. Heat the remaining water to 160°F. Add the heated water to the first two ingredients. Mix at 160°F. for 15 to 30 minutes. Cool the batch slowly with moderate agitation. Between 150 and 140°F., add the salt solution and the color. Incorporate the perfume at 120°F. Some increase in agitation may be necessary at approximately 90°F. at which point the composition tends to gel. Continue the agitation until the batch cools to room temperature. An opaque, viscous liquid is formed. Pearlescence develops in the formulation after standing for a few days.

*Directions for Use*

Pour approximately 1 oz. of cream rinse into an 8-oz. container. Fill the container with tepid water. Stir well and pour through the hair. Subsequent rinsing with plain water is optional depending upon the amount of conditioning desired.

## Variations in the Formula

The salt concentration may be varied between 0.2 and 0.5%. Increases in the salt concentration will result in increases in the viscosity of the formulation.

Further increases in the viscosity of the rinse may be obtained by increases in the concentration both of the polyethylene glycol 400 distearate (up to 3%) and of the "Pendit" CA (8-9%). If such increases are made, more than 0.5% salt may be required.

### No. 2
### (Acid)

| | |
|---|---|
| Cetyl Alcohol | 0.5 |
| "Emcol" E-607S | 2.0 |
| Sodium Chloride | 0.8 |
| Water | 96.7 |
| Perfume | To suit |

### No. 3

| | |
|---|---|
| "Cetrimide" | 2 |
| Perfume | To suit |
| Water | To make 100 |

### No. 4

| | |
|---|---|
| "Triton" X 400 | 7.0 |
| Polyethyleneglycol 400 Distearate | 1.0 |
| Cetyl Alcohol NF | 0.5 |
| Color and Perfume | 0.5 |
| Water | 91.0 |

The emulsion is made at a temperature of 70 to 75°C. with constant stirring, and the perfume is added as the temperature drops at not higher than 40°C. Stirring should be continued until the emulsion reaches room temperature.

---

### Odor Stable Shampoo
### U. S. Patent 2,773,834

| | Formula No. 1 (Cream) | No. 2 (Paste) |
|---|---|---|
| Sodium Lauryl Sulfate | 20.55 | 30.00 |
| Sodium Chloride | 0.89 | 12.00 |
| Lanolin | 0.47 | 0.50 |
| Sodium Benzoate | 0.24 | 0.25 |
| Methyl-p-Hydroxy Benzoate | 0.14 | 0.15 |
| Stearic Acid | 6.58 | — |
| Potassium Hydroxide (34.2%) | 4.04 | — |
| Lauric-Myristic Diethanolamide | 1.50 | — |
| Monomethyloldimethyl Hydantoin | 0.10 | 0.50 |
| Disodium Phosphate | — | 1.50 |
| Titanium Dioxide | — | 0.40 |
| Perfume | 0.30 | 0.40 |
| Water | To make 100.00 | To make 100.00 |

## Temporary Hair Coloring

### Formula No. 1
#### (Blond)

| | |
|---|---|
| Citric Acid | 80.00 |
| Modified Lanolin | 0.50 |
| Perfume Oil | 0.25 |
| Alcohol or Acetone | 25.00 |
| Butyl or Ethyl Citrate | 3.00 |
| F.D. & C. Yellow No. 5 | 7.50 |
| F.D. & C. Yellow No. 6 | 7.50 |

### No. 2
#### (Brown)

| | |
|---|---|
| Crystalline Citric Acid | 80.00 |
| Modified Lanolin | 0.50 |
| Perfume Oil | 0.25 |
| Alcohol or Acetone | 25.00 |
| Citrate Ester or Butyleneglycol | 3.00 |
| F.D. & C. Yellow No. 6 | 3.00 |
| D. & C. Brown No. 1 | 1.00 |

### No. 3
#### (Black)

| | |
|---|---|
| Citric Acid | 80.00 |
| Modified Lanolin | 0.50 |
| Perfume Oil | 0.25 |
| Alcohol or Acetone | 25.00 |
| Butyl or Ethyl Citrate | 5.00 |
| D. & C. Brown No. 1 | 4.00 |
| F.D. & C. Green No. 2 | 2.00 |
| F.D. & C. Violet No. 1 | 3.00 |
| D. & C. Black No. 1 | 10.00 |
| F. D. & C. Red No. 1 | 1.50 |

Manufacture of the crystal hair rinse consists of coating the crystalline citric acid with an alcoholic solution of the other ingredients, followed by drying.

Because of the hygroscopic nature of the product, waterproof or water-repellant packaging should be used. Materials include laminated glassine, saran, saran-coated foil, polyethylene or polyethylene bonded to glassine, foil or paper. The film is formed into packets or envelopes on fully automatic or semi-automatic equipment which fills 8 to 15 gr. of the crystals into the container and crimps or heat seals both ends. Packets should be tested by immersing a number of them in warm water and then making the required changes in the sealing operation.

---

### Liquid Tinting Hair Rinse

| | |
|---|---|
| "Ammonyx" 4 | 25.0 |
| "Polawax" | 0.5 |
| "Tegacid" Regular | 6.5 |
| "Lanogel" 41 | 1.0 |
| Perfume | 0.5 |
| Urea | 5.0 |
| F.D. & C. Violet No. 1 | 5.0 |
| Water | 156.0 |
| F.D. & C. Blue No. 2 | 2.0 |
| Nigrosine | 10.0 |

The formula is given as parts by weight; by the use of proper color combinations, a full range of shades may be developed.

The "Ammonyx," "Tegacid," "Lanogel," and "Polawax" are heated together to 80°C. Half of the water is heated to 85°C. with the urea. The rest of the water is mixed with the certified colors and heated to 70°C. The urea solution is slowly poured into the melted fats with constant agitation and allowed to mix until the temperature drops to 70°C. The dye solution is added and mixing continued until the temperature reaches 45°C., when the perfume is added. Continue agitation with external cooling until room temperature is reached. Fill cold, using low vacuum in the filler.

### White Henna

| | |
|---|---|
| Magnesium Peroxide | 50 |
| Magnesium Carbonate | 50 |

### Pine-Oil Bubble Bath

| | |
|---|---|
| Pine-Needle Oil | 20 |
| Isopropyl Alcohol | 10 |
| "Texapon" Q (Surfactant) | 70 |

### Paste Cuticle Remover
*German Patent 939,048*

| | |
|---|---|
| Methylcellulose | 5.0 |
| Triethanolamine | 15.0 |
| Potassium Hydroxide | 0.1 |
| Alcohol | 25.0 |
| Water | 54.9 |

### Preshave Lotion

#### Formula No. 1

| | |
|---|---|
| Diisopropyl Sebacate | 20 |
| Ethyl Alcohol | 80 |

#### No. 2

| | |
|---|---|
| Isopropyl Myristate | 20 |
| Ethyl Alcohol | 80 |

#### No. 3

| | Concentrate A | Concentrate B |
|---|---|---|
| Cetyl Alcohol | 2.4 | 1.0 |
| Hydrous Lanolin | 1.5 | 2.0 |
| Heavy Mineral Oil | 4.8 | 5.0 |
| "Avitex" C | 16.2 | 30.0 |
| "Avitex" SF | 27.0 | 15.0 |
| "Triton" X-100 | 3.0 | 3.0 |
| Stearic Acid | 1.0 | — |
| Glycerin | — | 5.0 |
| Perfume | 0.2 | 0.2 |
| Water | 43.9 | 38.8 |
| Concentrate (A or B) | 90 | |
| Dichlorodifluoromethane | 4 | |
| Dichlorotetrafluoroethane | 6 | |

The stability of the emulsions is improved by using perfume oils soluble in the propellants.

To prepare this shaving cream, the cetyl alcohol, lanolin, mineral oil, and the stearic acid or glycerin are heated together to

70°C. The foaming and wetting agents are mixed in half the water and are also heated to 70°C. The oil phase is added to the water solution with stirring, then the remainder of the water is added, mixed well, and the perfume added. The cream is then stirred again until cool.

### No. 4

| | |
|---|---|
| *a* Stearic Acid | 1.0 |
| Cetyl Alcohol | 2.4 |
| Hydrous Lanolin | 1.5 |
| Heavy Mineral Oil | 4.8 |
| *b* "Avitex" C | 16.2 |
| "Avitex" SF | 27.0 |
| "Igepal" CO-630 | 3.0 |
| *c* Water | 43.9 |

Heat *a* to 71°C., mix *b* with half of *c* and heat to 71°C. Add *a* to *b* with stirring. Add the remainder of *c* and mix the batch thoroughly. Then the mixture is allowed to cool to 50°C., perfumed, and then cooled to room temperature. It is reworked at room temperature and packaged with suitable propellants.

### No. 5

| | | |
|---|---|---|
| *a* Deionized Water | | |
| To make 500 gal. | | |
| "Sorbo" | 160 lb. | |
| Triethanola- | | |
| mine | 169 lb. | |
| KOH Pellets | 7 lb. | 8 oz. |
| *b* Polyvinyl- | | |
| pyrrolidone | 10 lb. | |
| *c* Stearic Acid | 262 lb. | |
| Coconut | | |
| Fatty Acid | 75 lb. | |
| Cetyl Alcohol | 61 lb. | |
| Petrolatum | 94 lb. | |
| Perfume | 5 lb. | 8 oz. |

---

### Aerosol Shave Creams

| | Formula No. 1 | No. 2 |
|---|---|---|
| "Modulan" | 0.8 | 0.8 |
| Myristic Acid | 2.0 | — |
| Stearic | | |
| Acid, TP | 6.0 | 6.0 |
| Oleic Acid | — | 2.0 |
| Cetyl Alcohol | 0.5 | 0.5 |
| "Sorbo" | 3.0 | 3.0 |
| Triethano- | | |
| lamine | 4.0 | 4.0 |
| "Tween" 20 | 2.5 | 2.5 |
| "Tween" 80 | 2.5 | 2.5 |
| Neutral Soap | 1.0 | 1.0 |
| Borax | 0.1 | 0.1 |
| Perfume | 0.3 | 0.3 |
| Distilled | | |
| Water | 77.3 | 77.3 |

Above concentrate
92.0 %by wt.
"Genetron" 114A/12
(43:57) 8.0% by wt.

### No. 3

| | |
|---|---|
| *a* Triple-Pressed | |
| Stearic Acid | 6.00 |
| Diglycol Stearate S | 1.00 |

Polyethyleneglycol
   600 Monostearate   2.25
Isopropyl Myristate   1.25
b "Sorbo" (70%
   Solution)   3.00
Triethanolamine   2.75
Distilled Water   83.20
c Coconut Oil   0.25
Perfume   0.30

Heat a and B separately to 80°C. Add b to a slowly with constant agitation, cool to room temperature, and add c to mix.

For fill:   Above
        concentrate 92%
   "Genetron" 12/114a
       (57:43)   8%

Heat the water to 160°F. Add the triethanolamine, potassium hydroxide, "Sorbo" and maintain temperature. Mix well. Dissolve b. in 25 gal. of water and add to a. Heat c to 165°F. and mix. Then add the oils to the water mixture and stir for twenty minutes. Begin to cool the emulsion and agitate at the same time. Add perfume at 105 to 110°F. Cool the shaving cream to room temperature.

The above product is charged with 10% propellant-dichlorodifluoromethane or a mixture of dichlorodifluoromethane and tetrafluorodichloroethane.

## Brushless Shaving Cream

### Formula No. 1

| | |
|---|---|
| Glycerol | 1.7 |
| White Mineral Oil* | 2.0 |
| Butyl Stearate | 2.0 |
| Stearic Acid | 13.4 |
| Triethanolamine | 0.7 |
| Glycerol<br>  Monostearate S.E. | 3.5 |
| Preservative | 0.1 |
| Water | 76.6 |

* Saybolt Viscosity at 100°F. = 65 to 75 sec.

Heat all the ingredients, in one container, to 90 to 95°C. Allow the mixture to cool to 35°C. with stirring.

### No. 2

| | |
|---|---|
| Diglycol Stearate S.E. | 10.0 |
| Glycerol | 2.5 |
| White Mineral Oil* | 2.0 |
| Butyl Stearate | 2.5 |
| Stearic Acid | 10.0 |
| Spermaceti | 5.0 |
| Preservative | 0.1 |
| Water | 67.9 |

* Saybolt Viscosity at 100°F. = 65 to 75 sec.

Heat all the ingredients, in one container, at 90 to 95°C. Allow the mixture to cool to 35°C. with stirring.

## No. 3

a Triple-Pressed

| | |
|---|---|
| Stearic Acid | 17.5 |
| "Emulphor" VT 679 | 2.5 |
| "Igepal" CO-710 | 1.0 |
| White Petrolatum | 5.0 |

b
| | |
|---|---|
| Propyleneglycol | 5.0 |
| Water | 67.0 |

c 28% Ammonia Water 2.0

Melt a and heat to 80°C. Heat b to 82°C. Charge c to b quickly and add to a with moderate agitation. Stir until the batch cools to 55°C. Perfume and stir slowly until the temperature is 45°C. Stir for 2 to 3 days at room temperature. Remix and pack in suitable containers.

## No. 4

a
| | |
|---|---|
| Stearic Acid | 24.0 |
| "Arlacel" 165 (Acid-Stable g.m.s.) | 3.0 |
| Glycerol | 6.0 |
| Stearyl Alcohol | 1.0 |
| Petrolatum | 8.0 |
| Lanolin | 2.0 |

b
| | |
|---|---|
| Triethanolamine | 1.1 |
| Borax | 1.0 |
| Water | 53.9 |
| Preservative | To suit |

Heat a to 80°C. Heat b to 82°C. Add b to a with agitation. Stir until cream has set. Allow to stand overnight. Mix well. Pack.

## Aftershave Cream

### Formula No. 1

| | |
|---|---|
| "Cetrimide" | 0.5 |
| Cetostearyl Alcohol | 4.5 |
| Cetoxypolyethylene Oxide Condensate | 1.0 |
| Pure Glyceryl Monostearate | 10.0 |
| Isopropyl Myristate or Mineral Oil | 10.0 |
| Menthol | 0.1 |
| Perfume | To suit |
| Water | To make 100.0 |

Similar preparations used also as preshave creams enable exceptionally smooth, clean shaves to be obtained.

### No. 2

| | |
|---|---|
| "Cetrimide" | 1 |
| Aluminum Lactate | 1 |
| Urea | 3 |
| Ethyl Alcohol | 25 |
| Menthol | To suit |
| Water | To make 100 |

---

### Rose Bouquet Base

| | |
|---|---|
| Geraniol | 30 |
| Citronellol | 25 |
| Phenethyl Alcohol | 25 |
| Linalool | 5 |

This skeleton formula can be varied and modified to suit a

wide variety of requirements, as for example a superfine rose for extracts or cosmetics can be obtained by replacing citronellol by rhodinol, geraniol ex Java citronella by geraniol palmarosa, given added strength by traces of nonyl and lauryl aldehydes and tone by including geranium bourbon and alpha ionone, and so forth. Interesting variations may be obtained by the inclusion of the esters (acetates, butyrates, etc.) of the rose alcohols—geraniol, rhodinol, citronellol and phenylethyl alcohol, and numerous others will come to mind as the experiments proceed. Excellent fixation can be achieved by using phenylacetic acid, musk ketone, sandalwood and trichlormethyl phenyl carbinyl acetate in small quantities. The results obtained by the addition of 1 to 2% of genuine otto or rose or a high-grade imitation will easily justify the extra cost involved. For the less expensive soap perfume, special "soap" grades of the various aromatics are readily available and these together with a fairly generous percentage of diphenyl oxide and citronella oil Java or Ceylon will provide a good basis for further work.

## Rose Bulgarian-Type Soap Perfume

| | |
|---|---|
| Bulgaryol I | 450 |
| Tetrahydro Geraniol | 100 |
| Phenyl Ethyl Alcohol Supreme | 100 |
| I-Rhodinol | 100 |
| Rosatol Extra | 50 |
| Essence Absolue Reseda 10% | 10 |
| Cognac Oil Rect. 10% | 10 |
| Geraniol | 50 |
| Citronellol | 50 |
| Cananga Oil | 10 |
| Aldehyde C 9 (Nonyl) 10% | 10 |
| Methyl Iridone 100% | 40 |
| Oil of Geranium Bourbon Ia | 10 |
| Dimethyl Hydroquinone | 10 |

## Tuberose Perfume Base

| | |
|---|---|
| Linalool | 25 |
| Tuberose Absolute | 5 |
| Methyl Salicylate | 5 |
| Isobutyl Salicylate | 3 |
| Balsam Peru | 10 |
| Methyl Anthranilate | 3 |
| Benzyl Propionate | 5 |
| Benzyl Alcohol | 5 |
| Methyl Laurate | 5 |
| Coumarin | 3 |
| Isoeugenol | 1 |
| Cananga Oil | 10 |
| Geraniol | 10 |

| | |
|---|---|
| Siam Benzoin | 5 |
| Ylang-Ylang Oil | 5 |

---

## Tuberose Flowers Soap Perfume

| | |
|---|---|
| Amyl Cinnamic Aldehyde | 25 |
| Phenylethyl Alcohol | 20 |
| Oil Tolu Distilled | 20 |
| Mimosa Absolute | 20 |
| Rhodinol | 10 |
| Methyl Ionone | 10 |
| Ylang Bourbon Extra | 12 |
| Tuberyl Acetate | 6 |
| Methyl Salicylate | 4 |
| Oil Celery ex Seeds | 2 |
| Hydroxycitronellal | 10 |
| Methyl Anthranilate | 10 |
| Amyl Salicylate | 4 |
| "Palatone" (5% in Phenylethyl Alcohol) | 2 |
| Isoeugenol | 2 |
| Cumin Ketone | 2 |
| Cuminic Aldehyde (10% in DEP) | 2 |
| Hexyl Cinnamic Aldehyde | 5 |
| Tolyl Acetate | 4 |
| Phenylpropyl Alcohol | 2 |
| Civettiane | 6 |
| Heliotropine | 8 |
| Phenylethyl Isobutyrate | 2 |
| "Glycidiol" | 4 |
| Aldehyde, C-14 (10% in DEP) | 1 |
| "Veronel" Aldehyde (10% in DEP) | 2 |

| | |
|---|---|
| Cinnamyl Isobutyrate | 4 |
| Aldehyde C-12 MNA (10% in DEP) | 1 |

---

## Jasmine Base

| | |
|---|---|
| Benzyl Acetate | 40 |
| Linalool | 10 |
| Amyl Cinnamic Aldehyde | 15 |
| Cinnamyl Alcohol | 5 |
| Phenethyl Alcohol | 5 |

Numerous "build ups" of this basic formula will soon be found with a little experimentation. A fine sweet note is that most generally favored in jasmine but various shadings toward fruity and rosy-lilac are also quite acceptable. A few suggestions for the elaboration of the above theme would include the addition of small quantities of other benzyl esters, in particular the butyrate and propionate, together with linalyl acetate or bergamot oil to add a sweet freshness. Traces of methyl anthranilate and indole (not more than 1%, generally less) may be used if desired but great care should be observed with a view to possible discoloration at a later stage. Once again a small percentage of natural jasmin or a good imitation will appreciably improve the composition providing the costs will permit. The sweet note can be further strengthened by the use of musk ketone as a fixative and fruity background can be introduced by traces of so called aldehydes C-14 and C-16 in 10% solution. Other products from which interesting modifications may be obtained include, hydroxy citronellal, ylang ylang, p-cresyl esters, sweet orange oil and many more. For soaps, good fixation is essential and the resinoids of styrax and benzoin will be found particularly useful.

Fougere perfumes are based on the powerful tones of oakmoss blended with the pleasing freshness of a lavender citrus complex and find appeal especially in the more masculine type of preparations.

### Jasmine Perfume Base

| | |
|---|---|
| Benzyl Acetate | 15.0 |
| Jasmine Absolute | 10.0 |
| Amyl Cinnamic Aldehyde | 6.0 |
| Indol (100%) | 0.3 |
| Linalool | 8.0 |
| Benzyl Alcohol | 15.0 |
| Alpha Ionone | 5.0 |
| Benzyl Salicylate | 5.0 |
| Dimethyl Benzylcarbinyl Acetate | 4.0 |

| | |
|---|---|
| Rhodinol | 5.0 |
| Rose de Mai | 2.0 |
| Methyl Anthranilate | 4.0 |
| Phenethyl Alcohol | 10.0 |
| Paracresylphenyl Acetate | 2.0 |
| Isojasmone | 1.0 |
| Rhodinol Isobutyrate | 0.5 |
| Siam Benzoin (50%) | 7.2 |

### Jasmine Soap Perfume

| | |
|---|---|
| Indol | 5 |
| Civettal | 15 |
| Linalool ex Bois de Rose | 50 |
| Linalyl Acetate | 60 |
| Hydroxycitronellal "Naarden" | 30 |
| Aldehyde C-14 | 10 |
| Aldehyde C-18 | 5 |
| Benzyl Isobutyrate | 15 |
| Benzyl Proponiate | 425 |
| Alpha Hexyl-Cinnamic Aldehyde | 150 |
| Nerol | 50 |
| Petitgrain Oil Nardenised | 40 |
| Mandarin Oil Nardenised | 20 |
| Ylang-Ylang Oil Nardenised | 60 |
| Cypress Oil Nardenised | 10 |
| Angelica Root Oil Nardenised | 5 |
| Basil Oil Nardenised | 5 |

African Geranium Oil
Nardenised ........................ 25
Spanish Rosemary Oil
Nardenised ........................ 20

---

Golden Jasmine Soap Perfume

Musk Ketone ...................... 30
Benzyl Propionate ............... 30
Tolyl Acetate .................... 48
Linalool ......................... 60
Ylang Bourbon Extra ............. 60
Geranyl Acetate ................. 60
Phenylethyl Alcohol ............. 60
Terpineol ........................ 60
Hydroxycitronellal .............. 90

---

Amber for Chypres E.054

Amyl Salicylate .................. 6
Aldehyde C-12 MNA
(10% in DEP) .................... 5
Isobutyl Phenylacetate .......... 6
Phenylacetic Acid ............... 10
Tuberyl Acetate (Verona
Chemical) ....................... 10
Civettiane (Perfumery
Associates) ..................... 14
Benzyl Propionate ............... 15
Ylang Bourbon ................... 20
Alpha Ionone .................... 20
Yara Yara ....................... 20
Heliotropine .................... 20
Hydratropyl Alcohol ............. 20
Absolute Castoreum .............. 20
Aldehyde "Veronol"
(10% in DEP) .................... 20
Oil Myrrh Distilled ............. 20

Musk Xylol ...................... 25
Musk Ketone ..................... 30
South American Petit-
grain ........................... 30
Citronellyl Oxyacetalde-
hyde ............................ 30
California Oil Lemon ............ 40
Amyl Cinnamic Aldehyde .......... 42
Musk Ambrette ................... 45
Oil Bitter Orange ............... 60
Methyl Cinnamate ................ 60
Coumarin ........................ 65
Methyl Ionone ................... 55
Oil Labdanum Abso-
lute (10% in DEP) ............... 70
Ethyl Vanillin .................. 80
Phenylethyl Phenylace-
tate ............................ 87

The amber E.054 is useful in all branches of the LFC group, and particularly for the preparation of men's cologne fragrances. It adds a touch of "luxury" to various floral fragrances such as carnation and jasmine.

---

Violet Base

Methyl Ionone ................... 30
Alpha Ionone .................... 15
Bergamot ........................ 10
Heliotropin ..................... 4
Methyl Heptin Carbonate ......... ½

If cost is of primary importance ordinary 100% ionone may be used, together with a good synthetic bergamot either as an extender or a complete replacement for the natural oil. To obtain a sweet floral background about 3 to 5% of ylang-ylang oil and smaller amounts of the natural absolutes of cassie, mimosa, jasmine and violet leaves will be most beneficial. Again if cost prohibits the inclusion of these costly natural materials, many excellent specialty replacements are available and can be used to great advantage. For cheaper versions use may be made of such aromatics as anisic aldehyde, benzyl acetate, terpineol and its esters, phenylethyl alcohol and esters, also traces of lauryl aldehyde in 10% solution. Fixation will largely be taken care of by the natural absolutes but in their absence and also to supplement them, about 1 to 5% of sandalwood, orris resinoid, musk ambrette and ketone, or vanillin will prove of value. Violet soap perfumes may contain cananga synthetic in place of ylang ylang, "soap" grades of the ionones along with fairly generous quantities of the resinoids of benzoin, Peru balsam and styrax; also musk xylol and sandalwood and vetivert residues.

The perfume of lily of the valley is probably the most exquisite of all of the florals and care should be taken to produce a compound which in dilution is both delicate and elusive but yet at the same time sweet and persistent.

---

### Cyclamen des Alpes

| | |
|---|---|
| "Alpine Violet" Standard | 15 |
| Hydroxycitronellal | 30 |
| Hydroxycitronnellyl Dimethyl Acetal | 10 |
| Cumin Ketone | 2 |
| Rhodinol | 8 |
| Linalool ex Bois de Rose | 15 |
| Ylang Bourbon Extra | 5 |
| Italian Oil Lemon | 6 |
| Benzyl Acetate | 10 |
| Benzyl Propionate | 2 |
| Amyl Cinnamic Aldehyde | 2 |
| Coumarin | 1 |
| Musk Ketone | 4 |
| Methyl Ionone | 5 |
| Alpha Ionone | 4 |
| Alpha Terpineol (Best Grade) | 10 |
| Algerian Oil Geranium | 5 |
| Phenylethyl Propionate | 1 |
| Hydratropyl Alcohol | 3 |
| Cinnamic Alcohol | 4 |
| Tolyl Alcohol | 5 |
| Linalyl Acetate | 5 |
| South American Petitgrain | 4 |

| | |
|---|---|
| Aldehyde C-14 (10% in DEP) | 1 |
| Phenylacetaldehyde Trimethylene Glycol Acetal | 4 |
| Benzyl Alcohol | 14 |
| Amyl Cinnamic Aldehyde | 90 |
| Benzyl Acetate | 90 |
| Benzyl Salicylate | 174 |
| Diethylphthalate | 4 |

Amber Perfume Base

| | |
|---|---|
| Labdanum Absolute | 10 |
| Musk Ketone | 10 |
| Vanillin | 10 |
| Methyl Ionone | 20 |
| Rose Base | 10 |

The basic formula given above will be found most suitable for the heavier types but by alteration and substitution all tones will suggest themselves as experiments proceed. Lighter compositions will result from the use of bergamot oil, linalyl acetate or traces of lemon or tangerine oil together with an increased quantity of the rose base or such rose alcohols as rhodinol, geraniol or phenylethyl alcohol. If the finished compound is required as a base and a semisolid substance is acceptable, then the deeper tones could be considerably strengthened by further quantities of musk and vanillin, plus a fairly heavy percentage of coumarin, benzyl isoeugenol, vetivenol and sandalwood. A liquid product can be obtained by adding quite large quantities of 3% tincture of musk, civet or castoreum. For a variety of interesting compositions, experiments with the following are recommended, oil of Peru balsam, clary sage and oakmoss, also some of the many amber type specialties readily available on the market today.

Magnolia

| | |
|---|---|
| Hydroxylcitronellal | 40 |
| Methyl Ionone | 30 |
| Phenylethyl Alcohol | 30 |
| Musk Ketone | 50 |
| Ylang Bourbon Extra | 25 |
| Amyl Cinnamic Aldehyde | 10 |
| Benzyl Acetate | 15 |
| Oil Bergamot | 18 |
| Isoeugenol | 12 |
| "Alpine Violet" Standard | 20 |
| Sandalwood Oil | 10 |
| Vetyverol | 15 |
| Musk Ambrette | 10 |
| Linalyl Acetate | 25 |
| Hydroxycitronellal-Methyl Anthranilate Base | 20 |
| Cinnamic Alcohol | 10 |
| Dimethyl Anthranilate | 5 |

| | |
|---|---|
| Ethyl Vanillin | 5 |
| Rhodinol | 30 |
| Aldehyde C-14 (10% in DEP) | 5 |
| Phenylethyl Propionate | 5 |
| Lemon Oil Italian | 15 |
| Oak Moss Resin | 1 |
| Aldehyde C-12 MNA (10% in DEP) | 2 |
| Indol (10% in DEP) | 8 |
| Styrallyl Propionate | 4 |
| Rose Otto | 4 |
| Citronellyl Oxyacetaldehyde | 3 |
| Diethyl Phthalate | 23 |

## Mignonette

| | |
|---|---|
| Methyl Ionone Standard | 110 |
| "Resedalia" (Diphenylethyl Acetal) | 40 |
| Benzyl Acetate | 10 |
| Oil Bergamot Natural | 50 |
| Phenylpropyl Alcohol | 30 |
| East Indian Oil Sandalwood | 50 |
| South American Oil Petitgrain | 20 |
| Ylang Bourbon Extra | 10 |
| Oil Sweet Basil | 12 |
| Isobutyl Salicylate | 8 |
| Galbanum Absolute | 1 |
| Labdanum Resin | 1 |
| Lavender Absolute | 1 |
| Methyl Octine Carbonate (100%) | 2 |

| | |
|---|---|
| Amyl Cinnamic Aldehyde | 50 |
| Phenylpropyl Acetate | 30 |
| Cinnamic Alcohol | 20 |
| Heliotropine | 20 |
| Hydroxycitronellal | 50 |
| Ethyl Decine Carbonate (100%) | 5 |
| Benzyl Isoeugenol | 30 |

## Muguet Extra

| | |
|---|---|
| Hydroxycitronellal | 35 |
| Cumin Ketone | 5 |
| Amyl Salicylate | 2 |
| Phenylacetaldehyde Propyleneglycol Acetal | 6 |
| Phenylethyl Alcohol | 20 |
| Hydratropyl Alcohol | 5 |
| Linalool | 15 |
| Benzyl Acetate | 8 |
| Rhodinol | 70 |
| Indol (10% in DEP) | 10 |
| Cuminic Aldehyde (10% in DEP) | 2 |
| Amyl Cinnamic Aldehyde | 8 |
| Phenylethyl Propionate | 3 |
| Algerian Geranium | 6 |
| Alpha Ionone | 5 |
| Musk Xylol | 10 |
| Musk Ketone | 2 |
| Diethyl Phthalate ("DEP") | 13 |

## Muguet de Mai

| | |
|---|---|
| Hydroxycitronellal | 30 |
| Rhodinol | 15 |
| Heliotropine | 10 |
| Phenylethyl Alcohol | 25 |
| Hydratropyl Alcohol | 5 |
| Linalool | 5 |
| Linalyl Acetate | 12 |
| Benzyl Acetate | 5 |
| Tolyl Acetate | 8 |
| Terpineol | 3 |
| Cumin Ketone | 2 |
| Ylang Absolute | 1 |
| Italian Oil Lemon | 1 |
| Dimethyl Octanyl Acetate | 2 |
| Anisic Aldehyde | 1 |

## Lavender

| | |
|---|---|
| Coumarin | 50 |
| Tonka Absolute | 20 |
| Orris Concrete | 30 |
| Linalyl Acetate | 100 |
| Geranyl Acetate | 30 |
| Nerol | 40 |
| African Geranium Oil Nardenised | 30 |
| Lemon Oil Nardenised | 10 |
| Mandarin Oil Nardenised | 40 |
| Bay Oil Nardenised | 30 |
| Basin Oil Nardenised | 10 |
| Petitgrain Oil Nardenised | 40 |
| Clary Sage Oil Nardenised | 60 |

Spanish Spike Oil

| | |
|---|---|
| Nardenised | 60 |
| Dalmatian Rosemary Oil Nardenised | 80 |
| Lavandin Oil Nardenised | 300 |

## English Type Lavender

| | |
|---|---|
| Ambroma "Naarden" | 30 |
| Musk Ketone | 20 |
| Nardipanol "Naarden" | 20 |
| Coumarin | 60 |
| Peru Balsam | 30 |
| Tolu Balsam | 30 |
| Labdanum Clair | 10 |
| Citronellyl Formate | 30 |
| Linalyl Acetate | 50 |
| Cinnamic Alcohol | 30 |
| Patchouli Oil Nardenised | 20 |
| Bergamot Oil Nardenised | 40 |
| Clary Sage Oil Nardenised | 60 |
| Petitgrain Oil Nardenised | 30 |
| African Geranium Oil Nardenised | 30 |
| Thyme Oil Nardenised | 10 |
| Basil Oil Nardenised | 10 |
| Dalmatian Rosemary Oil Nardenised | 60 |
| Sandalwood Oil E. I. Nardenised | 40 |
| Lemongrass Oil Nardenised | 10 |

| | |
|---|---|
| Mandarin Oil Narden- | |
| ised | 20 |
| Spanish Spike Oil Nar- | |
| denised | 50 |
| Bay Oil Nardenised | 40 |
| Lavandin Oil Nardin- | |
| ised | 70 |
| Lavender Oil Narden- | |
| ised | 200 |

---

### Lavender for TW E.048

| | |
|---|---|
| Lavender 40/42 | 300 |
| Lavendin | 100 |
| Lavender Spike | 100 |
| French Rosemary Oil | 100 |
| Oil Red Thyme | 50 |
| Oak Moss Resin | 40 |
| Terpinyl Acetate | 100 |
| Terpinyl Propionate | 50 |
| Linalool ex Bois de Rose | 70 |
| Styrax Resin | 40 |
| Labdanum Resin | 10 |
| Oil Patchouli | 20 |
| Musk Ambrette | 30 |
| "Lignyl" Acetate | 50 |
| Methyl Ionone | 5 |
| Methyl Coumarin | 10 |
| Oil Bergamot | 50 |
| Distilled Oil Myrrh | 10 |
| Tincture Musk 4/128 | 10 |
| Tincture Ambergris | |
| 4/128 | 5 |
| Ethyl Vanillin | 2 |
| Oil Estragon | 2 |
| Oil Clary Sage | 3 |

| | |
|---|---|
| Phenylacetaldehyde | |
| Dimethyl Acetal | 3 |
| Diethyl Phthalate | 40 |

This E.048 composition differs from E.047 in that it is specifically a lavender fragrance for colognes and toilet waters. It is not suitable for use as a base, nor is it especially desirable for extracts and high-class work. It is not practical in cosmetics. Where a lavender fragrance only is requested, this is the type of composition that should be set up.

---

### Lavender-Bouquet

| | |
|---|---|
| Oil of Lavender Mont | |
| Blanc 40/42 | 630 |
| Extractol Lavender 2099 | 50 |
| Melonia | 80 |
| Ambrofix | 60 |
| Tonkarol | 50 |
| Base Poivree | 10 |
| Lavender Extract | 20 |
| Musk Ketone | 50 |
| Oil of Geranium Bour- | |
| bon | 30 |
| Centifiol E/P 6088 | 20 |

---

### Lavender Blossoms E.047

| | |
|---|---|
| Ylang Bourbon | 10 |
| Oil Peppermint Hotch- | |
| kiss | 10 |
| South American Oil | |
| Petitgrain | 15 |

| | |
|---|---|
| Dimethyl Hydroquinone | 15 |
| Coumarin | 15 |
| Geranyl Acetate | 25 |
| Phenylethyl Alcohol | 25 |
| Cinnamic Alcohol | 25 |
| Ethyl Decylate | 25 |
| "Tepyl" Acetate | 25 |
| Musk Xylol | 25 |
| African Oil Geranium | 25 |
| Isobornyl Propionate | 25 |
| Brazilian Bois de Rose | 50 |
| Citronellol | 50 |
| Lavender Absolute | 50 |
| Linalyl Acetate | 75 |
| Natural Oil Bergamot | 75 |
| Italian Oil Lemon | 35 |
| Oil Lavender 40/42 | 200 |
| Benzyl Salicylate | 200 |

### Fougere

| | |
|---|---|
| Oil of Lavender Mont Blanc 40/42 | 250 |
| Extractol Lavender Green | 50 |
| Lavandin | 50 |
| Centifiol E/P | 100 |
| Jasminet | 100 |
| Jasmonon Decolorized | 50 |
| Coumarin | 80 |
| Brazilian Bois de Rose Oil | 100 |
| Ylangol | 50 |
| Iromuskon DP | 50 |
| Oakmoss Extract Green | 50 |
| Nerolyol | 20 |

| | |
|---|---|
| Oil of Geranium Bourbon | 10 |
| Oil of Coriander | 10 |
| Helioflor Concrete | 10 |
| Lavender Extract | 20 |

### Fougere Base

| | |
|---|---|
| Oakmoss Absolute | 4 |
| Bergamot Oil | 15 |
| Lavender Oil | 15 |
| Coumarin | 7 |

On the basis of the above formula, depth of odor can be improved with the addition of 1 to 3% of patchouly, vetivert and musk ambrette or ketone. For floral tones fairly liberal quantities of good quality rose and jasmine bases are recommended, with possibly a trace of orange blossom. Modifications of the fresh top note can be brought about by the use of such products as lemon oil, linalyl acetate or citronellyl formate and in certain cases, traces of tangerine oil. Many other materials may be used to provide varying creations, examples of these include geranium oil, clary sage, petitgrain, ylang-ylang, sandalwood, amyl salicylate and ionone. Fixation is adequately taken care of by the basic ingredients already mentioned but, if desired, increased quantities of the musks

along with vanillin may be incorporated. Cheaper versions
and soap grades can easily be
produced by using oakmoss resinoid and synthetic geranium
and bergamot oils and by replacing the high priced lavender oil
by lavandin.

---

### Fougere E.049

| | |
|---|---|
| Natural Oil Bergamot | 35 |
| Lavender 40/42 | 15 |
| California Oil Lemon | 5 |
| Ionone AB | 8 |
| Amyl Salicylate | 8 |
| Oil Vetyvert Bourbon | 5 |
| Oil Geranium Bourbon | 8 |
| Oil Patchouli | 5 |
| Citronellol | 5 |
| Ethyl Salicylate | 5 |
| Oak Moss Resin | 3 |
| Oil Carrot Seed | 1 |
| French Oil Estragon | 2 |
| Citronellyl Oxyacetal- | |
| dehyde | 1 |
| East Indian Oil Sandal- | |
| wood | 5 |
| Hydroxycitronellal | 8 |
| Linalyl Acetate | 6 |

This type of composition is the
foundation for many men's colognes and lotions. It can be varied extensively by manipulating
it with lavender components,
and by the use of floral bases and
floral adjuvants.

### Fougere d'Espagne

| | |
|---|---|
| Amyl Salicylate | 60 |
| Heliotropine | 45 |
| Tonkarol | 20 |
| Mouscalia | 20 |
| Mousse EME Extra De- | |
| colorized | 30 |
| Coumarin | 120 |
| Patchouly Oil | 45 |
| Lavandin Oil | 75 |
| Musk Ketone | 15 |
| Jasminet 6019 | 6 |
| Genuine Peru Balsam | 15 |
| "Standard" Benzoin Ex- | |
| tract | 25 |
| Vetivert Oil Bourbon | 35 |
| Olibanum Oil | 6 |
| Civet Synth. Liquid | |
| 6263 | 3 |
| Elemi Oil | 9 |
| Cumin Oil | 3 |
| Bergamot Oil Reggio | 66 |
| Vanillin | 6 |
| Santalol | 14 |
| Rosevertol | 6 |
| Ambron | 6 |
| Mousse de Perse | 150 |
| Cedarwood Oil | 140 |
| East Indian Sandal- | |
| wood Oil | 80 |

---

### Fougere Perfume for Soap

| | |
|---|---|
| Musk Ketone | 30 |
| Coumarin | 80 |
| Vanillin | 20 |
| Mousse Hyperessence | 20 |

| | | | |
|---|---|---|---|
| Peru Balsam | 30 | Tonkarol | 40 |
| Tolu Balsam | 40 | Muscocine | 20 |
| Cinnamic Aldehyde | 20 | Cinnamic Alcohol | 40 |
| Cinnamic Alcohol | 40 | Ambrofix | 20 |
| Methylionone (100%) | 80 | "Standard" Oakmoss | |
| Geraniol Extra | 80 | Extract | 50 |
| Sandalwood Oil W. I. | | Nerolyol | 50 |
| Nardenised | 50 | Sophodor | 50 |
| Patchouli Oil Nardenised | 70 | Vetivenon | 30 |
| Clove Oil Nardenised | 20 | Oil of Bergamot Reggio | 80 |
| Cananga Oil Nardenised | 20 | Carnatin 1045 | 10 |
| Bourbon Geranium Oil | | Iso-Butyl Salicylate | 10 |
| Nardenised | 80 | East Indian Oil of San- | |
| Petitgrain Oil Narden- | | dalwood | 20 |
| ised | 40 | Vial | 20 |
| Lavender Oil Nardenised | 60 | Bouvardia 3161 | 30 |
| Spanish Rosemary Oil | | Musk Ambrette | 20 |
| Nardenised | 40 | Dimethyl Hydroquin- | |
| Spanish Spike Oil | | one (10%) | 20 |
| Nardenised | 30 | Oil of Coriander | 5 |
| Cassia Oil Nardenised | 40 | Aldehyde G-12 (Lauric) | |
| Cinnamon-Leaf Oil | | (10%) | 5 |
| Nardenised | 40 | Aldehyde EdC | 10 |
| Clove-Leaf Oil Narden- | | | |
| ised | 70 | | |

## Modern Chypre

| | | | |
|---|---|---|---|
| | | Heliotropine | 25 |
| ## Chypre | | Musk Ketone | 100 |
| | | Vetivert Oil | 60 |
| Methyl Iridone | 200 | Majoram Oil | 3 |
| Colonal | 100 | Patchouly Oil | 20 |
| Linalool | 30 | Santalol | 30 |
| California Oil of Lemon | 30 | Vetiveryl Acetate | 20 |
| Citryl | 20 | Mousse de Perse | 97 |
| Jasminet | 40 | "Standard" Labdanum- | |
| Phenyl Ethyl Alcohol | 20 | Extract Discol. Extra | 30 |
| Heliotropine | 30 | | |

"Standard" Oakmoss Abs.

| | |
|---|---|
| Liq. Discol. | 15 |
| Methyl Iridon (100%) | 25 |
| Jasminet 6019 | 200 |
| Jasmador P | 30 |
| Bulgaryol III | 15 |
| Centifiol E/P 6088 | 125 |
| Bergamot Oil Reggio | 30 |
| Hydroxycitronellal | 40 |
| Carnatin 1045 | 15 |
| Alpha Amyl Cinnam. Ald. | 20 |
| Aldehyde C-12 (10%) | 30 |
| Aldehyde C-10 (10%) | 15 |
| Carbatin | 15 |
| Ambrofix | 10 |
| Benzylbenzoate | 30 |

### Chypre-Bouquet

| | |
|---|---|
| Chypre 2926 | 400 |
| Jacinthaflor 6268 | 90 |
| Methyl Iridone | 200 |
| Colonal | 50 |
| Carnatin 1045 | 30 |
| Jasminet | 60 |
| Ambrofix | 20 |
| Agrumenal | 40 |
| Bulgaryol III | 30 |
| Civethopon Extra 1712 (10%) | 20 |
| Noval | 20 |
| "Standard" Oakmoss Extract | 40 |
| Oil of Bergamot Reggio | 30 |
| Florida Sweet Oil of Orange | 30 |

### Chypre-Fougere for Soaps

| | |
|---|---|
| Hydroxal S | 100 |
| Florida Oil of Cedarwood Crude | 100 |
| Terpinyl Acetate | 100 |
| Oil of Lavender Synth. 9863 | 100 |
| Oil of Spike | 50 |
| Mousse Ambree | 50 |
| Benzyl Acetate | 50 |
| Tonka B | 50 |
| Amyl Salicylate | 50 |
| Eau de Cologne 26745 | 50 |
| Paraguay Oil of Petitgrain | 30 |
| Ylangol | 30 |
| Methyl Iridone for Soaps | 40 |
| Musk Xylol | 80 |
| Dimethyl Resorcin | 20 |
| Extractol Oakmoss SK Green 2735 a | 20 |
| Benzyl Benzoate | 60 |
| Isobutyl Salicylate | 20 |

### Chypre Oriental E.051

| | |
|---|---|
| Natural Oil Bergamot | 185 |
| Oak Moss Resin | 100 |
| Coumarin | 120 |
| Vanillin | 25 |
| Musk Ambrette | 18 |
| Musk Ketone | 35 |
| Phenylacetic Acid | 4 |
| Vetivert Bourbon | 8 |
| Methyl Ionone | 8 |
| Hydratropyl Alcohol | 7 |
| Rhodinol | 45 |

| | |
|---|---|
| Castoreum Absolute | 20 |
| Ylang Bourbon | 18 |
| Methyl Salicylate | 4 |
| Rose Otto | 4 |
| Oil Estragon | 4 |
| Oil Patchouli | 8 |
| Lavender 40/42 | 5 |
| South American Petit-grain | 10 |
| Cinnamic Aldehyde | 3 |
| Isoeugenol | 6 |
| Oil Bitter Orange | 15 |
| California Oil Lemon | 35 |
| Aldehyde "Veronol" (10% in DEP) | 7 |
| Aldehyde C-14 (10% in DEP) | 9 |
| Civettiane (Perfumery Associates) | 8 |
| Amyl Cinnamic Aldehyde | 15 |
| Benzyl Propionate | 10 |
| Tolyl Acetate | 15 |
| Hydroxycitronellal | 30 |
| Benzyl Acetate | 10 |
| Diethyl Phthalate | 9 |

The foregoing formula represents an elaborate chypre and demonstrates an extensive use of the various components previously noted. It is a useful material as a specialty for the variation of other compositions. Shown below is a very elemental chypre base. This is not a finished effect. It is merely a "Skeleton" that must be elaborated on by the perfumer. It is the chypre effect in its simplest form, and should be floralized and sweetened.

---

### Chypre Base E.052

| | |
|---|---|
| Natural Oil Bergamot | 40 |
| Linalool | 40 |
| East Indian Sandalwood | 30 |
| Oil Bitter Orange | 10 |
| Ylang Bourbon | 12 |
| California Oil Lemon | 5 |
| Methyl Ionone | 8 |
| Ionone AB | 4 |
| Heliotropine | 15 |
| Ethyl Vanillin | 2 |
| Oak Moss Resin | 8 |
| Musk Xylol | 6 |
| Oil Star Anise | 2 |
| Coumarin | 4 |
| Civettiane | 2 |
| Oil Clary Sage | 2 |
| Benzyl Propionate | 8 |
| Amyl Cinnamic Aldehyde | 6 |
| "Lignyl" Acetate | 20 |
| Diethyl Phthalate | 1 |

---

### Methyl Ionone Parma Type

| | |
|---|---|
| Labdanum Oil, Distilled | 3 |
| Aldehyde C-12 MNA (100%) | 3 |
| "Veronol" Aldehyde | 5 |
| Iris Concrete | 10 |
| Vetiverol | 50 |

| Santalol | 50 |
| Tetrahydro Methyl Ionone | 50 |
| Methyl Ionone Standard | 819 |

### Methyl Ionone Bouquet "A"

| Vitiverol | 8 |
| Santalol | 6 |
| Oil Labdanum Distilled | 1 |
| "Veronol" Aldehyde (10% in Methyl Ionone) | 1 |
| Standard Methyl Ionone | 984 |

### "Ultraionone"

| Cumin Ketone | 2 |
| Isobutyl Quinoleine (5% in DEP) | 4 |
| Labdanum Oil Distilled (10% in DEP) | 5 |
| "Palatone" | 5 |
| Vetiverol | 8 |
| Isopropyl Quinoleine (5% in DEP) | 8 |
| Hydratropyl Alcohol | 10 |
| "Veronol" Aldehyde. (10% in DEP) | 10 |
| Oil Carrot Seed (10% in DEP) | 20 |
| Hydroxycitronellal | 34 |
| Eugenyl Phenylacetate | 34 |
| Citronellyl Oxyacetaldehyde | 30 |
| Alpha Ionone | 100 |

| Standard Methyl Ionone | 190 |
| Dimethyl Ionone | 240 |

### Lily of the Valley Base

| Hydroxycitronellal | 30 |
| Linalool | 15 |
| Terpineol | 15 |
| Rhodinol | 10 |

Experiments may proceed on the basis of the above, blended with such products as ylang-ylang, phenylethyl alcohol, geranyl, and citronellyl esters, bergamot and benzyl acetate in relatively small proportions. Traces of phenylacetaldehyde dimethyl acetal are particularly useful for obtaining a green freshness, but this must not be taken to excess, 1% of a 10% solution being considered ample. The floral note can be further strengthened with good rose and jasmin bases and minute traces of cyclamen aldehyde. As fixatives, use can be made of musk ketone, methyl naphthyl ketone and trichlormethyl phenyl carbinyl acetate. In the case of cheaper compounds and perfumes for toilet soaps, bois de rose oil may be used to replace linalool, together with "soap" grades of hydroxycitronellal, phenylethyl alcohol and so forth.

## White Clover

| | |
|---|---|
| Orchid Base E.020 | 250 |
| Coumarin | 16 |
| Ylang Bourbon Extra | 15 |
| Dimethyl Hydroquinone | 3 |
| Musk Ketone | 6 |
| Oil Bergamot | 7 |
| Lavender 40/42 | 3 |
| Oak Moss Resin | 2 |
| Benzyl Acetate | 3 |
| "Palatone" (5%) in Phenylethyl Alcohol | 3 |
| Heliotropine | 5 |
| Amyl Salicylate | 10 |
| Vanillin | 2 |
| Acetophenone | 2 |
| Isobutyl Phenylacetate | 3 |
| Tuberyl Acetate | 2 |
| Benzyl Alcohol | 13 |

## New Mown Hay (Foin Coupe)

| | |
|---|---|
| Coumarin | 60 |
| Tonkarol | 50 |
| Acanthon | 20 |
| Anisic Aldehyde | 20 |
| Mousse de Perse | 130 |
| Patchouly Oil | 5 |
| Clary Sage Oil | 10 |
| Vetivert Oil | 10 |
| Linalool | 50 |
| Lavender Oil Montblanc | 50 |
| Foin Coupe G 3510 | 180 |
| Aldehyde MNA 10% Sol. | 5 |
| Bergamot Oil Reggio | 65 |

| | |
|---|---|
| Dihydrocoumarin | 20 |
| Peru Balsam Oil | 15 |
| Bulgaryol III | 15 |
| Amyl Salicylate | 60 |
| Vetiveryl Acetate | 30 |
| Vanillin | 5 |
| East Indian Sandalwood Oil | 60 |
| Ylang-Ylang 8258 or Ylangol | 50 |
| Phenyl Acetic Acid | 10 |
| Jasmador P | 80 |
| Tuberosal Extra D | 20 |
| Mimosaflor | 30 |
| Cassie 11793 | 10 |
| Hydroxycitronellal | 40 |

## Fleurs D'Orange for Soaps

| | |
|---|---|
| California Oil Sweet Orange | 300 |
| California Oil Lemon | 200 |
| Citronellol | 100 |
| Geranyl Acetate | 50 |
| Amyl Cinnamic Aldehyde | 60 |
| Benzyl Acetate | 60 |
| Methyl Naphthyl Ketone | 30 |
| Citral | 5 |
| Phenylacetic Acid | 5 |
| Lignyl Acetate | 90 |

## Fleurs d'Orange Bigarade for Fine Perfumery

| | |
|---|---|
| Rhodinyl Acetate | 20 |
| Rhodinyl Butyrate | 10 |

| | |
|---|---|
| Phenylethyl Dimethyl Carbinyl Acetate | 15 |
| Italian Oil of Mandarin | 10 |
| Petitgrain ex Citronnier | 25 |
| Methyl Naphthyl Ketone | 5 |
| Civettiane | 1 |
| Castoreum Absolute (10% in Benzyl Alcohol) | 1 |
| Musk Ketone | 5 |
| Rhodinol | 10 |
| Linalool ex Bois de Rose | 5 |
| Oil Bitter Orange | 10 |
| Oil Bergamot | 15 |
| Methyl Ionone | 5 |
| Phenylacetic Acid | 2 |
| Amyl Phenylacetate | 1 |
| Cinnamyl Butyrate | 2 |
| Hydroxycitronellal-Methyl Anthranilate | 20 |
| Cumin Ketone | 5 |
| Ylang Bourbon Extra | 3 |
| Phenylethyl Alcohol | 10 |
| Benzyl Phenylacetate | 10 |
| Phenylethyl Phenyl-acetate | 10 |

### Wood Note

| | |
|---|---|
| Noval | 160 |
| Sophonia | 80 |
| Sophodor | 80 |
| East Indian Sandalwood Oil | 70 |
| Vetivenon | 50 |
| Methyl Salicylate | 50 |
| Muguet P 06218 | 100 |

| | |
|---|---|
| Centifiol E/P | 90 |
| Coumarin | 50 |
| Iromuskon DP | 70 |
| Carnatin 1045 | 50 |
| Bergamot Oil 38% | 80 |
| California Lemon Oil | 50 |
| Ylang Oil Extra | 20 |

### Note Boisee

| | |
|---|---|
| Noval | 160 |
| Sophonia | 80 |
| Sophodor | 80 |
| East Indian Oil of Sandalwood | 100 |
| Amyl Salicylate | 50 |
| Convial 6122 | 40 |
| Centifiol E/P | 90 |
| Coumarin | 50 |
| Erodon | 70 |
| Carnatin 1045 | 50 |
| Oil of Bergamot Reggio | 80 |
| Oil of Lemon Messina | 50 |
| Oil of Ylang-Ylang Extra | 20 |
| Vetivenon | 50 |
| Singapore Oil of Patchouly | 20 |
| Oil of Vetivert Ia | 10 |

### Precious Wood Type

| | |
|---|---|
| Sophodor | 200 |
| Methyl Iridone | 100 |
| Colonal | 90 |
| Oil of Bergamot Reggio | 80 |
| Coumarin | 80 |
| California Oil of Lemon | 70 |

| | |
|---|---|
| East Indian Oil | |
| Sandalwood | 80 |
| Lignol | 80 |
| "Standard" Oakmoss | |
| Extract | 20 |
| Musk Ambrette | 30 |
| Bulgaryol III | 20 |
| Amyl Salicylate | 30 |
| Singapore Oil of | |
| Patchouly | 40 |
| Linalyl Acetate | 20 |
| Linalool | 10 |
| Aldehyde C–12 | |
| (Lauric) (10%) | 30 |
| Aldehyde C–11 | |
| (Undecylenic) (10%) | 20 |

### Gardenia

| | |
|---|---|
| Musk Ambrette | 30 |
| Vanillin | 10 |
| Heliotropin | 20 |
| Benzyl Proponiate | 160 |
| Linalyl Acetate | 80 |
| Jasmiral | 80 |
| Cinnamic Alcohol | 60 |
| Citronellol (Levoro- | |
| tatory) | 50 |
| Geraniol | 50 |
| Geranyl Formate | 20 |
| Citronellyl Formate | 30 |
| Hydroxycitronellal | |
| "Naarden" | 100 |
| Styralyl Acetate | 10 |
| Amyl Salicylate | 20 |
| Ylang-Ylang Oil | |
| Nardenised | 60 |

| | |
|---|---|
| Calamus Oil Nardenised | 5 |
| African Geranium Oil | |
| Nardenised | 45 |
| East Indian Sandalwood | |
| Oil Nardenised | 50 |
| Bay Oil Nardenised | 40 |
| Mandarine Oil | |
| Nardenised | 30 |
| Petitgrain Oil | |
| Nardenised | 50 |

### Gardenia for Soap

| | |
|---|---|
| Hydroxycitronellal | 20 |
| Cinnamic Alcohol | 5 |
| Geraniol | 8 |
| Ionone AB | 2 |
| Methyl Ionone | 1 |
| Amyl Salicylate | 1 |
| Dimethyl Hydroquinone | 2 |
| Heliotropine | 4 |
| Terpineol | 20 |
| Benzyl Acetate | 10 |
| Benzyl Propionate | 4 |
| Linalool | 5 |
| Styrallyl Acetate | 5 |
| Hydratropyl Acetate | 1 |
| Aldehyde C-14 | |
| (10% in DEP) | 2 |
| Phenylacetic Acid | 1 |
| Benzyl Phenylacetate | 9 |

### Gardenia for Cosmetics

| | |
|---|---|
| Benzyl Acetate | 45 |
| Benzyl Propionate | 15 |
| Amyl Cinnamic | |
| Aldehyde | 10 |

| | |
|---|---|
| Hexyl Cinnamic Alde- | |
| hyde | 4 |
| "Astrotone" BR (100%) | 2 |
| Cinnamyl Acetate | 5 |
| Ylang Bourbon Extra | 12 |
| Methyl Naphthyl Ketone | 6 |
| Phenylethyl Alcohol | 12 |
| Hydratropyl Alcohol | 4 |
| Italian Oil Lemon | 12 |
| Oil Petitgrain | 15 |
| Hydroxycitronellal | 9 |
| Linalyl Acetate | 15 |
| Styrallyl Propionate | 12 |
| Hydrotropyl Acetate | 2 |
| Coumarin | 2 |
| Sandalwood Oil | 6 |
| Ionone Alpha | 5 |
| Rhodinol | 8 |
| Tuberyl Acetate | 1 |
| Propylene Carbonate | 23 |

### Cheap Neroli for Soap

| | |
|---|---|
| Orange Terpenes (or | |
| Limonene Technical) | 500 |
| Geranyl Acetate | |
| (Soap Grade) | 200 |
| Bois de Rose Oil | 100 |
| Methyl Anthranilate | 100 |
| Amyl Cinnamic | |
| Aldehyde | 60 |
| Benzyl Acetate | 80 |
| Citral | 10 |
| Nerolin | 10 |
| Lignyl Acetate | 40 |

### Neroli Bigarade Type for Fine Perfumery

| | |
|---|---|
| South American | |
| Petitgrain | 100 |
| Linalyl Acetate | |
| (90-92%) | 100 |
| Phenylethyl Alcohol | 100 |
| Hydratropyl Alcohol | 50 |
| Linalool ex Bois de | |
| Rose | 50 |
| Geranyl Acetate | 50 |
| Hydroxycitronellal | 50 |
| Hydroxycitronellal- | |
| Methyl Anthranilate | 30 |
| Oil Coriander | 10 |
| Immortelle Absolute | 10 |
| Ylang Bourbon Extra | 10 |
| Ethyl Decylate | 150 |
| Benzyl Phenylacetate | 50 |
| Oil of Celery Seeds | 5 |
| Cuminic Aldehyde | 3 |
| Indol | 1 |
| Aldehyde C-12 MNA | |
| (10% in DEP) | 2 |
| Citronelyl | |
| Oxyacetaldehyde | 1 |
| Ethyl Anthranilate | 10 |
| Diethylphthalate | 18 |

### Narcissus

| | |
|---|---|
| French Sweet Basil Oil | 4 |
| Phenylacetic Acid | 10 |
| Hydratropyl Acetate | 10 |
| Cumin Ketone | 10 |
| Oil Clary Sage | 12 |

| | | | | |
|---|---|---|---|---|
| Anisic Aldehyde | 16 | Oil Bergamot | 10 |
| Heliotropine | 16 | Benzyl Acetate | 8 |
| Amyl Salicylate | 16 | Hydroxycitronellal | 6 |
| Musk Xylol | 16 | Hydratropyl Alcohol | 10 |
| Methyl Ionone | 16 | Heliotropine | 40 |
| Rosacetol | 20 | Benzyl Butyrate | 6 |
| Amyl Cinnamic | | Alpha Ionone | 2 |
| Aldehyde | 20 | Indole 10% (in DEP) | 2 |
| Benzyl Acetate | 20 | Coumarin | 2 |
| Para Cresyl | | Methyl Benzoate | 2 |
| Phenylacetate | 20 | Musk Ketone | 4 |
| Ylang Bourbon Extra | 20 | Phenylethyl Alcohol | 8 |
| Paracresyl Caprylate | 20 | Rhodinol | 8 |
| Tolyl Acetate | 20 | Para Methyl | |
| Musk Ketone | 40 | Hydratropic Aldehyde | 2 |
| South American | | Amyl Salicylate | 4 |
| Oil Petitgrain | 40 | Amyl Cinnamic | |
| Rhodinol | 40 | Aldehyde | 2 |
| Phenylethyl Alcohol | 40 | South American | |
| Nerolin | 50 | Petitgrain | 4 |
| Sandalwood Oil | 50 | | |
| Hydroxycitronellal | 60 | | |
| Oil Bitter Orange | 10 | Jonquille | |
| Oil Labdanum Distille | | Aldehyde C-12 | |
| (10% in DEP) | 10 | (MNA 10%) | 5 |
| Vetivert Acetate | 10 | Oil Estragon | 10 |
| Oil Patchouli | 10 | Phenylacetate | 20 |
| Benzyl Salicylate | 24 | Heliotropine | 20 |
| | | Phenylacetic Acid | 30 |
| Pink Hyacinth | | Musk Ketone | 30 |
| Phenylacetaldehyde | | Paracresyl Caprylate | 30 |
| (50% in Benzyl | | Ylang Bourbon Extra | 30 |
| Alcohol) | 70 | Paramethyl Hydratropic | |
| Cinnamic Alcohol | 50 | Aldehyde | 30 |
| Linalool | 10 | Standard Alpine Violette | 30 |
| Terpineol | 25 | South American Oil | |
| | | Petitgrain | 30 |

| | |
|---|---|
| Anisic Aldehyde | 30 |
| Phenylethyl Alcohol | 40 |
| Citronellol | 40 |
| Benzyl Acetate | 40 |
| Eugenyl Phenylacetate | 40 |
| Vanillin | 50 |
| Nerolin | 50 |
| Oil Sandalwood | 50 |
| Amyl Cinnamic Aldehyde | 60 |
| Benzyl Salicylate | 60 |
| Paracresyl Phenylacetate | 75 |
| Tolyl Acetate | 100 |
| Hydroxycitronellal | 100 |

### Jonquille Fleurs

| | |
|---|---|
| Aldehyde C-12 MNA (10% in DEP) | 5 |
| French Oil Estragon | 10 |
| Phenylethyl Acetate | 10 |
| Heliotropine | 20 |
| Phenylacetic Acid | 30 |
| Musk Ketone | 30 |
| Para Cresyl Caprylate | 30 |
| Ylang Bourbon Extra | 30 |
| Para Methyl Hydratropic Aldehyde | 30 |
| "Alpine Violet" | 30 |
| South American Petitgrain | 30 |
| Anisic Aldehyde | 30 |
| Phenylethyl Alcohol | 40 |
| Rhodinol | 40 |
| Eugenyl Phenylacetate | 40 |
| Vanillin | 50 |
| Nerolin | 50 |

| | |
|---|---|
| Sandalwood Oil | 50 |
| Amyl Cinnamic Aldehyde | 60 |
| Benzyl Salicylate | 60 |
| Para Cresyl Phenylacetate | 75 |
| Tolyl Acetate | 110 |
| Hydroxycitronellal | 100 |
| Benzyl Acetate | 40 |

### Acacia Blossoms

| | |
|---|---|
| Anisic Aldehyde | 45 |
| Methyl Anthranilate | 40 |
| Paramethyl Acetophenone | 3 |
| Cinnamic Alcohol | 6 |
| Isobutyl Phenylacetate | 6 |
| Benzyl Acetate | 3 |
| Citronellol | 8 |
| Linalool | 4 |
| Hydroxycitronellal | 4 |
| Hydratropyl Alcohol | 8 |
| Phenylacetaldehyde (50% in Benzyl Alcohol) | 6 |
| Vanillin | 3 |
| Phenylacetic Acid | 3 |
| Methyl Naphthyl Ketone | 2 |
| Heliotropine | 10 |
| Aldehyde C-16 (100%) | 2 |
| Musk Ketone | 4 |
| Aldehyde C-14 (10% in DEP) | 1 |
| Benzophenone | 4 |
| South American Petitgrain | 5 |
| Rose Otto | 2 |

| | |
|---|---|
| Coumarin | 2 |
| Benzyl Salicylate | 10 |
| Diethyl Phthalate | 22 |

---

### Synthetic Verbena Oil

| | |
|---|---|
| Violet Leaves, Absolute (10%) | 40 |
| Spearmint Oil Nardenised | 40 |
| Lemon Oil Nardenised | 100 |
| Basil Oil Nardenised | 20 |
| Ceylon Citronella Oil Nardenised | 20 |
| Lemongrass Oil Nardenised | 780 |

---

### Tabac Fleurs

| | |
|---|---|
| Tabac 1314 | 350 |
| Coumarin | 80 |
| Methyl Iridone | 80 |
| Tonkarol | 40 |
| Bulgaryol I | 40 |
| Oil of Bergamot Reggio | 20 |
| Oil of Ylang-Ylang | 20 |
| Cinnamic Alcohol | 20 |
| Hydroxy Citronellal | 20 |
| Phenyl Ethyl Alcohol | 50 |
| Citronellol | 20 |
| Allyl Phenyl Acetate | 20 |
| Geraniol | 50 |
| Sophodor | 40 |
| Vetivenon | 20 |
| Labdanum Extract | 40 |
| Light Storax Extract | 20 |

| | |
|---|---|
| Cirosia | 20 |
| Oil of Birch Tar (10%) | 10 |
| Mousse Ambree | 10 |
| Ambropur (10%) | 10 |
| "Standard" Orris Extract | 10 |
| Dimethyl Hydroquinone (10%) | 5 |
| Isobutyl Quinoline | 5 |

---

### Tabac Bosnia

| | |
|---|---|
| Tabak 1314 | 50 |
| Tabak 6531 | 90 |
| Vanillin | 15 |
| East Indian Sandalwood Oil | 120 |
| Jasmador P 3668 | 78 |
| Mousse Ambree 6067 | 25 |
| Coumarin | 25 |
| Dihydrocoumarin | 10 |
| Amberone C | 170 |
| Ambrofix | 24 |
| Extractol Olibanum | 30 |
| "Standard" Castoreum Extract | 3 |
| "Standard" Labdanum Extract Discol. | 12 |
| Olibanum Oil | 12 |
| Civet Synth. Liquid 6263 | 5 |
| Nerolyol | 5 |
| Musk Ambrette | 40 |
| Musk Ketone | 5 |
| Hydroxal E | 5 |
| Mousse de Perse | 106 |
| Madagascar Ylang Ylang Oil | 70 |
| Ylangol | 50 |

| Chypre Extra 3294 | 120 |
| Bulgaryol III | 12 |
| Mandarin Oil | 24 |
| Tilia 6227 | 24 |
| Tuberosal Extra D 13586 | 12 |
| Jasminal P | 10 |
| Evernol | 48 |

### Tobacco Odor for Soaps

| Phenyl Ethyl Alcohol | 100 |
| Terpinyl Acetate | 100 |
| Mousse Ambree | 100 |
| Tabac 1314 | 50 |
| Paraguay Oil of Petitgrain | 50 |
| Eugenol | 20 |
| Vanillin | 20 |
| Tonka B | 50 |
| Coumarin | 50 |
| Amyl Salicylate | 40 |
| Iridone | 50 |
| Methyl Iridone for Soaps | 20 |
| Dimethyl Hydroquinone | 20 |
| Florida Oil of Cedarwood Crude | 50 |
| "Standard" Extract Labdanum, Green | 50 |
| "Standard" Extract Storax | 30 |
| Musk Xylol | 60 |
| Helioflor Concrete | 20 |
| Singapore Oil of Patchouly | 20 |
| West Indian Oil of Sandalwood | 30 |
| Brazilian Oil of Rosewood | 20 |

| Aldehyde C-12 (Lauric) (10%) | 40 |
| Cirosia | 10 |

### Russian Leather

| Cuirisal | 100 |
| Colonal | 100 |
| Messina Oil of Lemon | 70 |
| Oil of Bergamot Reggio | 70 |
| Cuir Saffian | 12 |
| Mousse Ambree | 80 |
| Ambropur (10%) | 20 |
| Musc D | 10 |
| "Standard" Extract Labdanum Green | 20 |
| Peru Extractol | 20 |
| Oil of Birch Tar (10%) | 20 |
| "Standard" Extract Tolu, Dark | 30 |
| Iridone | 70 |
| Methyl Iridone | 60 |
| Cedryl Acetate | 30 |
| Cedrol | 20 |
| Bulgaryol I | 13 |
| Nerolin Yara Yara | 30 |
| Jasminet | 50 |
| Vanillin | 15 |
| Cinnamic Alcohol | 160 |

### Orchid Base

| Amyl Salicylate | 55 |
| Anisic Aldehyde | 25 |
| Rhodinol | 30 |
| Isobutyl Salicylate | 40 |
| Heliotropine | 10 |
| Dimethyl Hydroquinone | 2 |

| | | | | |
|---|---|---|---|---|
| Acetophenone | 5 | Mousse de Perse | 120 |
| Linalool | 15 | Vetivert Oil Bourbon | |
| Phenylethyl Alcohol | 16 | (Reunion) | 56 |
| Cumin Ketone | 5 | East Indian | |
| Coumarin | 4 | Sandalwood Oil | 50 |
| Phenylacetaldehyde (50% | | Orange Oil Sweel A I | |
| in Benzyl Alcohol) | 10 | Guinea | 12 |
| Hydroxycitronellal | 10 | Peru Balsam Oil | 40 |
| "Veronol" Aldehyde | | Ambrofix | 30 |
| (10% in DEP) | 2 | Carbatin | 20 |
| Phenylacetic Acid | 4 | Bergamot Oil Reggio | 64 |
| Ylang Bourbon | 5 | California Lemon Oil | 12 |
| Vanillin | 2 | Java Cananga Oil | 12 |
| Methyl Ionone | 4 | Eugenol | 132 |
| Oak Moss Resin | 1 | Jasmador P | 50 |
| Alpha Ionone | 1 | Borenia | 10 |
| Citronellyl | | | |
| Oxyacetaldehyde | 2 | | |
| "Palatone" (5% in | | **Georgette** | |
| Phenylethyl Alcohol) | 2 | Noval | 200 |
| Diethyl Phthalate | 6 | Oil of Bergamot | |
| | | Reggio | 200 |
| | | Colonal | 150 |
| **Fleurs de Mousse** | | Centifiol E/P 6088 | 50 |
| Nerolyol 1608 | 26 | Sophonia | 50 |
| Musk Ambrette | 65 | California Oil of Lemon | 50 |
| Methylisoeugenol | 48 | East Indian Oil of | |
| "Standard" Labdanum- | | Sandalwood | 50 |
| Extract Discol. | 20 | Sophodor | 50 |
| Mouscalia 1392 | 16 | Coumarin | 50 |
| Orange-Flower Extract | | "Standard" Oakmoss | |
| (30%) | 5 | Extract | 40 |
| Civet Synth. Liquid 6263 | 48 | Methyl Iridone | 40 |
| Pepper Oil | 4 | Vetivenon | 13 |
| Vanillin | 47 | Aldehyde C-11 (Unde- | |
| Heliotropine | 75 | cylenic) (100%) | 5 |
| Methyl Iridon (100%) | 38 | | |

| Aldehyde C-12 (Lauric) | |
|---|---|
| (100%) | 2 |
| Oil of Geranium | 10 |
| I-Rhodinol | 20 |
| Citronellol | 20 |

### Manola Exotique

| | |
|---|---|
| Musk Ambrette | 20 |
| Coumarin | 30 |
| Vanillin | 90 |
| Borenia 1395 | 140 |
| Patchouly Oil | 10 |
| East Indian | |
| Sandalwood Oil | 40 |
| Estragon Oil | 10 |
| Jasmin 6018 | 20 |
| Tabak 1314 | 10 |
| Decolorized Mousse | |
| EME Extra | 60 |
| Lemon Oil Messina | 100 |
| Bergamot Oil Reggio | 200 |
| Mousse de Perse | 110 |
| Nerolyol | 20 |
| Neroli 8235 | 40 |
| Amber Synth. Liq. | |
| 1049 | 100 |

### Sweet Pea

| | |
|---|---|
| Terpineol | 30 |
| Oil Bergamot | 25 |
| Linalool ex Bois de Rose | 40 |
| Paramethyl Hydratropic | |
| Aldehyde | 5 |
| Citronellyl | |
| Oxyacetaldehyde | 7 |
| Phenylacetone | 4 |

| South American Oil | |
|---|---|
| Petitgrain | 5 |
| Italian Oil Orange | 8 |
| Alpha Ionone | 10 |
| Isobutyl Phenylacetate | 8 |
| Musk Ambrette | 6 |
| Methyl Naphthyl | |
| Ketone | 12 |
| Heliotropine | 8 |
| Musk Ketone | 4 |
| Ylang Bourbon | 4 |
| Anisic Aldehyde | 6 |
| Cumin Ketone | 2 |
| Hydroxycitronellal | 6 |
| Rhodinol | 4 |
| Hydratropyl Alcohol | 4 |
| Dimethyl Octanyl | |
| Acetate | 4 |
| Benzyl Salicylate | 8 |

### Silvaflor

| | |
|---|---|
| Hydroxal S | 200 |
| Tonka B | 100 |
| Terpinyl Acetate | 100 |
| Lavandin | 100 |
| Siberian Synth. 1871 | |
| Pine-Needle Oil | 50 |
| Mousse Ambree | 50 |
| Florida Oil of | |
| Cedarwood Ia | 50 |
| Base Champignon | 50 |
| Iridone | 50 |
| Terpineol | 50 |
| Jasminet | 50 |
| Helioflor Concrete | 30 |

| Methyl Iridone for | |
|---|---|
| Soaps | 20 |
| Musk Xylol | 50 |
| Pseudo Ionone | 20 |
| Borenia | 20 |
| Extractol Oakmoss SK | |
| Green | 10 |

## Parfume Français

| | |
|---|---|
| Orgina 1355 | 380 |
| Coumarin | 90 |
| Sophonia | 80 |
| East Indian Oil of | |
| Sandalwood | 70 |
| Cassis 11793 | 50 |
| Iso-Butyl Salicylate | 20 |
| Singapore Oil of | |
| Patchouli | 20 |
| Musk Ketone | 20 |
| Musk Ambrette | 30 |
| Hydroxy Citronellal | 20 |
| Methyl Iridone | 50 |
| Bouvardia 3161 | 20 |
| Ambropur (10%) | 10 |
| Tonkarol | 10 |
| Light Storax Extract | 10 |
| Linalyl Acetate | 20 |
| Light Oil of | |
| Ylang-Ylang | 25 |
| California Oil of Lemon | 40 |
| Oil of Vetivert | 10 |
| Aldehyde C-12 MNA | |
| (10%) | 20 |
| Aldehyde C-10 (Decyl) | |
| (10%) | 5 |

## French Perfume

| | |
|---|---|
| Tuberosal Extra D 13586 | 70 |
| Jasmonon Decolorized | 90 |
| Jasminet | 80 |
| Mimosaflor 6293 | 60 |
| Citronellol | 50 |
| Ylangol | 100 |
| Hydroxycitronellal | 50 |
| Musc D | 40 |
| Phenyl Ethyl Alcohol | 100 |
| Bergamot Oil (38%) | 40 |
| Musk Ambrette | 30 |
| Vetivenon | 30 |
| Sandalwood Oil E | 20 |
| California Lemon Oil | 20 |
| Methyl Ionone | 30 |
| Ylang Oil Extra | 20 |
| Aldyl 2451 | 60 |
| Phenyl Acetaldehyde | |
| (50%) | 10 |
| Bulgaryol III | 20 |
| Vetiveryl Acetate | 60 |
| Phenyl Ethyl Alcohol | 20 |

## Lilac

| | |
|---|---|
| Hydroxycitronellal | 45 |
| Terpineol | 40 |
| Cinnamic Alcohol | 25 |
| Anisic Alcohol | 15 |
| Phenylethyl Alcohol | 20 |
| Hydratropyl Alcohol | 5 |
| Indol (10% in DEP) | 27 |
| Isoeugenol | 8 |
| Oil Styrax | 3 |
| Linalool | 8 |

| | |
|---|---|
| Benzyl Acetate | 15 |
| Benzyl Propionate | 3 |
| Cumin Ketone | 3 |
| Ylang Absolute | 2 |
| Tolyl Alcohol | 5 |
| Amyl Salicylate | 1 |
| Benzyl Alcohol | 10 |
| Benzyl Salicylate | 15 |

### Industrial Lilac Mask

| | |
|---|---|
| Terpineol | 600 |
| Phenylethyl Alcohol (Soap Grade) | 200 |
| Benzyl Acetate | 100 |
| Paracresyl Acetate | 5 |
| Paracresyl Methyl Ether | 5 |
| Ionone AB (Soap Grade) | 10 |
| Amyl Cinnamic Aldehyde | 80 |

### Lilac Royale

| | |
|---|---|
| Linalool | 3 |
| Terpineol | 15 |
| Heliotropine | 12 |
| Phenylacetaldehyde | 3 |
| Benzyl Acetate | 12 |
| Ylang Absolute | 9 |
| Hydroxycitronellal | 18 |
| Cinnamic Alcohol | 12 |
| Phenylethyl Alcohol | 15 |
| Rose Otto | 1 |
| Natural Rose Absolute | 5 |
| Tuberose Absolute | 1 |
| Isoeugenol | 2 |
| Tolu Concrete | 5 |

| | |
|---|---|
| Tincture Ambergris | 5 |
| Cumin Ketone | 2 |
| Jasmine Absolute | 5 |

### Lilac for Cosmetics

| | |
|---|---|
| Hydroxycitronellal | 14 |
| Cumin Ketone | 4 |
| Terpineol | 40 |
| Isoeugenol | 1 |
| Phenylethyl Alcohol | 22 |
| Benzyl Acetate | 10 |
| Phenylacetaldehyde | 1 |
| Heliotropine | 4 |
| Amyl Cinnamic Aldehyde | 4 |
| Hexyl Cinnamic Aldehyde | 2 |
| Anisic Aldehyde | 4 |
| Benzyl Dimethyl Carbinyl Acetate | 1 |
| Paramethyl Acetophenone (10% in DEP) | 2 |
| Cinnamic Alcohol | 14 |
| Oil Styrax | 3 |
| Linalool | 4 |
| Paracresyl Caprylate | 10 |
| Benzyl Isoeugenol | 10 |

### Medium-Priced Soap Lilac

| | |
|---|---|
| Terpineol | 600 |
| Amyl Cinnamic Aldehyde | 140 |
| Benzyl Acetate | 100 |
| Linalool | 100 |
| Anisic Aldehyde | 75 |
| Paracresyl Phenylacetate | 35 |
| Isoeugenol | 10 |

| | |
|---|---|
| Paracresyl Acetate | 5 |
| Phenylethyl Alcohol | 35 |
| Benzyl Salicylate | 100 |

### Cheap Lilac for Soap

| | |
|---|---|
| Linalool | 40 |
| Ionone AB for Soap | 40 |
| Bromstyrol | 40 |
| Benzyl Acetate | 50 |
| Heliotropine | 200 |
| Aldehyde C-14 (10% in DEP) | 10 |
| Paramethyl Benzaldehyde | 10 |
| Terpineol | 110 |

### Classical Compound

| | |
|---|---|
| Musk Ambrette | 20 |
| Musk Ketone | 50 |
| Vanillin | 5 |
| Coumarin | 30 |
| Styrax Resinoid | 40 |
| Labdanum Resinoid | 30 |
| Civette Paste "Naarden" | 5 |
| Orris Concrete | 15 |
| Isobutyl Salicylate | 20 |
| Methylionone Delta | 80 |
| Linalyl Acetate | 50 |
| Styralyl Acetate | 15 |
| Hydroxycitronellal "Naarden" | 80 |
| Para-Methyl Acetophenone | 10 |
| Vioparmone "Naarden" | 50 |
| Muguet Nardarome | 50 |

| | |
|---|---|
| Jasmine Supreme | 45 |
| Floranger | 60 |
| Opoponax "N" | 60 |
| Aldehyde C-9 (1%) | 20 |
| Aldehyde C-12 (1%) | 10 |
| Methyl Nonyl Acetaldehyde (1%) | 20 |
| Undecylenic Aldehyde (1%) | 20 |
| Bourbon Vetiveryl Acetate | 50 |
| Petitgrain Oil Nardenised | 40 |
| Clove Oil Nardenised | 40 |
| Ylang-Ylang Oil Nardenised | 50 |
| Patchouli Oil Nardenised | 40 |
| Bourbon Vetivert Oil Nardenised | 40 |
| East Indian Sandalwood Oil of E. I. Nardenised | 50 |
| Bergamot Oil Nardenised | 10 |
| Cassia Oil Nardenised | 5 |

### Bouquet de Fleurs
### Formula No. 1

| | |
|---|---|
| Bulgaryol III | 50 |
| Tuberosal Extra D | 100 |
| Jasmonon Decolorized | 90 |
| Jasminet | 80 |
| Jacinthaflor 6268 | 60 |
| I-Citronellol | 50 |
| Ylangol | 100 |
| Hydroxy Citronellal | 50 |
| Musc D | 40 |

| | |
|---|---|
| Phenyl Ethyl Alcohol | |
| Supreme | 100 |
| Oil of Bergamot | |
| Reggio | 40 |
| Musk Ambrette | 30 |
| Vetivenon | 30 |
| East Indian Oil of | |
| Sandalwood | 20 |
| California Oil of Lemon | 20 |
| Methyl Iridone | 30 |
| Oil of Ylang-Ylang | 20 |
| Aldyl 2451 | 60 |
| Phenyl Acetaldehyde | |
| (50%) | 10 |
| Cinnamic Alcohol | 20 |

### No. 2

| | |
|---|---|
| Mimosaflor 6293 | 100 |
| Convial 6122 | 150 |
| Lilac 06000 | 350 |
| Cinnamic Alcohol | 20 |
| Ylang Oil Extra | 50 |
| Phenyl Ethyl Alcohol | 50 |
| Jasminet 6019 | 100 |
| Isoeugenol | 10 |
| Hydroxycitronellal | 50 |
| Benzyl Benzoate | 40 |
| Fixon | 40 |
| Methyl Jonone | 40 |

---

### Bouquet de Fleurs for Soaps

| | |
|---|---|
| Terpineol | 180 |
| Phenyl Ethyl Alcohol | 100 |
| Helioflor Concrete | 100 |
| Centifiol S | 100 |
| Hydroxal S | 100 |

| | |
|---|---|
| Cinnamic Alcohol | 50 |
| Tonka B | 50 |
| Jasminet | 50 |
| Convial | 50 |
| Anisic Aldehyde | |
| (Aubepine) | 30 |
| Linalyl Acetate Supreme | 20 |
| Oil of Cananga | 20 |
| African Oil of Geranium | 20 |
| Musk Xylol | 50 |
| Eugenol | 40 |
| Isoeugenol | 20 |
| Oil of Cloves (Buds) | 10 |
| Isoeugenol Methylether | 10 |

---

### Flowery Perfume for Soap

| | |
|---|---|
| Terpineol | 150 |
| Coumarin | 50 |
| Methyl Naphthyl | |
| Ketone | 40 |
| Trirosol | 30 |
| Para Methoxy | |
| Acetophenone | 40 |
| Dimethyl Benzyl | |
| Carbinol | 50 |
| Geraniol Extra | 100 |
| Isopropyl Hydroxal | 60 |
| Cinnamic Alcohol | |
| Synth. | 100 |
| Methylionone (100%) | 40 |
| Linalool ex Bois de | |
| Rose | 100 |
| Bourbon Geranium Oil | |
| Nardenised | 40 |
| Petitgrain Oil | |
| Nardenised | 50 |

| | |
|---|---|
| Ceylon Citronella Oil | |
| Nardenised | 20 |
| Cananga Oil Nardenised | 80 |
| Lavender Oil Nardenised | 50 |

---

**Fancy Flower Perfume**

| | |
|---|---|
| Musk Ketone | 80 |
| Amyl Salicylate | 30 |
| Isobutyl Salicylate | 40 |
| Linalool ex Bois de | |
| Rose | 50 |
| Alcohol C-12 (100%) | 15 |
| Cervolide "Naarden" | 25 |
| Benzyl Proponiate | 50 |
| Methylionone Gamma | 150 |
| Hydroxycitronellal | |
| "Naarden" | 70 |
| Rhodinyl Acetate | 30 |
| Geranyl Acetate | 40 |
| Phenyl Acetaldehyde | |
| (10%) | 10 |
| Lilacs Blanc Nardarome | |
| "Naarden" | 80 |
| Methyl Para | |
| Cresol (1%) | 5 |
| Para Cresyl Acetate | |
| (1%) | 5 |
| Jacinthe Nardarome | |
| "Naarden" | 70 |
| Petitgrain Oil | |
| Nardenised | 50 |
| Ylang-Ylang Oil | |
| Nardenised | 60 |
| Coriander Oil | |
| Nardenised | 20 |
| Bergamot Oil Nardenised | 60 |

| | |
|---|---|
| Clove Oil Nardenised | 40 |
| African Geranium Oil | |
| Nardenised | 10 |
| Carrotseed Oil | |
| Nardenised | 10 |

---

**Synthetic Ambergris Base**

| | |
|---|---|
| "Ambropur" (50%) | 100 |
| Peru Resinoid, | |
| Decolorized | 100 |
| Santalol | 50 |
| Nerol Extra | 50 |
| Labdanum Absolute | 50 |
| Terpeneless Clary | |
| Extra | 550 |
| Farnesol | 30 |
| Cyclic Ethylene | |
| Glycol Brassylate | 20 |
| Methylnonyl Acetalde- | |
| hyde (10% in | |
| Civet Tincture) | 20 |
| Decylaldehyde (10%) | 10 |
| Undecylene Aldehyde | |
| (10%) | 10 |
| Terpeneless Angelica | |
| Root Oil (5-10%) | 10 |

---

**Synthetic Ambergris Fixative**

| | |
|---|---|
| Decolorized Resinoid | |
| Tolu | 200 |
| Extra Light Resinoid | |
| Labdanum | 100 |
| Terpeneless Cypress Oil | 100 |
| "Ambropur" in Diethyl | |
| Phthalate (10%) | 300 |

| | | | | |
|---|---|---|---|---|
| Absolute Tonka | 50 | Ambron | 18 |
| Ethyl Vanillin | 30 | Civet Synth. Liquid | |
| Decolorized Resinoid | | 6263 | 10 |
| Vanilla | 20 | Cassis 11793 | 8 |
| Extra Clary | 100 | Messinal | 10 |
| Heliotropin | 50 | "Standard" Opoponax- | |
| Absolute Castoreum | 5 | Extract | 30 |
| Civet Absolute | 10 | "Standard" Castoreum- | |
| Musk D (100%) | 10 | Extract | 20 |
| Olibanum Oil | 5 | Carbatin | 10 |
| Extra Resinoid Benzoin | 20 | Peau d'Espagne 6132 | 40 |
| | | Russofixol | 30 |
| | | Colonal | 30 |
| Eau de Cologne | | Hydroxycitronellal | 79 |
| Colonal | 250 | Methyl Iridone (100%) | 90 |
| Oil of Bergamot Reggio | 150 | Jasmador P 3668 | 90 |
| Messina Oil of Lemon | 150 | Mouscalia 1392 | 30 |
| Florida Oil of Orange | | Ocolofix | 38 |
| Sweet | 130 | Eau de Cologne Oil | |
| Citryl | 100 | K 8600 | 150 |
| Nerolyol | 100 | Rose B 3553 | 29 |
| Messinal | 50 | Ambrofix | 30 |
| Agrumenal | 50 | Decolorized Mousse | |
| Aldehyde Eau de | | EME Extra | 20 |
| Cologne | 20 | Musk Ketone | 30 |

Eau de Cologne "Cuir"

| | | | |
|---|---|---|---|
| Coumarin | 4 | | |
| Vanillin | 12 | | |
| Heliotropine | 50 | | |
| East Indian Sandalwood | | | |
| Oil | 5 | | |
| Mace Oil | 7 | | |
| Mousse de Perse | 100 | | |
| Orange Flower Oil 8248 | 15 | | |
| Nerolyol | 15 | | |

Eau de Cologne-Chypre

| | |
|---|---|
| Colonal | 200 |
| Oil of Bergamot Reggio | 150 |
| Florida Oil of Orange | |
| Sweet | 100 |
| California Oil of Lemon | 90 |
| Nerolyol | 50 |
| Messinal | 10 |
| Benzyl Acetate | 20 |
| Jasminet | 50 |

| | | | |
|---|---|---|---|
| Tonka B | 30 | German-Type Cologne | |
| Vetivenon | 10 | Linalyl Acetate | 120 |
| Iromuskon DP | 20 | Bigarade Neroli Oil | 40 |
| "Standard" Oakmoss | | Absolute Orange Flowers | 30 |
| Extract Green | 45 | Lime Oil Nardenised | 20 |
| Methyl Iridone | 40 | Lemongrass Oil | |
| Aldehyde C-11 (Unde- | | Nardenised | 20 |
| cylenic) (100%) | 10 | Spanish Rosemary Oil | |
| Aldyl 2451 | 5 | Nardenised | 50 |
| Sophonia | 10 | Lavender Oil Nardenised | 60 |
| Cuminic Aldehyde | 10 | Lemon Oil Nardenised | 80 |
| Resinoid Opoponax | 10 | Orange Oil Nardenised | 100 |
| Phenyl Ethyl Alcohol | 50 | Petitgrain Oil | |
| I-Rhodinol | 50 | Nardenised | 120 |
| Geraniol | 20 | Bergamot Oil | |
| Citronellol | 20 | Nardenised | 360 |

### Cooling Eau de Cologne Stick

| | Formula No. 1 | No. 2 | No. 3 | No. 4 |
|---|---|---|---|---|
| Sodium Stearate | 60 | 70 | 80 | 80 |
| Glycerin | 30 | 30 | — | 30 |
| Propylene Glycol | 30 | — | 50 | — |
| Isopropyl Myristate | — | — | — | 100 |
| Water | 50 | — | — | — |
| Alcohol | 810 | 870 | 855 | 765 |
| Perfume Oil | 20 | 30 | 15 | 25 |

For production, sprayed sodium stearate purissima is used. If there is water in the batch, the soda soap is moistened with it beforehand, as the soap dissolves better and more rapidly in alcohol if swollen. For small batches the soap is dissolved along with the other components in a water bath, using a reflux condenser. When everything is dissolved, the batch is left to cool somewhat, after which the perfume is mixed in and the mixture at once filled into the shells used for packaging in so far as they are suitable for direct filling. The faster the mass congeals after pouring, the more homogeneous it looks. For this reason it is best to pour at as low a temperature as possible, since the mass hardens amorphously when so manipulated. These sticks can also be prepared with the formation of soda soap taking place during the manufacturing process itself.

|  | No. 5 | No. 6 |
|---|---|---|
| Stearic Acid | 45.0 | 55.5 |
| Glycerin | 30.0 | 30.0 |
| Propylene Glycol | — | 30.0 |
| Sodium Hydroxide | 7.5 | — |
| Water | 73.0 | — |
| Sodium Hydroxide Solution (Containing 840 mg. NaOH) | — | 42.0 |
| Alcohol | 829.5 | 822.5 |
| Perfume Oil | 15.0 | 20.0 |

The stearic acid is dissolved in a part of the alcohol at the same time as the plasticizer (glycerin, propylene glycol, etc.) is added. The sodium hydroxide solution is then warmed up in water with the remaining alcohol and added to the alcoholic stearic acid solution. Saponification proceeds rapidly under the influence of heating and stirring. The perfume oil is added next. Instead of sodium stearate, chips of a well-dried, white toilet soap may also be used. Before being dissolved in alcohol they are moistened with water, speeding up the process of dissolution under heat.

|  | No. 7 | No. 8 |
|---|---|---|
| Dry Soap Chips | 110 | 140 |
| Glycerin | 30 | 50 |
| Propylene Glycol | 20 | — |
| Isopropyl Myristate | 10 | 40 |
| Water | — | 10 |
| Alcohol | 800 | 745 |
| Perfume Oil | 20 | 15 |

## Perfume Stick

| | |
|---|---|
| Acetanilide | 125 |
| Magnesium Carbonate | 15 |
| Wax Bodies | 5 |
| Musk Xylol | 50 |
| Heliotropin | 25 |
| Perfume Oil | 12 |
| Benzyl Alcohol | 6 |

---

## Water Solubilized Perfume

| | |
|---|---|
| Perfume Oil | 0.5-2 |
| "Solulan" 98 | 3-12 |
| Water | To make 100 |

---

## Deodorant Foot Powder

| | |
|---|---|
| Impalpable Chlorhydrol | 10.00 |
| Starch | 3.00 |
| "G-4" | 0.25 |
| "G-11" | 0.25 |
| Talcum | 86.50 |
| Essential Oils | To suit |

Mix ingredients thoroughly, add perfume gradually, until thoroughly blended.

---

## Antiseptic Body Powder

| | |
|---|---|
| Impalpable Chlorhydrol | 10.0 |
| Calcium Carbonate | 3.0 |
| "G-11" | 0.5 |
| Talcum | 86.5 |
| Perfume | To suit |

## Ion-Exchanger Body Powder

| Formula | No. 1 | N. 2 |
|---|---|---|
| "Amberlite" XE-64 (Micropowder) | 10.00 | — |
| Oxyquinoline Sulfate | 0.15 | — |
| Zinc Stearate | 5.00 | 5 |
| Talc | 25.00 | 50 |
| Cornstarch | 60.00 | — |
| Carboxyl Resin Ion-Exchanger | — | 30 |
| Zinc Oxide | — | 10 |
| Precipitated Chalk | — | 5 |

---

## Aerosol Body Powder

| | |
|---|---|
| Glycerin | 1.0 |
| Perfume Oil | 1.0 |
| Hexachlorophene USP | 0.4 |
| Isopropyl Myristate | 1.0 |
| Dipropyleneglycol | 5.0 |
| Anhydrous Alcohol SDA40 | 91.6 |

Filling Charge:

| | |
|---|---|
| Concentrated Formula (as above) | 70% |
| Propellent Mixture | 30% |
| "Freon" 114 or "Genetron" 320 | 60% |
| "Propellent" 12 | 40% |

## Styptic Stick

| | Formula No. 1 (Transparent) | No. 2 (Blue Transparent) | No. 3 (Opaque) |
|---|---|---|---|
| Potash Alum | 350 | 960 | 330 |
| Aluminum Sulfate | 640 | — | 620 |
| Borax | 5 | — | — |
| Glycerin | 5 | 20 | 40 |
| Pet. Ferrocyanide | — | 10 | — |
| Zinc Oxide | — | — | 10 |

## Washable Pharmaceutical Ointment Bases
### (Oil-in-Water)

| | Formula No. 1 | No. 2 | No. 3 |
|---|---|---|---|
| "Amerchol" L-101 | — | — | 10.0 |
| "Amerchol" CAB | — | 18.8 | — |
| "Amerchol" H-9 | 5.0 | — | — |
| "Modulan" | 4.0 | — | 3.5 |
| "Carbowax" 4000 | — | 15.1 | — |
| Stearyl Alcohol | 1.5 | 19.6 | — |
| Glyceryl Monostearate (Self-Emulsifying) | 1.5 | — | — |
| Glyceryl Monostearate (Neutral) | — | — | 13.5 |
| Spermaceti | — | — | 1.5 |
| "Tween" 61 | — | — | 8.5 |
| Petrolatum | 37.5 | — | — |
| Glycerin | — | 10.0 | — |
| Water | 50.0 | 36.2 | 63.0 |
| Sodium Lauryl Sulfate | 0.5 | 0.3 | — |
| Preservative | To suit | To suit | To suit |

Add the water containing the water-soluble ingredients to the melted fats with the temperature of both phases at 85°C. Mix at moderate speed until cool and remix the following day for a smoother ointment.

## No. 4

| Octadecyl Alcohol | 2 |
| "Lantrol" (Liquid Fraction of Lanolin) | 8 |
| White Petrolatum | 90 |

## No. 5

| Magnesium Oleate | 25 |
| Mineral Oil | 75 |

Warm together and mix until uniform. This can take up 50 of water and spreads easily on skin.

## No. 6

| "Carbowax" 4000 | 34 |
| "Carbowax" 400 | 50 |
| 1,2,6-Hexanetriol | 16 |

Melt the "Carbowax" 4000 in the 1,2,6-hexanetriol at 60 to 70°C. on a water bath. Add the melt to the "Carbowax" 400 and stir.

Most major ointment ingredients are soluble in this base in the following generally-used concentrations.

| Medicinal Ingredient | Weight Per Cent of Active Ingredient in Ointment Base |
|---|---|
| Ammoniated Mercury | 5.00 |
| Benzoic Acid and | 12.00 |
| Salicylic Acid | 6.00 |
| Boric Acid | 10.00 |
| Calamine | 15.00 |
| Coal Tar | 3.00 |
| Zinc Oxide and | 24.25 |
| Starch | 24.25 |
| Ethyl Aminobenzoate | 5.00 |
| Iodine and | 4.00 |
| Potassium Iodide | 4.00 |
| Ichthammol | 10.00 |
| Yellow Mercuric Oxide | 1.00 |
| Phenol | 3.00 |
| Resorcinol | 10.00 |
| Precipitated Sulfur | 15.00 |
| Tannic Acid and | 20.00 |
| Exsiccated Sodium Sulfite | 0.20 |
| Zinc Oxide | 20.00 |
| Zinc Oxide and | 25.00 |
| Starch | 25.00 |

## No. 7

| Glycerol | 3.0 |
| "Kessco" Wax A-21 | 4.0 |
| Stearic Acid | 19.0 |
| Propylene Glycol | 4.5 |
| Isopropyl Myristate | 5.0 |
| Sodium Alginate Solution (1%) | 64.5 |

Heat all of the ingredients to 90°C. in one container. Allow the mixture to cool to 30°C. with continual stirring. This formulation may be poured at temperatures as low as 25°C. It sets to a firm cream after 24 hours.

## No. 8

| | |
|---|---|
| Pure Glyceryl Mono- | |
| stearate | 4.5 |
| Stearic Acid | 7.2 |
| Preservative | 0.1 |
| Isopropyl Myristate | 4.3 |
| Polyethyleneglycol | |
| 1000 Monostearate | 6.0 |
| Lanolin | 1.0 |
| Propyleneglycol | 2.5 |
| Water | 74.4 |

Heat all of the ingredients to 95°C. in one container. Stir the mixture uniformly until it cools to 35°C. Perfume may be added below 40°C.

## No. 9

| | |
|---|---|
| Diglycol Stearate S. E. | 8.5 |
| Spermaceti | 2.0 |
| Water | 74.9 |
| Stearic Acid | 5.0 |
| White Mineral Oil° | 2.0 |
| Butyl Stearate | 4.5 |
| Propyleneglycol | 3.0 |
| Preservative | 0.1 |

Heat all of the ingredients to 95°C. in one container. Stir the mixture slowly until it cools to 35°C.

* Saybolt Viscosity at 100°F. = 65 to 75 sec.

## No. 10

| | |
|---|---|
| *a* Glycerol Mono- | |
| stearate 24 S. E. | 10.0 |
| Water | 62.9 |

| | |
|---|---|
| Borax | 2.0 |
| Preservative | 0.1 |
| *b* Isopropyl Palmitate | 3.0 |
| Petrolatum | 7.0 |
| Paraffin Wax | |
| (M.P. 135°F.) | 15.0 |

Heat all the ingredients under *a* at 90 to 95°C. in one container. Stir the mixture until it cools to 65°C. In another container, heat the ingredients under *b* to 65°C. Add *b* to *a* slowly, with stirring. Stir until the mixture cools to 35°C. Pour at 33 to 35°C. The cream will become firm after standing overnight.

## No. 11
### (Silicone)

| | |
|---|---|
| Pure Glyceryl Mono- | |
| stearate | 4.3 |
| Stearic Acid | 6.8 |
| "Kesscomir" | 4.1 |
| Polyethyleneglycol | |
| 1000 Monostearate | 5.7 |
| Lanolin | 1.0 |
| Propyleneglycol | 2.4 |
| Water | 70.6 |
| Silicone Oil° | 5.0 |
| Preservative | 0.1 |

Heat all of the ingredients to 95°C. in one container. Stir at this temperature for several minutes. Allow the mixture to cool to 35°C., with continual stirring. A fine emulsion does not form until the temperature is approximately 50°C.

\* Dimethyl Polysiloxane-Viscosity at 25°C. equals 1000 cps.

## No. 12
### (Cationic)

| | |
|---|---|
| Stearic Acid | 6.5 |
| Spermaceti | 5.0 |
| Propyleneglycol | 2.0 |
| Polyethyleneglycol | |
| 600 Distearate | 6.5 |
| White Mineral Oil° | 2.0 |
| Butyl Stearate | 2.0 |
| "Pendit" C A | 0.2 |
| Preservative | 0.1 |
| Water | 75.7 |

Heat all the ingredients to 95°C. in one container, with continual agitation. Allow the mixture to cool to 35°C., with continual stirring.

\* Saybolt Viscosity at 100°F. = 65 to 75 sec.

## No. 13

| | |
|---|---|
| Water | 61.0 |
| Ethyl Alcohol | 29.0 |
| Polyethyleneglycol | |
| 4000 Distearate | 5.0 |
| Lanolin USP | 0.5 |
| "Carbopol" 934 | 1.2 |
| Diisopropanolamine | To pH 6 |

Dissolve the "Carbopol" in the water at room temperature. Add the remaining ingredients except diisopropanolamine, warm to 60°C., and mix well. Add diiso-

propanolamine with agitation until a pH of 6 is reached. Allow the mixture to cool to 35°C., with gentle stirring.

## No. 14

| | |
|---|---|
| Isopropyl Palmitate | 3.0 |
| Glycerol Monostearate | |
| 24 S. E. | 12.0 |
| Spermaceti | 3.0 |
| Propyleneglycol | 5.0 |
| Preservative | 0.1 |
| Water | 76.9 |
| Perfume | As desired |

In one container, heat all of the ingredients, except the perfume, to 95°C. Allow the mixture to cool to 35°C., with stirring, and then add the perfume.

## No. 15

| | | |
|---|---|---|
| a | White Mineral Oil° | 20.0 |
| | Butyl Stearate | |
| | (Cosmetic Grade) | 1.0 |
| | Cetyl Alcohol | 1.0 |
| | Beeswax | 10.0 |
| | Polyethylene Glycol | |
| | 300 Monostearate | 14.0 |
| | Preservative | 0.1 |
| b | Triethanolamine | 1.5 |
| | Water | 52.4 |

Heat the ingredients under a to 75°C. Slowly add b to a while stirring and continue to stir until the cream has cooled to 30°C.

\* Saybolt Viscosity at 100°F. = 65 to 75 sec.

## No. 16

| | |
|---|---|
| Water | 52.0 |
| Ethyl Alcohol | 37.8 |
| Polyethyleneglycol | |
| 600 Monostearate | 7.0 |
| Isopropyl Myristate | 1.0 |
| "Carbopol" 934 | 2.2 |
| Diisopropanolamine | To pH 6 |

Dissolve the "Carbopol" in the water at room temperature. Add the remaining ingredients, except the diisopropanolamine, warm to 45°C. and mix well. Add diisopropanolamine with agitation until a pH of 6 is reached. Allow the mixture to cool to 35°C., with gentle stirring.

---

## Hydrocortisone Ointment Base

### Formula No. 1

| | |
|---|---|
| Glycerin | 24.8 |
| Cetyl Alcohol | 14.9 |
| Sesame Oil | 20.0 |
| Polyoxyl 40 Stearate | 5.0 |
| Glyceryl Monostearate | 5.0 |
| Distilled | |
| Water, | To make 100 |

### No. 2

| | |
|---|---|
| Glycerin | 24.8 |
| Cetyl Alcohol | 14.9 |
| Sesame Oil | 20.0 |
| "Myrj" 52 | 5.0 |
| Glyceryl Monostearate | 5.0 |
| Water | 30.3 |

## Infant's Wound Ointment

| | |
|---|---|
| Zinc Oxide | 10 |
| Talc | 10 |
| Balsam Peru | 3 |
| "Emulgade" F | 12 |
| "Cetiol" V | 6 |
| White Petrolatum | 18 |
| Mineral Oil | 6 |
| Water, | to make 100 |
| Preservative | To Suit |

"Emulgade" F, "Cetiol" V, the white petrolatum, and the mineral oil are melted in a water-jacketed vessel at 158 to 176°F. The water, in which the preservative has been dissolved, is heated to the same temperature and then stirred into the fatty components. The zinc oxide is mixed with the talc and then blended with the balsam of Peru. This mixture is stirred into the ointment base at about 86°F. and then the finished ointment is passed through a roller mill.

---

## Cod-Liver Oil Ointment

| | |
|---|---|
| Cod-Liver Oil | 30 |
| "Dehydag" Wax SX | 10 |
| White Petrolatum | 15 |
| Distilled Water | 45 |
| Preservative | To suit |

The fish-liver oil, "Dehydag" wax SX and the white petrolatum are melted together on a water-jacketed vessel at about 70°C. (158°F.). The water is also warmed to about 70°C. (158°F.) and stirred in. Emulsification results while stirring until cold.

Ointments as oil-in-water emulsion systems are subject to water evaporation. The finished preparations are, therefore, best supplied in tubes. These should be coated with a suitable inside protective coating so as to avoid any corrosion.

---

### Glyceryl Ointment

| | |
|---|---|
| Potato Starch | 6 |
| Sorbitol (60% Solution) | 20 |
| Glycerol | 40 |

| | |
|---|---|
| Water | 40 |
| β-Methylhydroxy Benzoate | ⅛ |
| Alcohol | 4 |
| Tragacanth Powder | 2 |

---

### Antipruritic Ointment

| | |
|---|---|
| Dyclonine Hydrochloride | 1.0 |
| Stearic Acid | 18.0 |
| Polyethyleneglycol 1000 Monostearate | 8.0 |
| Polyethyleneglycol 400 Monostearate | 5.0 |
| Beeswax (White) | 2.0 |
| Hydrous Chlorobutanol | 0.3 |
| Distilled Water | To make 100.0 |

CHAPTER V

# EMULSIONS

Look through all other chapters for emulsion formulation. For example, under Cosmetics and Polishes there are many oil, wax, and other emulsions. In the Paint section there are emulsions of oils, synthetic resins, plasticizers, pigments, etc. Other sections contain many varied types of emulsions.

## Cosmetic Water-in-Oil Emulsions

| | Formula No. 1 | No. 2 | No. 3 | No. 4 | No. 5 | No. 6 |
|---|---|---|---|---|---|---|
| "Amerchol" L-101 | 10 | 15 | — | — | — | 5.00 |
| "Amerchol" CAB | — | — | 20 | — | — | — |
| "Amerchol" H-9 | — | — | — | 19 | 10 | — |
| Lanolin | 10 | 10 | 15 | 3 | 20 | 3.00 |
| Mineral Oil (70 viscosity) | 25 | 20 | 8 | 16 | — | 33.50 |
| Vegetable Oil | — | — | — | — | 15 | — |
| Microcrystalline Wax | 5 | 15 | — | 4 | — | — |
| Beeswax | — | — | 7 | — | — | 2.00 |
| Cetyl Alcohol | — | — | — | — | 3 | — |
| Petrolatum | — | — | — | 5 | — | 5.00 |
| Sorbitan Sesquioleate | — | — | — | — | — | 3.00 |
| "Carbowax" 1500 | — | — | — | — | 5 | — |
| Water | 50 | 40 | 50 | 53 | 47 | 47.75 |

| | | | | | | |
|---|---|---|---|---|---|---|
| "Veegum" | — | — | — | — | — | 0.25 |
| Borax | — | — | — | — | — | 0.50 |
| Consistency | cream | cream | cream | cream | cream | lotion |
| Perfume and Preservative to suit | | | | | | |

Heat the "Amerchol," waxes, and oils to a temperature just above their melting point. Heat the water and water-soluble materials to the same temperature, then add the water solution to the oils, with constant rapid agitation and continue mixing until the batch is cold. For best results, these emulsions should be homogenized while the batch is slightly warm.

### Turbine-Oil Emulsion
### (For Corrosion Protection)

| | |
|---|---|
| Turbine Oil | 64 |
| Oleic Acid | 23 |
| Triethanolamine | 13 |
| Water | To suit |

### Drying-Oil Emulsion

| | |
|---|---|
| Water | 100.0 |
| "Carbopol" | 0.2 |
| 10% Sodium Hydroxide | 0.6 |
| "Ethomeen" C-25 | 0.2 |
| Linseed Oil | 100.0 |

### Oil-Well Drilling-Fluid Emulsion
### U. S. Patent 2,862,881

| | |
|---|---|
| Vegetable Pitch | 15 |
| Petroleum Sulfonate | 5 |
| Wood-Rosin Extract | 5 |
| Calcium Chloride | 10 |
| Lime | 10 |
| Adsorptive Clay | 25 |

Use 20 to 70 parts of above with water to make a barrel of drilling fluid.

### Tar Emulsion

| | |
|---|---|
| Water | 48.90 |
| "Carbopol" | 0.25 |
| 10% Sodium Hydroxide | 0.75 |
| "Ethomeen" C-25 | 0.10 |
| Tar TR-12 | 50.00 |

### Cationic Asphalt Emulsion

#### Formula No. 1

| | |
|---|---|
| Asphalt | 70.0 |
| Water | 30.0 |
| Tallow Diamine | 0.5 |
| Acetic Acid | 0.2 |
| Calcium Chloride | 0.1 |

#### No. 2

| | |
|---|---|
| Asphalt | 60.0 |
| Water | 40.0 |

"Duomeen" T　　　　0.2
Acetic Acid　　　　0.5
Calcium Chloride　　0.1

Using a two-phase system, "Duomeen" T, the fatty diamine, is added to preheated asphalt at 190 to 250°F., while the acetic acid and calcium chloride are added to preheated water at 180°F. The two phases are then combined by high-speed mixing, producing a stable cationic emulsion that is brownish and has a viscosity of 88 SSF at 77°F.

## Transparenting Oils in Water
### German Patent 876,308

Menthyloxacetic Acid　　8.0
Triethanolamine　　　　4.6
Polyglycol Monolaurate　3.5
Water　　　　　　　　20.0

The above will dissolve

Peppermint Oil　　　　1.5
or Cincole　　　　　　2.0
or p-Tolymethyl-
　carbinol　　　　　　4.0
or Camphor　　　　　1.5

These can be diluted with water to 100 and further to give clear dispersions.

## Solubilizing Oils

| Oil | Parts by Weight | Solubilizing Agent | Parts by Weight |
|---|---|---|---|
| Pine | 1 | "Ethofat" C/25 | 3 |
| Mineral | 1 | "Ethofat" O/20 | 2 |
| Peppermint | 1 | "Ethofat" C/25 | 5 |
| Peppermint | 1 | "Ethofat" 242/25 | 7 |
| Lemon | 1 | "Ethofat" C/25 | 5 |
| Orange | 1 | "Ethofat" C/25 | 3 |
| Wintergreen | 1 | "Ethofat" C/25 | 7 |
| Hexachlorophene | 1 | "Ethofat" C/25 | 9 |

## Hydrogen Peroxide Emulsion
### U. S. Patent 2,886,532

Saturated Aromatic
　Hydrocarbons (B.P.
　100-300°C.)　　　　30-50
Saturated Nonionic
　Emulsifier　　　　　15-45
Hydrogen Peroxide　　55-5

This gives a water-in-oil emulsion.

## Silicone Emulsion

### Formula No. 1

| Union Carbide L-45 | |
|---|---|
| Silicone Oil | 35.0 |
| "Tergitol" Nonionic NPX | 3.5 |
| Water | 61.5 |

About 0.15 pounds of sodium nitrite can be added per 100 lb. of emulsion as a corrosion inhibitor.

### No. 2
### German Patent 910,340

| Silicone | 1.00 |
|---|---|
| Acetone | 0.13 |
| Isopropanol | 0.12 |
| Alcohol | 0.75 |

Then bring to *pH* 8 with triethylamine.

## Allyl Starch Emulsion
### U. S. Patent 2,740,724

| Allyl Starch | 80 |
|---|---|
| "Flexol" TWS | 20 |
| "Triton" X-100 | 5 |
| Water | 170 |
| Ammonium Hydroxide | 3 |

Stability is improved by addition of a little "Alkanol" S.

## Polytriflurochloroethylene ("Teflon") Dispersion
### German Patent 935,330

| | Formula No. 1 | No. 2 |
|---|---|---|
| "Teflon" | 1.00 | 1.0 |
| Water | 3.00 | 2.5 |
| Cyclohexylamine | 0.15 | — |
| Pyridine | — | 0.5 |

## Nylon Dispersion
### German Patent 957,883

| Nylon | 30 |
|---|---|
| Ethylenechlorohydrin | 70 |

Warm under reflux and mix until dissolved. Knead water into above to give <25% solids. This may be diluted for use in coatings.

## "Vistanex" Emulsion

### Formula No. 1

| "Vistanex" LM | 100 |
|---|---|
| Stearic Acid | 3 |
| Triethanolamine | 2 |
| Water | 80 |

Temperature 90 to 95°C.

### No. 2

| "Vistanex" LM | 100.0 |
|---|---|
| Potassium Stearate | 5.4 |
| Water | 72.7 |

Temperature 70°C.

## Paraffin Wax Emulsion (Nonionic)

### Formula No. 1

| | |
|---|---|
| Paraffin Wax (138-140°F., Refined) | 72 |
| "Ethomid" HT/15 | 21 |
| "Ethofat" 60/20 | 7 |

Melt the paraffin wax and "Etho" chemicals together and add any desired quantity of water heated above the melting point of the mixture. The above formula may be prepared as a concentrate if desired and can then be further diluted.

### No. 2

| | |
|---|---|
| Paraffin Wax (138-140°F.) Refined | 20.00 |
| "Ethofat" 60/25 | 4.00 |
| "Ethomeen" 18/12 | 2.00 |
| Water | 73.65 |
| Glacial Acetic Acid | 0.35 |

The paraffin, "Ethofat" 60/25 and "Ethomeen" 18/12 are heated to 150°F. and mixed. Water at 150°F. is added slowly with paddle agitation. After the emulsion cools to 140°F. the acetic acid is added and agitation is continued until the temperature drops to 110 to 120°F.

The emulsion prepared in this manner has an extremely fine particle size and appears translucent with a bluish fluorescence. The fluid emulsion may be diluted with water to any concentration. The wax film deposited from the emulsion is continuous, transparent and has good gloss without polishing. Using the same ratio of emulsifier to paraffin wax, the emulsion concentration can be increased to 40% solids and still remain fluid. Also 125°F. refined paraffin may be substituted for the 138 to 140°F. paraffin with equal results. Phosphoric acid (½%) may also be used in place of the acetic acid.

---

## Nonyellowing Substantive Paraffin (Waterproofing) Emulsion

| | |
|---|---|
| Paraffin Wax (139°F. Refined) | 17 |
| "Ethomeen" 18/12 | 2 |
| Glacial Acetic Acid | 1 |
| Water | 80 |

For "renewable" water repellents utilizing waxes and aluminum salts, "Ethomeen" 18/12 acetate is one of the best nonyellowing, nonrewetting emulsifiers.

The paraffin and "Ethomeen" 18/12 are melted together and poured with vigorous stirring into hot water and the acetic acid. This emulsion should be put through a colloid mill to obtain optimum particle size. Aluminum salts can also be incorporated into this formulation for improved water repellency.

---

### "Armowax" Emulsion

| | |
|---|---|
| "Armowax" (M.P. 132°C.) | 10-20 |
| "Ethomeen" T/25 | 1.54-3.08 |
| Stearic Acid ("Neo Fat" 18) | 0.46-0.92 |
| Water | 88.0-76.0 |

Heat the "Armowax," "Ethomeen" and stearic acid together to approximately 140°C. Add in a thin stream to water heated to the boiling point and continue as above.

---

### Nonionic Carnauba Wax Emulsion

#### Formula No. 1

| | |
|---|---|
| Carnauba Wax | 10 |
| "Ethomid" HT/25 | 3 |
| Water | 87 |

The wax, emulsifier and 25% of the water are charged into a kettle and heated until the wax has completely melted. Boiling of the mixture will often produce a good emulsion, but mechanical agitation is preferably applied at this time. The emulsification is completed as soon as uniform mixing has taken place. The remainder of the cool water is then added, and after mixing, no further processing is required. The above formulation may be extended with 10% shellac solubilized with ammonia, if desired.

When mixtures of waxes or other components are to be incorporated, the oily materials can be melted and mixed together in the kettle. Water can then be added as in the above procedure.

#### No. 2
#### (Cationic)

| | |
|---|---|
| Carnauba Wax | 10 |
| "Ethomeen" C/12 acetate | 4 |
| Water | 86 |

Melt the Carnauba wax and "Ethomeen" C/12 acetate and raise the temperature to 95°C. Add boiling water in small increments, thoroughly incorporating each addition with a paddle type stirrer before making the next water addition. A viscous emulsion may form at first, which will thin out as more water is added. After the emulsion has reached the low-viscosity stage, cooler water may be added to complete the dilution if desired.

This emulsion has excellent

leveling and dry-bright properties. The dried film has good resistance to water spotting.

Extenders such as shellac, synthetic resins, etc., which are often used, cannot be used in this formula because of the cationic nature of the emulsifiers.

## Antifoam

### Formula No. 1
### U. S. Patent 2,727,009

| | |
|---|---|
| Lecithin | 25.0 |
| "Ucon" LB1715 | 10.0 |
| Polyglycol 400 Dilaurate | 8.7 |
| Kerosene | 13.0 |
| Mineral Seal Oil | 47.8 |

### No. 2
### U. S. Patent 2,773,041

| | |
|---|---|
| Kerosene | 50-70 |
| Stearic Acid | 10-15 |
| Polyoxyethylene Tallate | 2-6 |
| Alumina Hydrate (Fine Powder) | 5-10 |
| or | |
| Magnesium Carbonate | 10-30 |
| Water | To suit |

## Yeast Process Defoamer

### Formula No. 1

Aqueous emulsions of oleic acid (5 to 10%) are effective.

### No. 2

| | |
|---|---|
| Mineral Oil | 95 |
| Oleic Acid | 5 |

## Decreasing Foaming and Increasing Stability of Foaming Beer
### Danish Patent 85,551

The addition of 0.1 to 1 mg. of cobalt per liter of beer works well. The cobalt is added in the form of nitrate or chloride salt.

## Sugar-Refining Antifoam
### U. S. Patent 2,727,009

| | |
|---|---|
| Lecithin | 21.8 |
| "Ucon" LB1715 | 8.7 |
| Polyethyleneglycol 400 Dilaurate | 8.7 |
| Kerosene | 13.0 |
| Mineral Seal Oil | 47.8 |

## Sewage Defoamer
### British Patent 793,737

| | |
|---|---|
| Mineral Oil | 98.3 |
| Cyclohexylamine Oleate | 1.5 |
| Polyethyleneglycol Bis Triricinoleate | 0.2 |

Use at the rate of 5 to 10 p.p.m.

## Protein Antifoam Agent

| | |
|---|---|
| "Triton" A-20 | 0.75 |
| Silicone | 0.01 |
| Glycerol | 1.00 |

Potassium Bicarbonate 1.00
Water          To make 100.00
Superior to octyl alcohol on proteins.

---

Antifreeze Foam Inhibitor
*U. S. Patent 2,777,821*

Borax                    2
Tritolyl Phosphate 0.01-0.07
Based on ethyleneglycol.

---

Ringless Orange-Oil Emulsion
Orange Oil              6.00 l.
Brominated
    Vegetable Oil        3.75 l.
    (Sp. Gr. 1.33-1.34)
Gum Arabic              4.50 kg.
Water    To make 100.00 l.
FD&C Yellow No. 6  2.25 kg.
Sodium Benzoate  120.00 g.

The orange oil and the brominated vegetable oil are mixed thoroughly together. The specific gravity of the mixture is taken with the aid of a special hydrometer that can be read to the third decimal point. The specific gravity of the mixture would be about 1.025 at 70°F. (21°C.). If it is not, then the specific gravity must be adjusted by adding brominated vegetable oil or conversely orange oil, depending on whether the specific gravity is less than or more than 1.025. The gum arabic is stirred in and stirring is continued until all the lumps have disappeared. Sufficient water is added to bring the total volume to 100 liters. The mixture is stirred thoroughly and is passed through a homogenizer.

If coloring matter is to be used, it is added and the mixture stirred until the color is dissolved; then it may be passed through the homogenizer again. It is preferable to dissolve the color and sodium benzoate in a portion of the water, adding it before the first homogenizer pass.

A small volume of the orange oil can be replaced by fivefold or terpeneless orange oil to improve the flavor or the flavor can be modified by replacing some of the orange oil with mandarin oil, tangerine oil, lemon oil, or other citrus oils.

---

Ringless Lemon-Flavor
Emulsion

Lemon Oil              12 fl.oz.
Brominated
    Vegetable Oil       10 fl.oz.
Gum Arabic              12 oz.
32° Bé Syrup            1 pt.
FD&C Yellow No. 5   2 oz.
Sodium Benzoate      10 g.
Water        To make 2 gal.

The oils in this formulation should be adjusted to a specific gravity of 1.028. This formulation is used in the ratio of 2 oz. of emulsion flavor to 1 gal. of syrup.

---

Ringless Lime-Flavor Emulsion

| | |
|---|---|
| Distilled Lime Oil | 14.0 fl.oz. |
| Brominated | |
| Vegetable Oil | 12.0 fl.oz. |
| 32° Bé Syrup | 1.0 pt. |
| FD&C Green No. 2 | 2.6 g. |
| Gum Arabic | 12.0 oz. |
| Sodium Benzoate | 10.0 g. |
| Water | To make 2.0 gal. |

The oils in this formulation are adjusted to a specific gravity of 1.028 and the emulsion is used in the ratio of 2 oz. per gal. of syrup.

---

Low-Viscosity Wax Emulsions

Formula No. 1

| | |
|---|---|
| Paraffin Wax | 45.0 |
| Carnauba Wax | 3.0 |
| Triethanolamine | 6.0 |
| Stearic Acid | 13.5 |
| Water | 225.0 |

No. 2

| | |
|---|---|
| Paraffin Wax | 45.0 |
| Carnauba Wax | 3.0 |
| Monoethanolamine | 2.6 |
| Stearic Acid | 9.0 |
| Water | 225.0 |

No. 3

| | |
|---|---|
| Paraffin Wax | 45.0 |
| Carnauba Wax | 3.0 |
| Morpholine | 3.7 |
| Stearic Acid | 9.0 |
| Water | 225.0 |

No. 4

| | |
|---|---|
| Paraffin Wax | 45.0 |
| Beeswax | 3.0 |
| Triethanolamine | 6.0 |
| Stearic Acid | 13.5 |
| Water | 225.0 |

No. 5

| | |
|---|---|
| Paraffin Wax | 45.0 |
| Beeswax | 3.0 |
| Monoethanolamine | 2.6 |
| Stearic Acid | 13.5 |
| Water | 225.0 |

No. 6

| | |
|---|---|
| Paraffin Wax | 45.0 |
| Beeswax | 3.0 |
| Morpholine | 3.7 |
| Stearic Acid | 13.5 |
| Water | 225.0 |

No. 7

| | |
|---|---|
| Paraffin Wax | 40.0 |
| Casein | 5.0 |
| Monoethanolamine | 2.6 |
| Stearic Acid | 13.5 |
| Water | 225.0 |

No. 8

| | |
|---|---|
| Paraffin Wax | 40.0 |
| Casein | 5.0 |

| | |
|---|---|
| Morpholine | 3.7 |
| Stearic Acid | 13.5 |
| Water | 225.0 |

### No. 9

| | |
|---|---|
| Paraffin Wax | 40.0 |
| Carnauba Wax | 4.0 |
| Triethanolamine | 6.0 |
| Oleic Acid | 11.0 |
| Water | 225.0 |

### No. 10

| | |
|---|---|
| Paraffin Wax | 40.0 |
| Carnauba Wax | 4.0 |
| Monoethanolamine | 2.6 |
| Oleic Acid | 9.0 |
| Water | 225.0 |

### No. 11

| | |
|---|---|
| Paraffin Wax | 40.0 |
| Carnauba Wax | 4.0 |
| Morpholine | 3.7 |
| Oleic Acid | 9.0 |
| Water | 225.0 |

### No. 12

| | |
|---|---|
| Carnauba Wax | 15.0 |
| Triethanolamine | 1.5 |
| Oleic Acid | 2.3 |
| Water | 90.0 |

### No. 13

| | |
|---|---|
| Carnauba Wax | 15.0 |
| Monoethanolamine | 1.0 |
| Oleic Acid | 1.7 |
| Water | 90.0 |

### No. 14

| | |
|---|---|
| Carnauba Wax | 15.0 |
| Morpholine | 1.3 |
| Oleic Acid | 1.7 |
| Water | 90.0 |

### No. 15

| | |
|---|---|
| Beeswax | 15.0 |
| Triethanolamine | 1.5 |
| Oleic Acid | 2.6 |
| Water | 100.0 |

### No. 16

| | |
|---|---|
| Japan Wax | 15.0 |
| Triethanolamine | 1.5 |
| Oleic Acid | 3.0 |
| Water | 100.0 |

### No. 17

| | |
|---|---|
| Ester Gum | 23.0 |
| Triethanolamine | 0.8 |
| Oleic Acid | 4.2 |
| Water | 100.0 |

The emulsions of paraffin wax with carnauba wax and stearic acid are the most stable of all those suggested and show good stability even in extremely high dilutions. Casein can be used in place of carnauba wax or beeswax, although the resulting emulsions are less stable than those containing carnauba wax. For best results, oleic acid is not recommended with paraffin wax when either casein or beeswax is used. A mold inhibitor should be added to the hot melted wax mixture just before adding the water, when casein is used in the formulation. About 5% phenol, based on the casein, may be used.

Two different methods may be used for the preparation of these emulsions:

### METHOD 1

Melt the waxes and fatty acid in a hot-water or steam-jacketed kettle and heat to 95°C. When casein is used, heat the paraffin wax–stearic acid mixture to only 75°C. before stirring in the dry casein.

Add the amine slowly, stirring constantly, until the solution clears.

Heat the water to boiling and add it slowly, a little at a time, thoroughly incorporating each small portion before another addition is made.

When the water is first added, the mixture becomes increasingly viscous. When from one third to two thirds of the water has been added, it begins to thin out again. The emulsions containing casein, however, do not become so viscous during the addition of water as those made with the waxes. The water can be added more rapidly to emulsions without casein, but the rate of addition should be adjusted by experiment.

After the mixture has become definitely thinned and is of *smooth, even* consistency, the rest of the water may be added slowly, but continuously, with constant stirring. The emulsion should be stirred slowly, but continuously, until it has cooled to room temperature.

### METHOD 2

Melt the waxes and fatty acid and add the amine as directed in method 1. A temperature of 95°C. should be maintained.

Heat the water to the boiling point and pour it quickly, all at one time, into the melted wax–soap mixture. Stir thoroughly until a smooth, homogeneous mixture is obtained and then stir slowly, but continuously, until the emulsion has cooled to room temperature. If it is more convenient, the hot wax–soap mixture can be added quickly, all at one time, to the heated water.

Stearic acid can be substituted for oleic acid in all the formulae. While the resulting emulsions may be slightly more viscous, the stability should be as good as or better than when oleic acid is used.

## "Thiokol" Liquid Polymer Emulsion

### Formula No. 1

OIL PHASE

| | |
|---|---|
| "Thiokol" LP–2 | 32.8 |
| "Indopol" H–50 | 1.8 |
| "Indopol" L–100 | 1.8 |
| Xylene | 15.0 |
| Toluene | 15.0 |

WATER PHASE

| | |
|---|---|
| Emulsifier 160* | 1.4 |
| "Permosol" Base | 0.7 |
| Water | 31.5 |

### No. 2

OIL PHASE

| | |
|---|---|
| "Thiokol" LP–2 | 27.7 |

WATER PHASE

| | |
|---|---|
| 5% "Methocel" (15 cp) Solution | 33.1 |
| "Veegum" | 1.1 |
| Water | 38.1 |

The emulsions are prepared by first mixing the oil phase and warming at 110 to 120°F. The same procedure is followed with the water phase. Both solutions are then fed simultaneously into a colloid mill with 0.003-in. clearance. The emulsion is

---

* Emulsifier 160

| | |
|---|---|
| Glycerol Monostearate ("Aldo" 28) | 40 |
| "Permosol" Base | 10 |
| Sulfonated Neatsfoot Oil | 28 |
| "Tergitol" Penetrant No. 4 | 22 |

passed once more through the mill and, after standing for 24 hours at room temperature, it is given a final pass.

The "Thiokol" liquid polymer in the emulsion can be cured to an elastomer by means of suitable oxidizing agents, such as:

"Thiokol" Curing Agent

| | |
|---|---|
| Cumene Hydroperoxide (70%) | 3.3 |
| Diphenylguanidine | 1.2 |
| "Permosol" Base | 1.1 |
| 5% "Methocel" (15 cps) Solution | 4.0 |
| Water | 90.4 |

The emulsion is prepared by adding the cumene hydroperoxide to a solution of the water, "Permosol" base and "Methocel." The resultant mixture is emulsified by passing it once through a colloid mill with 0.003-in. clearance after which the diphenylguanidine is added to the solution. The final mixture is then given two more passes through the colloid mill. After standing at room temperature for 24 hours, the emulsion is passed once again through the colloid mill. The stability of the emulsion is limited to approximately 1 month.

Use 6 parts of curing agent in 100 parts of "Thiokol."

### Phenyl-2-Naphthylamine Dispersion
*U. S. Patent 2,510,882*

| | |
|---|---|
| Phenyl-2-Naphthylamine | 25.0 |
| Bentonite | 0.2 |
| Gum Arabic | 0.5 |
| Isopropanol (90%) | 5.0 |
| Water | 69.3 |

Ball mill for 24 hours.

---

### Potassium Cresylate Dispersion

| | |
|---|---|
| Cresol | 100.0 |
| Potassium Hydroxide | 34.7 |
| Water | 199.0 |

---

### Bacteria-Stable Casein Dispersion

| | |
|---|---|
| Casein | 100 |
| Potassium Hydroxide (25% Solution) | 15 |
| Potassium Cresylate (30% Solution) | 4 |
| "Moidol" (25% Solution) | 12 |
| Water | 869 |

---

### Calcium Silicate Dispersion

| | |
|---|---|
| Calcium Silicate | 50 |
| Sodium Alkyl Sulfonate | 2 |
| Water | 48 |

Ball mill for 3 hours.

### Calcium Carbonate Dispersion

| | |
|---|---|
| Calcium Carbonate | 30 |
| Sodium Alkyl Sulfonate | 2 |
| Water | 68 |

Mix the last two ingredients before adding the calcium carbonate. Stir for 10 minutes at 60V speed in an Eppenbach Homomixer or in a ball mill for 4 hours.

---

### Calcined Kaolin Dispersion for "Geon"

| | |
|---|---|
| Calcined Kaolin | 55.0 |
| Sodium Pyrophosphate | 0.5 |
| "Aerosol" OT | 0.5 |
| "Santomerse" S | 0.5 |
| Water | 43.5 |

---

### Clay Dispersion

| | |
|---|---|
| Clay | 50 |
| Sodium Alkyl Sulfonate | 2 |
| Water | 48 |

Mix the last two ingredients before adding the clay. Ball mill for 2 hours.

---

### Nonsettling Colloidal Clay
*French Patent 880,827*

| | |
|---|---|
| Colloidal Clay | 1 kg. |
| Water | 2 liters |
| Sodium Hexametaphosphate | 4.2 g. |

## Ft Carbon-Black Dispersion

| | |
|---|---|
| "Ft Black" | 60.0 |
| Sodium Alkyl Sulfonate | 2.4 |
| Water | 37.6 |

Ball mill for 16 hours.

---

### Sulfur Dispersion

| | |
|---|---|
| Sulfur | 99 |
| "Antaron"N–185 or "Antaron"L–135 or "Antaron"R–275 | 1 |

This will give a dispersion in hard or soft water.

---

## Titanium Dioxide Dispersion for Latex

### Formula No. 1

| | |
|---|---|
| Titanium Dioxide | 30.0 |
| Low-Viscosity Sodium Carboxymethyl-cellulose | 0.9 |
| Water | 69.1 |

Titanium dioxide, whiting and various clays can be dispersed, using a similar formula, by ball milling 3 to 8 hours. The amount of sodium carboxymethylcellulose required will generaly be 3 to 8% of the solid content of the slurry.

Where a low-viscosity dispersion is preferred, the following formula may be used:

### No. 2

| | |
|---|---|
| Titanium Dioxide | 45.00 |
| Sodium Metaphosphate | 1.35 |
| Water | 53.65 |

Sometimes the use of a combination of dispersing agents may be advantageous. Combination dispersions may be prepared by ball milling a compound similar to the following:

### No. 3

| | |
|---|---|
| Titanium Dioxide | 15.0 |
| Whiting | 20.0 |
| Low-Viscosity Sodium Carboxymethyl-cellulose | 1.2 |
| Water | 63.8 |

The dispersions should be alkaline before they are added to latex. The alkalinity can be adjusted by the addition of ammonia if the dispersion is to be added to a plastic latex or by sodium carbonate if the dispersion is to be added to "Geon" Latex 31X.

---

## Antimony Oxide Dispersion for "Geon"

| | |
|---|---|
| Antimony Oxide | 50.00 |
| Sodium Pyrophosphate | 0.25 |
| "Dresinate" 731 | 2.00 |
| "Santomerse" S | 0.50 |
| Water | 47.25 |

## CHAPTER VI

# FARM AND GARDEN PRODUCTS

Summer-Foliage Preservative

A solution made from 1 part glycerin and 2 parts water serves as the preservative for magnolia, beech, eucalyptus leaves and many other tree and shrub materials. Stems are split and placed to a depth of 3 to 5 in. in the glycerin-water solution. Allow to stand for approximately 2 weeks or until they have turned a warm brown in color. Foliage preserved in this way is soft, pliable, and attractive, lasts indefinitely, and can be used in either fresh or dry arrangements.

---

Cut-Flower Preservative

Formula No. 1
*Australian Patent 165,666*
2,2¹-bis (4,6-Dichlorophenol) Sulfide water solution. Use up to 5 p.p.m.

No. 2
*German Patent 952,753*

| | |
|---|---|
| Hydrazine Sulfate | 12 |
| Sodium Chloride | 6 |
| "Pril" (Detergent) | 1 |

Add 1 g. of above to 1 liter of water.

---

Fruit and Vegetable Peeling Liquid

| | |
|---|---|
| "Miranol" C2M Conc. | 1 |
| "Carbitol" | 1 |
| Caustic Soda | 20 |
| Water | 78 |

---

Peat Fertilizer
*Japanese Patent 8671 (1957)*

| | | |
|---|---|---|
| Peat | 100 | g. |
| Calcium Hydroxide | 2½ | g. |
| Water | 300 | g. |

Boil for 2 hours at 60 lb. pressure. Let stand for 20 hours. Dry.

## Gibberellin Aerosol

|  | Formula No. 1 (For Glass) | No. 2 (For Lacquered Metal) |
|---|---|---|
| Gibberellin | 0.00625 | 0.00625 |
| Surfactant | 0.20000 | 0.20000 |
| Alcohol (S.D. 40 Anhydrous) | 50.00000 | 20.00000 |
| Propellent | 50.00000 | 80.00000 |

---

### Fertilizing Compost
*German Patent 933,989*

| | |
|---|---|
| Penicillin Fermentation Waste (24% Dry Basis) | 1-13 |
| Peat (75% Dry Basis) | 1 |

Allow to ferment.

---

### Reducing Nicotine Content of Tobocco
*Swiss Patent 300,581*

| | |
|---|---|
| Hydrochloric Acid | 100 |
| Water | 100 |

Neutralize with dilute potassium silicate.

Immerse tobacco in solution from 1 minute to several hours, depending on the type of tobacco and amount of treatment desired. Wash quickly with water and dry.

Treatment for 14 hours reduces the nicotine content to 0.1 to 0.25%.

### Cotton-Plant Defoliant

| | |
|---|---|
| Magnesium Chlorate | 5 |
| Water | 95 |

---

### Protection of Seeds Against Crows
*German Patent 947,210*

| | |
|---|---|
| Diphenylguanidine | 70 g. |
| Oleic Acid | 430 g. |

Add the above to 50 kg. seeds and coat evenly.

---

### Potato Sprouting Stimulant

| | |
|---|---|
| Ethylene Chlorhydrin | 7 |
| Ethylene Dichloride | 3 |
| Carbon Tetrachloride | 1 |

---

### Tree-Wound Paint

The addition of ¼% of phenyl mercuric nitrate to an asphalt based paint or varnish prevents fungus growth.

## Dog Repellent
*German Patent 954,023*

| | |
|---|---|
| Citronella Oil | 8.5 |
| Anise Oil | 1.0 |
| Eucalyptus Oil | 0.5 |
| Polyacrylic Acid | 3.0 |
| Acetone-Alcohol | 95.0 |

This remains effective for 14 days after application.

---

## Feed Appetizer for Animals

The addition of small amounts of imitation anise oil (250 p.p.m.) or β-ionone (27 p.p.m.) to cattle and hog feeds stimulates them to eat more.

---

## Cattle Ringworm Control

Tincture of iodine mixed with equal parts of glycerin. This solution is applied after the body areas showing signs of ringworm infection (in most cases the head and the neck) are rubbed or scraped with a coarse brush or sandpaper.

---

## Poultry Drink Antiseptic

| | |
|---|---|
| "Cetrimide" | 25 g. |
| Water-Soluble Bacitracin | 25 g. |
| Dextrose | To make 1 lb. |

---

## Mastitis Udder Antiseptic Base

| | |
|---|---|
| "Cetrimide" | 0.5 |
| Neomycin (as Sulfate) | 0.4 |

## CHAPTER VII

# FOOD PRODUCTS

Imitation Mushroom Flavor

| | |
|---|---|
| Glucose | 1 |
| Sodium Glutamate | 1 |

Heat slowly with stirring for 1½ hr. at 195°C.

---

Food Flavor Intensifier

*U. S. Patent 2,790,719*

| | |
|---|---|
| Extract of Gentian | 1 |
| Glutamic Acid | 3-20 |

Dilute with salt or sugar for better distribution.

---

Horse-Radish Pungency Preserver

| | |
|---|---|
| Vitamin C | 0.005% |
| Isopropyl Gallate | 0.050% |

Add to horse radish.

---

Food Antioxidant

| | |
|---|---|
| Monoglycerides | 25 |
| Citric Acid | 5 |
| Propyl Gallate | 5 |
| Glycerol | ½ |
| Butylated Hydroxyanisole | 11 |
| Rice Oil | 53½ |

Use 8 oz. per 1000 lb. lard.

---

Food Antimold Coating

*U. S. Patent 2,856,294*

| | |
|---|---|
| Sorbic Acid | 2 |
| Sodium CMC (Medium Viscosity) | 3 |
| Water | 95 |

Mix until dissolved. Spray or dip and drain.

---

Meat Color Preservative

*U. S. Patent 2,863,777*

| | |
|---|---|
| Nicotinic Acid | 75 mg. |
| Ascorbic Acid | 100 mg. |
| Glucose | 1000 mg. |

Mix the above with

| | |
|---|---|
| Minced Meat | 1000 g. |

## Meat Tenderizer

### Formula No. 1
*Italian Patent 532,410*

| Papain | 6 |
| Rice Starch | 25 |
| Salt | 69 |

Use about 1.5 g. above per 100 g. meat.

### No. 2
*U. S. Patent 2,825,654*

| Fungal Enzymes | 5.000 |
| Salt | 77.000 |
| Dextrose | 15.000 |
| Monosodium Glutamate | 2.000 |
| Pepper | 0.125 |
| Celery Salt | 0.125 |
| Vegetable Oil | 0.750 |

## Champagne Clarifier

5 to 20 g. of Bentonite per 100 liters improves the clarity, aroma and bouquet.

## Noncaking Salt
*U. S. Patent 2,854,341*

Add a small amount of sodium aluminum silicate of particle size < 0.1 micron.

## Table Salt Substitute
*U. S. Patent 2,824,008*

| Potassium Chloride | 100.0 |
| Dipotassium Fumarate | 30.0 |
| Dipotassium Succinate | 70.0 |
| Sugar | 0.2 |

## Instant Coffee Tablet
*U. S. Patent 2,889,226*

| Instant Coffee Powder | 60-90 |
| Polyethyleneglycol (M.W. 3000-7500) | 4-10 |
| Sodium Bicarbonate | 2-10 |
| Alginic Acid | 4-20 |

## Baking Powder

### Formula No. 1
*U. S. Patent 2,793,121*

| Monobasic Calcium Phosphate | 8.7 |
| Aluminum-Sodium Sulfate | 20.0 |
| Calcium Sulfate | 14.2 |
| Sodium Bicarbonate | 30.0 |
| Cornstarch | 17.1 |

### No. 2
*U. S. Patent 2,870,017*

| Sodium Bicarbonate | 30 |
| Sodium Pyrophosphate | 42 |
| Calcium Sulfate | 5-28 |
| Filler (Starch) | 0-23 |

## Compressed Yeast Preservation
*East German Patent 9817*

Silver Nitrate 1 to 10 g. per ton of compressed yeast does not affect fermentation but will destroy harmful organisms.

## Hop Bitter Mix
(100 Gallon Batches)

| Formula | No. 1 | No. 2 | No. 3 | No. 4 | No. 5 | No. 6 | No. 7 | No. 8 | No. 9 | No. 10 | No. 11 | No. 12 |
|---|---|---|---|---|---|---|---|---|---|---|---|---|
| Hops | 6 lb. | 4 lb. | 3 lb. | 2 lb. | 2 lb. | 2 lb. | 3 lb. | 1½ lb. | 3 lb. | 2½ lb. | 1 lb. | 2½ lb. |
| Ginger | 2 lb. | 1 lb. | 2 lb. | 1½ lb. | 1 lb. | 2 lb. | ½ lb. | 1½ lb. | 2 lb. | 1 lb. | 1½ lb. | 2 lb. |
| Cayenne | 1 oz. | — | — | 2 oz. | — | — | — | — | — | — | — | 2 oz. |
| Horehound | — | 1 lb. | ¼ lb. | — | — | — | ½ lb. | — | — | — | 4 oz. | 4 oz. |
| Gentian | — | — | ½ lb. | — | ½ lb. | — | ½ lb. | — | — | — | — | 4 oz. |
| Comfrey | — | — | ¼ lb. | — | — | — | — | — | — | ½ lb. | — | — |
| Quassia | — | — | — | 4 oz. | — | — | — | — | — | — | 4 oz. | — |
| Chiretta | — | — | — | — | ¼ lb. | — | ¼ lb. | — | — | — | 4 oz. | — |
| Camomile | — | — | ¼ lb. | — | — | — | — | 1 lb. | — | — | — | 4 oz. |
| Sweet Flag | — | — | — | — | — | ½ lb. | — | — | — | — | — | 4 oz. |
| Dandelion | — | — | — | 4 oz. | — | — | — | — | — | — | 4 oz. | — |
| Liquorice | — | — | ½ lb. | 2 lb. | — | — | ¼ lb. | ½ lb. | — | ½ lb. | — | — |
| Nettle | — | — | — | — | — | — | ½ lb. | — | — | — | — | — |
| Sassafras | — | — | — | — | — | ¼ lb. | — | — | — | — | — | — |
| Alehoof | — | — | — | — | ½ lb. | — | — | — | — | — | 1 lb. | ¾ lb. |
| Buckbean | — | — | — | — | — | — | ½ lb. | 1 lb. | — | — | 1 lb. | 1 lb. |
| Orange Peel | — | — | — | — | — | 1 lb. | — | 2 lb. | — | — | — | 1½ lb. |
| Lemon Peel | — | — | — | — | — | ½ lb. | — | — | — | — | — | — |
| Centaury | — | — | — | — | — | — | — | — | ½ lb. | — | — | — |
| Zedoary | — | — | — | — | — | — | — | — | — | ½ lb. | — | — |
| Guaiac | — | — | — | — | — | — | — | — | ¼ lb. | — | — | — |
| Coltsfoot | — | — | — | — | — | — | — | — | ½ lb. | ½ lb. | — | — |

## Preventing Crystal Growth in Canned Marine Products

The presence of natural, hard, sharp, transparent crystals, glass-like in appearance in canned crab and canned lobster is not uncommon and has been the cause of many consumer complaints over recent years. These crystals can be prevented by the addition of 0.5% of sodium hexameta-phosphate. Such an addition appears to have no significance from a health angle.

## Cake Shortening
### U. S. Patent 2,882,167

| | |
|---|---|
| Vegetable Oil | 85-97 |
| Acetylated Monoglyceride | 3-15 |

## Liquid Shortening
### U. S. Patent 2,868,652

| | |
|---|---|
| Soybean Oil | 93 |
| Glyceryl Monostearate | 5 |
| Refined Stearic Acid | 2 |

Heat to 140°F., chill with stirring and run over "Votator."

## Swiss Meringue

| | |
|---|---|
| Egg Whites | 1 |
| Grandulated Sugar | 2 |

Flavor and color to suit the taste. Varied amounts may be made, but always in the proportion of twice as much sugar as egg whites.

Place sugar and whites in a grease-free mixing bowl. Mix until sugar is well moistened. Then set bowl in pan of simmering water and stir constantly until the mixture reaches 160°F. Remove bowl from water and put back on mixer. Beat at high speed until it becomes just a little warmer than the hand. Add flavor. Make up white pieces first. It is unnecessary to keep the machine running after the icing has reached a peak volume, but it is necessary to whip the icing well each time before filling the bag.

When coloring, keep it on the light side. Excess amounts will break down the icing. If a deeper color is desired, the pieces may be touched up with a medium stiff brush, after they are thoroughly dried. The brush must not be too damp or the pieces will become shiny, spoiling the entire effect. Proper shading will add new beauty and depth.

Fast Curing Meat Without Salt
U. S. *Patent 2,910,369*

Potassium Chloride 2,387 g.
Calcium Chloride 239 g.
Potassium Glutamate 460 g.
Potasium Nitrate 70 g.
Calcium Nitrate 35 g.

Sugar 525 g.
Ascorbic Acid 34 g.

Add this quantity per gal, of curing solution. This solution will cure meats in 48 hours against the usual one week to ten-day period.

## CHAPTER VIII

# INKS

Stencil Duplicating Ink

## Formula No. 1

### U. •S. Patent 2,771,372

| | |
|---|---|
| Nigrosine Dye | 5 |
| Polyethyleneglycol (6,000) | 47 |
| Glyoxal | 3 |
| "Aerosol" OT | 1 |
| Water | 44 |

## No. 2

### U. S. Patent 2,771,373

| | |
|---|---|
| Calcocid Scarlet M00 | 6 |
| "Aerosol" OT | 1 |
| Glyoxal | 1½ |
| Gelatin | 3 |
| Ethyleneglycol | 40 |
| Water | 45½ |
| Urea | 3 |

## No. 3

### U. S. Patent 2,771,374 ••

| | |
|---|---|
| Nigrosine J | 5 |
| Gum Arabic | 23 |
| Ethyleneglycol | 47 |
| Water | 25 |

## No. 4

### U. S. Patent 2,772,175

| | |
|---|---|
| Nigrosine H | 5.0 |
| Wetting Agent | 1.0 |
| Glyoxal (30% Sol) | 1.5 |
| Ethyleneglycol | 35.0 |
| Zinc Ammonium Poly-acrylate (2.5% Sol) | 57.5 |

---

Hectograph Transfer Sheet

### U. S. Patent 2,748,024

| | |
|---|---|
| Carnauba Wax | 8 |
| "Cardis" Wax | 11 |
| "Oronite" 128 | 3 |
| Petrolatum | 6 |
| Mineral Oil (100 sec.) | 17 |
| p-Diazoethylaniline Zinc Chloride | 55 |

Melt together and coat onto base sheet (15 to 20 lb. Per 3,000 ft.²).

## Duplicating Fluid

| | |
|---|---|
| Phloroglucinol | 0.5 |
| 2-Naphthol | 5.0 |
| Diethanolamine | 2.0 |
| Methanol | 92.5 |

## Hectograph Ink
### U. S. Patent 824,812

| | |
|---|---|
| Sugarcane Wax | 23 |
| Beeswax | 5 |
| Castor Oil | 18 |
| Lanolin | 24 |
| Mineral Oil | 30 |

Mix at 160 to 212°F.

## Hectograph Ink Base (Black)
### U. S. Patent 2,752,254

| | |
|---|---|
| Crystal Violet | 35-45 |
| Chrysoidine | 40-45 |
| Rhoduline Blue | 15-20 |

Disperse 100 g. of above in 80 to 200 cc. water and dry to moisture content of <2%.

## Improved Hectograph Solvent
### U. S. Patent 2,862,444

| | |
|---|---|
| Methanol | 80 |
| "Cellosolve" | 10 |
| Butyrolactone | 10 |

## Nontoxic Laundry Marking Ink

| | |
|---|---|
| Ethyleneglycol Phenyl Ether | 75 |
| Isopropanol | 5 |
| Nigrosine Base | 20 |

## Hard Finish Marking Ink
### U. S. Patent 2,836,573

| | |
|---|---|
| "Epon" 1,004 | 15.4 |
| Aluminum Nitrate | 15.4 |
| Rosin | 7.7 |
| Ethyl "Cellosolve" | 61.5 |

## Photographic Film Marking Ink
### U. S. Patent 2, 879,168

| | |
|---|---|
| 2-Furaldehyde | 80-2 |
| Cellulose Acetate Butyrate | 1-2 |
| Yellow Azo Dye | 10 |
| Blue & Orange Azo Dye | 2-6 |

## Erasable Drawing Ink
### U. S. Patent 2,833,736

| | |
|---|---|
| Graphite (2-5 microns) | 8-12 |
| Polyvinyl Alcohol | 0.2-0.3 |
| Water | 160-240 |

## Typewriter Ribbon Ink
### U. S. Patent 2,160,511

| | |
|---|---|
| Carbon Black | 6 |
| Tricresyl Phosphate | 15 |
| Nigrosine Base | 9 |
| Diglycol Laurate | 15 |

## Ball Point Pen Ink
### Formula No. 1
### U. S. Patent 2,882,172

| | |
|---|---|
| Carbon Black | 12 |

| | |
|---|---|
| Mineral Oil | 50 |
| Oleic Acid | 6 |
| Hydroabietyl Alcohol | 32 |
| Rosin | 3.5 |

### No. 2

| | |
|---|---|
| Victoria Blue Oleate | 56 |
| Victoria Blue Phospho-tungstic Toner | 27 |
| "Crill" S.6 | 2 |
| "Carbowax" 1,500, mixed with | 8 |
| Polyethyleneglycol 400 | 7 |

### Black Stand Oil Ink
### (Litho or Offset)

| | |
|---|---|
| Carbon Black | 19 |
| Bronze Blue | 9 |
| Linseed Stand Oil | 69 |
| Paste Driers | 3 |

It is common to partially substitute the stand oil with wax preparation (e.g. 10% wax pomade) and other additives, where improvement in rub resistance, gloss, appearance, or some special ink property is required.

### Co-Polymers Offset Ink

| | |
|---|---|
| Lake Red C | 80 |
| "Scopol" 41 | 100 |
| High Boiling Aliphatic Diluent | 50 |
| Lead Metal (As Naphthenate) | 1 |

| | |
|---|---|
| Cobalt Metal (As Naphthenate) | 0.12 |

### Blue Dry-Offset Ink

| | |
|---|---|
| Monastral Blue | 10 |
| Titanium Dioxide | 40 |
| Litho Varnish | 39 |
| Wax Pomade | 10 |
| Paste Driers | 1 |

### White Tin Printing Ink

| | |
|---|---|
| Titanium Dioxide Kronos Grade E | 50.6 |
| "Paralac" 15 | 46.3 |
| Terpineol | 2.3 |
| Cobalt Naphthenate (6%) | 0.8 |

### Letterpress Black Ink

| | |
|---|---|
| Carbon Black | 18 |
| Milori Blue | 5 |
| "Bentone" 34 | 3-4 |
| Resin (Maleic Rosin Modified) | 28.8 |
| Triethanolamine Resinate | 14.4 |
| Diethylene Glycol | 33.8 |

### White Cellophane L/P Ink

| | |
|---|---|
| Rutile | 60 |
| Wood Oil Phenolic | 30 |
| Refined Linseed Oil | 9 |
| Mild Driers | 1 |

## White Screen Ink

| | |
|---|---|
| Anastase | 26.5 |
| Precipitated Chalk | 15.0 |
| Blanc Fixe | 12.5 |
| "Beckosol" 9610 (50% Solid) | 46.0 |

Driers recommended are 0.5% lead, 0.025% cobalt, and 0.025% manganese.

---

## Rotary Press Ink

| | |
|---|---|
| Zein G200 | 100 |
| Gum Rosin | 100 |
| Oleic Acid | 20 |
| Aqueous Ammonia (28%) | 25 |
| Water | 600 |
| n-Butyl Alcohol | 60 |
| Pigment | 150 |

The alkaline dispersion of zein and resin in water is modified with 40 to 100 parts of organic solvent to make it less viscous and to produce a continuous film with high gloss when printing with the finished ink. The pigmented ink dries quickly and adheres to non-moistureproof cellophane, paper and paperboard.

---

## Flexographic Ink

| | |
|---|---|
| Zein G200 | 100 |
| Maleic or Fumaric Modified Rosin Ester | 100 |
| Alcohol, Denatured | 320 |

| | |
|---|---|
| Water | 20 |
| Iron Blue Pigment | 115 |
| Sodium Polyacrylate (Dry Basis) | 5 |

After the zein and resin are dissolved, the pigment is ground into the vehicle. This is done in a ball mill or with soft pigments in a colloid mill.

---

## Ink for Printing on Gelatin Sheets
### U. S. Patent 2,821,821

| | |
|---|---|
| Propyleneglycol | 500 |
| Water | 500 |
| FD & C Blue #1 (Dye) | 70 |
| Aerosol O.T. (70%) | 1 |

Allow to stand overnight before use.

---

## Fluorescent Printing Ink
### U. S. Patent 2,845,023
#### Formula No. 1
#### (Hot Melt)

| | |
|---|---|
| Fluorescent Pigment | 12 |
| Ethylcellulose | 1 |
| Beeswax, White | 9 |
| Boiled Linseed Oil | 2 |

#### No. 2
#### (Plastisol)

| | |
|---|---|
| Fluorescent Pigment | 45 |
| Polyvinyl Chloride, Powdered | 23 |
| Dioctyl Phthalate | 27 |
| Naphtha | 5 |

## Nonmisting Printing Ink

### Formula No. 1
### U. S. Patent 2,750,296

| | |
|---|---|
| Mineral Oil (Saybolt Viscosity 100-110 sec. at 100°F.) | 15.8 |
| Mineral Oil (Saybolt Viscosity 260-280 sec. at 130°F.) | 70.0 |
| Gilsonite | 1.2 |
| Carbon Black | 12.0 |
| Amine Bentonite | 1.0 |

### No. 2
### U. S. Patent 2,766,127

| | | |
|---|---|---|
| a | Mineral Oil | 87 |
| | Channel (Carbon) Black | 12 |
| | Gilsonite | 1 |
| b | Lecithin | 1 |
| | Water | 15 |
| c | Bentonite | 6 |

Mix a on 3 roller mill. Add b, mix vigorously and sift in c slowly. Stir slowly until uniform.

## Heat Set Ink

These inks set when the solvents are driven out by heating. The solvents, therefore, must be volatile, usually those boiling below 200°C. Inks of this type are made by first dissolving zein in low boiling glycols to produce the vehicle and then grinding the pigment into the vehicle. A typical heat set ink consists of a formulation such as this:

| | |
|---|---|
| Zein G200 | 100 |
| Propyleneglycol | 300 |
| Carbon Black | 100 |

## Nonsmudging Heat-Set Letterpress Black

| | |
|---|---|
| Carbon Black | 19 |
| Talc | 2 |
| Bronze Blue, In linseed oil (50%) | 4 |
| Reflex Blue, In linseed oil (40%) | 4 |
| "Bentone" 34 | 3-4 |
| Stand Oil | 5 |
| Varnish (Modified Phenol-Formaldehyde Ester Gum Varnish; 40% Solids) | 65 |

## Vapor Set Ink

### Formula No. 1

| | |
|---|---|
| Zein G200 | 100 |
| Maleic or Fumaric Modified Rosin Ester | 40 |
| Diethyleneglycol | 300 |
| Pigment | 150-300 |

In vapor set inks of this type, a resin of the dibasic acid modified condensate type is usually added with the zein. Rosin esters modified with maleic or fumaric acid are often used for this purpose.

The zein and modifying resin can be dissolved in the glycol at room temperature or at a higher temperature to make solution easier. Pigment is then ground into the resulting solution or vehicle on a conventional three or five roll mill.

### No. 2

| | |
|---|---|
| Zein G200 | 100 |
| Sodium Resinate | 20 |
| Diethyleneglycol | 230 |
| Hexyleneglycol | 50 |
| Water | 30 |
| Pigment | 50-150 |

Drying rate, viscosity and tack of such an ink can be controlled by varying the solvent composition.

### Water-Base Printing Ink

| | |
|---|---|
| Zein G200 | 100 |
| Hydrogenated Rosin | 50 |
| Isopropyl Alcohol | 225 |
| Aqueous Ammonia (28%) | 20 |
| Water | 375 |
| Titanium Dioxide | 180 |

The pigment is slurried into the vehicle which is then ball-milled or passed twice through a colloid mill.

### Water Color Ink Medium

| | |
|---|---|
| Glycerol | 36.50 |
| Water | 49.00 |
| Potassium Stearate | 1.00 |
| Formalin | 0.75 |
| Ammonium Alum | 0.25 |
| Gum Arabic | 9.00 |
| Dextrin | 3.50 |

### Waxless Carbon Paper Ink

#### Formula No. 1
*U. S. Patent 2,820,717*

| | | |
|---|---|---|
| a | "Vinylite" VYHH | 10.0 |
| | Mineral Oil | 27.5 |
| | Pigment | 7.5 |
| b | Ethyl Acetate | 45.0 |
| | Toluol | 15.0 |

Grind *a* in warm ball mill until uniform. Then add *b*.

### No. 2
*U. S. Patent 2,864,720*

| | |
|---|---|
| Nickel Sulfate | 55.08 |
| "Alkaterge" C | 0.92 |
| Mineral Oil | 18.23 |
| Lanolin | 2.76 |
| Gum Arabic | 0.92 |
| Beeswax | 1.83 |
| Carnauba Wax | 10.09 |
| Castor Oil | 3.49 |
| "Igepal" CA | 2.32 |
| Sodium p-Cymenesulfonate | 4.36 |

## No. 3
### (One-Time Black for Pencil and Machine)

| | |
|---|---|
| Raven Beads/"Spheron" 9 | 11 |
| Malacco/"Sterling" S | 9 |
| Methyl Violet Toner Oleate 30% | 2 |
| Calcium Carbonate | 5 |
| Paraffin Wax 143/148°F. | 25 |
| "Indracane" No. 517-A | 9 |
| Ink Oil (100 sec. Viscosity) | 32 |
| Petrolatum | 7 |

## No. 4
### (Blue One-Time Pencil for Form Use)

| | |
|---|---|
| Milori Blue | 20 |
| Methyl Violet Base | 1 |
| Montan Wax | 9 |
| "Indracane" No. 517-A | 7 |
| Paraffin Wax 143/148°F. | 28 |
| Green Petrolatum | 10 |
| Ink Oil (100 sec. Viscosity) | 25 |

Grind 3 hours in ball mill.

## No. 5
### (One-Time Blue)

| | |
|---|---|
| Iron Blue | 21.0 |
| Methyl Violet Oleate | 1.5 |
| Calcium Carbonate | 6.0 |
| Paraffin Wax 143/145°F. | 32.0 |
| "Indracane" No. 517-A | 9.0 |
| Ink Oil (1500 sec. Viscosity) | 30.5 |

Grind 3 hours in ball mill.

## No. 6
### (One-Time Blue)

| | |
|---|---|
| Milori Blue | 19.5 |
| Methyl Violet Oleate | 3.0 |
| Paraffin Wax 143/148°F. | 20.0 |
| No. 4517 Ink Wax | 26.0 |
| "Indracane" No. 517 | 8.0 |
| Ink Oil (100 sec. Viscosity) | 10.5 |
| Ink Oil (1500 sec. Viscosity) | 13.0 |

Grind 3 hours.

---

## Ceramic Ware Decorating Crayon
### U. S. Patent 2, 835,600

| | | |
|---|---|---|
| a | Lead Nitrate | 73 |
| | Antimony Oxide | 33 |
| | Alumina | 12 |
| | Sodium Chloride | 100 |
| b | Calcium Stannate | 3 |
| | Dextrin (10%) Solution | 1 |

Add 10 parts of a to b.

Mix to a paste. Press into form of crayon and heat to 400 to 1000°C.

This crayon is used to write on ceramic ware before firing, after which a yellow color develops.

## Chapter IX

# INSECTICIDES, FUNGICIDES, AND WEED KILLERS

## Insect Repellent

### Formula No. 1

| | |
|---|---|
| Diglycollaurate | 14 |
| 2-Ethylhexandiol | 8 |
| Dimethylphthalate | 16 |
| "Indalone" | 6 |
| Water | 55 |

### No. 2

| | |
|---|---|
| Sorbitol Monolaurate | 7.0 |
| Citronellal | 5.0 |
| Terpinylacetate | 2.0 |
| Ethylbutylpropandiol | 8.0 |
| "Nipagin" | 0.1 |
| Water | 78.0 |

### No. 3

| | |
|---|---|
| Stearin | 20.00 |
| Olein | 0.60 |
| Dimethylphthalate | 20.00 |
| "Indalone" | 2.00 |
| Triethanolamine | 1.10 |
| Sodium Hydroxide | 0.36 |
| Sodium Acetate | 0.50 |
| Water | 100.00 |

### No. 4

| | |
|---|---|
| Isopropyl Cinnamate | 4 |
| Dimethylphthalate | 20 |
| "Crag" Repellent | 16 |
| Alcohol (96%) | 60 |

### No. 5

| | |
|---|---|
| "Crag" Fly Repellent | 50.0 |
| Methoxychlor, technical (90%) | 5.0 |
| Emulsifier A | 4.4 |
| "Velsicol" AR-50 | 40.6 |

## Silicone Insect Repellent Emulsion
### U. S. Patent 2,684,878

| | |
|---|---|
| Mixture of 2-Phenylcyclohexanol (70%) and 2-Cyclohexylcyclohexanol (30%) | 59.2 |
| Dimethyl Polysiloxane Fluid, 1000 cs | 6.6 |
| Polyethylene Oxide Stearate | 10.5 |

Glyceryl Monostearate 10.5
Water 13.2

---

### Insect Repellent Sticks
*U. S. Patent 2,819,995*

| | |
|---|---|
| 2-Ethyl-1,3-Hexan-ediol | 65-75.0 |
| Stearic Acid, Triple pressed | 18-23.4 |
| Ozokerite | 7-11.6 |

Warm and mix until clear; pour into molds and cool slowly. Rub stick on skin to leave an insect repellent coating which is not washed off by water or perspiration. It can be removed by washing with water and soap or detergent.

### Clothing Insecticide Impregnant

| | |
|---|---|
| Benzyl Benzoate | 30 |
| N-Butylacetanilide | 30 |
| 2-Butyl-2-Ethyl-1,3 Propanediol | 30 |
| "Tween" 80 | 10 |

---

### Insecticide Cone (Burning)
*Italian Patent 505,718*

| | |
|---|---|
| Lindane | 58 |
| Rosin | 25 |
| Sugar | 12 |
| Mint Oil | 5 |

Melt together, mix and pour into forms.

---

### Aerosol Spray Concentrate

| | Formula No. 1 | | No. 2 | |
|---|---|---|---|---|
| | *gal.* | *lb.* | *gal.* | *lb.* |
| "Crag" Fly Repellent | 50.0 | 412.5 | 50.0 | 412.5 |
| "Pyrenone" 50-5 | 10.0 | 79.0 | — | — |
| "Pyrexcel" 50-5 Special | — | — | 10.0 | 82.0 |
| Petroleum Distillate | 40.0 | 260.0 | 40.0 | 260.0 |

The final formulation should be prepared to contain 70% fluorinated hydrocarbon propellent and 30% concentrate. The average pressurized sprayer, which holds 12 oz. (340 g.), will contain 3.6 oz. (102 g.) concentrate and 8.4 oz. (238 g.) of propellent.

---

### Fumigant Insecticidal Tablets
*Spanish Patent 208,955*

| | |
|---|---|
| Benzenehexachloride | 10 |
| Potassium Chlorate | 25 |
| Perfume (Incense) | 2 |
| Sugar | 6 |
| Talc or Kaolin | 47 |

| Wettable Powder Insecticide | | Emulsifier B | 8.0 |
|---|---|---|---|
| Dieldrin | 75.75 | Xylene | 27.8 |
| "Micro-Cel" 805 | 19.25 | | |
| "Polyfon" H | 4.00 | | |
| "Wetanol" | 1.00 | | |

**Lindane Emulsion**

| Lindane | 20.0 |
|---|---|
| Emulsifier C | 7.5 |
| "Velsicol" AR-50 | 32.5 |
| Isophorone | 40.0 |

**Insecticide Emulsion**

2-Ethylhexyl 2,4,5 Tri-
chlorophenoxyacetate 64.2

### Methyl Parathion Emulsion Concentrate
(100-gal. batch)

| Ingredients | % weight | lb. | gal. at 68°F. |
|---|---|---|---|
| Technical Methyl Parathion | 31.40 | 253.0 | 24.6 |
| Xylene | 63.60 | 511.8 | 70.8 |
| "Atlox" 3335 | 4.25 | 34.2 | 3.9 |
| "Atlox" 8916P | 0.75 | 6.0 | 0.7 |
| | 100.00 | 805.0 | 100.0 |

Typical Properties: Density (68°F.) = 8.05 lb. per gal. Methyl parathion contents = 2.02 lb. per gal.

### Methyl Parathion Dust Concentrate

| | 20% MEP Dust Concentrate | | 25% MEP Dust Concentrate | |
|---|---|---|---|---|
| Ingredients | % weight | Quantity to Make 1000 lb. | % weight | Quantity to Make 1000 lb. |
| Technical Methyl Parathion | 25.3 | 253 lb. | 31.6 | 316 lb. |
| "Attaclay" | — | — | 68.4 | 684 lb. |
| "Pike's Peak" 9T66 | 74.7 | 747 lb. | — | — |
| | 100.0 | 1000 lb. | 100.0 | 1000 lb. |

Typical Properties:

    Particle size; % through a 200-mesh screen = 95

    Methyl Parathion Content, % Weight

| | |
|---|---|
| 20% MEP Dust Concentrate | = 20.2 |
| 25% MEP Dust Concentrate | = 25.3 |
| Flowability | = Free flowing |

---

## "Strobane" Formulations for Liquid Sprays

Very effective space sprays for use against both flying and crawling insects are formulated by using approximately 2% "Strobane" with suitable knockdown agents. It is however, possible to formulate an effective spray for flying insects containing as little as 0.5% of this new insecticide.

Suitable knockdown agents include:

Pyrethrin concentrates

Pyrethrin concentrates plus synergists

Allethrin plus synergists

"Lethan" 384 regular

"Thanite"

Typical formulations based on active ingredients by weight are as follows. These formulations will far surpass minimum AA-Grade requirements.

### Formula No. 1

| | |
|---|---|
| Petroleum Distillate | 97.925 |
| "Strobane" | 2.000 |
| Pyrethrins | 0.075 |

### No. 2

| | |
|---|---|
| Petroleum Distillate | 97.775 |
| "Strobane" | 2.000 |
| Technical Piperonyl Butoxide | 0.200 |
| Pyrethrins | 0.025 |

### No. 3

| | |
|---|---|
| Petroleum Distillate | 97.663 |
| "Strobane" | 2.000 |
| Technical Piperonyl Butoxide | .281 |
| Allethrin | .056 |

### No. 4

| | |
|---|---|
| Petroleum Distillate | 97.06 |
| "Strobane" | 2.00 |
| B-Butoxy-B'-Thicyano-diethyl Ether | .94 |

### No. 5

| | |
|---|---|
| Petroleum Distillate | 95.6 |
| "Strobane" | 2.0 |
| Technical Isobornyl Thiocyanate | 2.4 |

## Aerosols

A concentration of 2% "Strobane" with knockdown agents and synergists will produce a highly effective aerosol formulation for use against both flying and crawling household pests.

The following are typical aerosol formulations.

### Formula No. 1

| | |
|---|---|
| Petroleum Distillate | 12.0 |
| "Strobane" | 2.0 |
| Technical Piperonyl Butoxide | 0.8 |
| Pyrethrins | 0.2 |
| Propellants | 85.0 |

### No. 2

| | |
|---|---|
| Petroleum Distillate | 9.8 |
| "Strobane" | 2.0 |
| Sulfox-cide-Technical | 1.0 |
| Pyrethrins | .2 |
| Propellants | 87.0 |

### No. 3

| | |
|---|---|
| Petroleum Distillate | 17.3 |
| "Strobane" | 2.0 |
| Technical Piperonyl Butoxide | .5 |
| Pyrethrins | .2 |
| Propellants | 80.0 |

### 65% "Toxaphene" Emulsifiable Concentrate (Cationic)

| | |
|---|---|
| "Toxaphene" | 65 |
| Xylol or kerosene | 25 |
| "Arquad" 2C | 5 |
| "Ethofat" 0/20 | 5 |

Combine all ingredients, heat and stir until a clear solution is formed. This concentrate disperses in water with very slight agitation. Any cream forming on long standing is easily redispersed by simply stirring.

---

### 45% Chlordane Emulsifiable Concentrate (Cationic)

| | |
|---|---|
| Technical Chlordane | 45.0 |
| "Arquad" 2C | 2.5 |
| "Ethofat" 0/20 | 2.5 |
| Kerosene | 50.0 |

The ingredients are simply mixed to form a clear solution. The concentrate disperses easily in water to form a stable emulsion.

---

### 85% Chlordane Emulsifiable Concentrate (Nonionic)

| | |
|---|---|
| Technical chlordane | 85 |
| "Ethofat" 0/20 | 15 |

The ingredients are mixed until homogeneous.

## Phosdrin Insecticide Dust

| Ingredients | % Weight | Ingredients to Make 1000-lb. Batch |
|---|---|---|
| Phosdrin Insecticide | 1.02 | 10.2 lb. |
| "Pyrax" ABB | 98.98 | 989.8 lb. |
| | 100.00 | 1000.0 lb. |

Typical Properties:

 Particle size, % through a 200-mesh screen = 95
 Phosdrin insecticide content, % weight = 1.02
 Bulk density, lb. per cu. ft. = 57 to 60
 Flowability = Free flowing

## Endrin Insecticide Emulsion Concentrate

| Ingredients | % Weight | Ingredients to Make 100-gal. Batch |
|---|---|---|
| Technical Endrin | 19.8 | 161.0 lb. |
| Xylene | 72.7 | 82.1 gal. |
| "Triton" X-155 | 5.0 | 40.3 lb. |
| "Triton" B-1956 | 2.5 | 20.5 lb. |

## Endrin Insecticide Dust

| Ingredients | Formula No. 1 % Weight | No. 2. % Weight |
|---|---|---|
| Technical Endrin | 25.2 | 25.2 |
| "Attaclay" | 71.8 | 35.9 |
| Barden Clay, AG | – | 37.0 |
| Hexamethylenetetramine | 3.0 | 1.9 |

## Dieldrin Emulsifiable Concentrate

| Ingredients | % Weight | Ingredients to Make 100-gal. Batch |
|---|---|---|
| Technical Dieldrin | 19.1 | 153.0 lb. |
| Xylene | 74.9 | 83.4 gal. |
| "Triton" X-155 | 4.0 | 32.4 lb. |
| "Triton" B-1956 | 2.0 | 16.4 lb. |

Dieldrin Insecticide Base

| Ingredients | % Weight |
|---|---|
| Technical Dieldrin | 50.5 |
| "Attaclay" | 22.4 |
| "Velvex" | 22.4 |
| Urea | 2.2 |
| "Tamol" 731 Dry | 2.0 |
| "Duponol" ME Dry | 0.5 |

Aldrin Wettable Powder
Insecticide

Technical Aldrin (85% minimum Purity).

Dry Ice.

"Hi-Sil" 101.

"Santomerse" D.

Note: Other wetting agents may be used but, before doing so, they must be known to be non-phytotoxic to the seed to be treated.

Reduce the technical aldrin to chunks of about 2 in. diameter or less. Add the "Hi-Sil" 101 to the blender and, while blender is operating, add the crushed aldrin to the blender as fast as possible without stalling the motor. Blend until the majority of the aldrin is reduced to about ¼ in. pieces or smaller. This step will usually require 15 to 20 minutes and its completion is readily apparent from the reduction of noise in the blender. Add the "Santomerse" D to the blender and blend for an additional 5 minutes. The amount of wetting agent used is held to a minimum (not more than 1%) because it replaces part of the necessary sorptive carrier.

Gross grind the preblended powder, add pulverized dry ice, and fine grind the mixture through a Mikro-Pulverizer or the equivalent. With some types of equipment, the gross grinding step can be eliminated and the preblended powder fed directly to the final grinding mill. It is necessary to have appreciable amounts of dry ice in the feed to the Mikro-Pulverizer, in order to prevent plugging of the screen. Transfer the finely ground material to the ribbon afterblender and blend for 10 to 15 minutes before packaging.

## Aldrin-DDT Emulsifiable Concentrate

| Ingredients | % Weight | Ingredients to Make 100-gal. Batch |
|---|---|---|
| Technical Aldrin | 11.6 | 102.2 lb. |
| Technical DDT | 23.2 | 203.0 lb. |
| Xylene (Petroleum type) | 58.2 | 71.1 gal. |
| "Triton" X-155 | 3.0 | 3.0 gal. |
| "Triton" B-1956 | 4.0 | 4.0 gal. |

## DDT Insecticide

| | |
|---|---|
| DDT | 25 |
| Emulsifier A or C* | 5 |
| Xylene | 70 |

## DDT Dispersion Concentrate
### German Patent 922,383

| | |
|---|---|
| DDT | 18 |
| Tetrahydronaphthalene | 72 |
| Cyclohexylaminedodecyl Sulfonate | 10 |

This stirred into water gives a stable white dispersion.

## 25% DDT Emulsifiable Concentrate (Cationic)

| | |
|---|---|
| DDT | 25.0 |
| "Arquad" 2C | 1.0 |
| "Ethofat" 0/20 | 0.5 |
| Xylol | 73.5 |

---

* A—2 parts "Tergitol" Nonionic NPX
  1 part "Tergitol" Nonionic XD
  2 parts Isopropanol (99%)
  C—95 parts "Tergitol" Nonionic XD
  5 parts Isopropanol (99%)

The DDT is dissolved in xylol and the emulsifiers added. The mixture is stirred until clear and uniform.

This concentrate is readily emulsified in water. The emulsion is unaffected by water hardness.

## DDT-Oil Larvicide

| | |
|---|---|
| DDT | 1.0 |
| "Ethofat" 0/20 | 0.5 |
| Kerosene | 98.5 |

## Self-emulsifying Agricultural Spray Oil

| | |
|---|---|
| Agricultural Spray Oil | 97.5 |
| "Ethofat" 0/15 | 2.0 |
| Mahogany Soap | 0.5 |

The ingredients are mixed until a clear solution is obtained. The concentrate is self-emulsifying and the emulsion can be used as an oil emulsion spray.

## Agricultural Spray Adhesive or "Sticker"

*Japan Patent 3999 (1954)*

|  | No. 1 | No. 2 |
|---|---|---|
| Sodium Alkyl-benzene Sul-fonate | 50 | 95 |
| Sodium Lignosul-fonate | 50 | — |
| Ammonium Citrate | — | 5 |

### No. 3

*German Patent 850,537*

| | |
|---|---|
| Copper Oxychloride | 3.0 |
| Calcium Oxide | 1.0 |
| Sodium Cellulose Glycollate | 0.4-0.6 |
| Water | To suit |

This gives a water and carbon dioxide resistant coating.

---

## Cattle & Space Fly Spray

| | |
|---|---|
| "Crag" Fly Repellent | 57.8 |
| Synergized Pyrethrins | 33.0 |
| Methoxychlor, (90% Technical Oil Conc.) | 7.6 |
| Petroleum Distillate | 572.0 |

---

## Aerosol Cattle Fly Spray

### Formula No. 1

| | |
|---|---|
| Pyrethrins | 0.2 |
| Piperonyl Butoxide | 2.0 |
| Methoxychlor | 3.0 |
| Methylated Naphtha-lenes | 5.0 |
| Deodorized Base Oil (Petroleum) | 9.8 |
| Propellent | 80.0 |

### No. 2

| | |
|---|---|
| Pyrethrins | 0.2 |
| Piperonyl Butoxide | 2.0 |
| Methoxychlor | 3.0 |
| "Thanite" | 2.0 |
| Methylated Naphtha-lenes | 5.0 |
| Deodorized Oil | 17.8 |
| Propellent | 70.0 |

### No. 3

| | |
|---|---|
| Pyrethrins | 0.2 |
| Piperonyl Butoxide | 2.0 |
| Methoxychlor | 3.0 |
| "Lethan" 384 | 4.0 |
| Methylated Naphtha-lenes | 5.0 |
| Deodorized Base Oil | 15.8 |
| Propellent | 70.0 |

### No. 4

| | |
|---|---|
| Pyrethrins | 0.25 |
| Piperonyl Butoxide | 2.50 |
| Methoxychlor | 3.00 |
| Methylated Naphtha-lenes | 5.00 |
| Deodorized Base Oil | 19.25 |
| Propellent | 70.00 |

### No. 5

| | |
|---|---|
| Pyrethrins | 0.25 |
| Piperonyl Butoxide | 2.50 |
| Methoxychlor | 3.00 |
| Methylated Naphtha-lenes | 5.00 |

| | |
|---|---|
| Deodorized Base Oil | 4.25 |
| Propellent | 85.00 |

### No. 6

| | |
|---|---|
| Pyrethrins | 0.5 |
| Piperonyl Butoxide | 4.0 |
| Base Oil, Deodorized | 10.5 |
| Propellent | 85.0 |

### No. 7

| | |
|---|---|
| Pyrethrins | 0.5 |
| Piperonyl Butoxide | 4.0 |
| Deodorized Base Oil | 25.5 |
| Propellent | 70.0 |

### No. 8

| | |
|---|---|
| Allethrin | 0.4 |
| Piperonyl Butoxide | 2.0 |
| Methoxychlor | 3.0 |
| Methylated Naphthalenes | 5.0 |
| Deodorized Base Oil | 19.6 |
| Propellent | 70.0 |

### No. 9

| | |
|---|---|
| Pyrethrins | 0.25 |
| Piperonyl Butoxide | 2.50 |
| Methoxychlor | 3.00 |
| Methylated Naphthalenes | 5.00 |
| Deodorized Base Oil | 4.25 |
| Propellent | 85.00 |

### Mothproofing Aerosol

| | |
|---|---|
| Methoxychlor | 5.0 |
| Chlordane | 1.5 |
| Odorless Kerosene | 18.8 |

| | |
|---|---|
| Methylene Chloride | 9.5 |
| Perfume | 0.2 |
| "Freon" 11 | 32.5 |
| "Freon" 12 | 32.5 |

### Peanut Storage Fumigant

### Formula No. 1

For use in mechanical generators such as the Microsol, Challenger, or Skilblower, or as a spray.

| | |
|---|---|
| Pyrethrins | 0.5 |
| Synergist (Piperonyl butoxide, sulfoxide, n-propyl isome) | 5.0 |
| Tetrachloroethylene | 50.0 |
| Deodorized Kerosene | 44.5 |

Mixing directions when formulated on the job:

| | |
|---|---|
| Concentrate containing 5% pyrethrins and 50% synergist | 2 pt. |
| Tetrachloroethylene | 6 pt. |
| Deodorized Kerosene | 8 pt. |

*Application rate:*

1 pt. per 10,000 cu. ft. of space above the load, or 2 gal. to an average warehouse 100 x 100 ft., with 15 to 20 ft. of space above the load.

Note: Where peanuts are piled almost to the roof, apply as a wet spray to the top surface of the load, at the rate of 2 gal. per 100 x 100 ft. of surface.

## No. 2

For use in thermal type generators, such as the Tifa, Swingfog, etc.

| | |
|---|---|
| Pyrethrins | 0.2 |
| Synergist | 2.0 |
| Tetrachloroethylene | 50.0 |
| Deodorized Kerosene | 47.8 |

*Mixing directions when formulated on the job:*

| | |
|---|---|
| Concentrate (5-50) | 1 pt. |
| Tetrachloroethylene | 6 pt. |
| Deodorized Kerosene | 9 pt. |

*Application rate:*

2½ pt. per 10,000 cu. ft. of space over the load, or 5 gal. in average warehouse 100 x 100 ft., with 15 to 20 ft. of space above the load.

---

## Grain Fumigant
### U. S. Patent 2,895,870

| | |
|---|---|
| Carbon Tetrachloride | 82-83 |
| Carbon Disulfide | 16½-17½ |
| Petroleum Ether | ½-1 |

---

## Nonflammable Fumigant
### U. S. Patent 2,803,581

| | |
|---|---|
| Carbon Bisulfide | 13 |
| Ethylene Dibromide | 7 |
| Methylene Chloride | 2-20 |
| Carbon Tetrachloride | 66-78 |

## Increased Flash Point Fumigant
### U. S. Patent 2,824,040

| | |
|---|---|
| Carbon Tetrachloride | 79 l. |
| Carbon Disulfide | 21 l. |
| Benzol | 3 l. |
| Sulfur Dioxide | 0.5-1.25 kg. |
| Petroleum Ether | 3 l. |

Flashpoint 120°F.

---

## Dog Repellent

| | |
|---|---|
| Lemongrass Oil | 13 |
| Denatured Alcohol | 87 |

---

## Rodenticide & Rat Repellent
### Formula No. 1

| | |
|---|---|
| Zinc Phosphide | 50 |
| Talc | 48 |
| DDT | 2 |

This is strewn around rat runs.

## No. 2
### U. S. Patent 2,862,849

2,3,5,6-Tetrachloronitrophenol butyl ether.

## No. 3
### British Patent 790,022

| | |
|---|---|
| Fluoroacetamide | 0.1 - 1 |
| Bran, Oats and Starch | To make 100 |

## For Cordage & Sacks

### Formula No. 1
*U. S. Patent 2,822,295-6*

Dodecyl alcohol or its acetate dissolved in a volatile solvent.

### No. 2
*U. S. Patent 2,864,727*

Treat twine with hydrocarbon solution of

| | |
|---|---|
| Quinaldine | 2 |
| Naphthenic Acid | 2 |

---

### Plant Defoliant
*U. S. Patent 2,749,227*

| | |
|---|---|
| Sodium Chlorate | 25 - 60 |
| Sodium Metaborate | 75 - 40 |

---

### Fruit Tree Defoliant

| | |
|---|---|
| Magnesium | |
| Chloride | 0.25 - 0.5 |
| "Endothal" | 0.075 - 0.1 |
| Water | To make 100 |

---

### Borax Weed Killer
*U. S. Patent 2,711,367*

| | |
|---|---|
| Aluminum | |
| Sulfate | ¼ - 2 lb. |
| Borax | ½ - 4 lb. |
| Water | 1 - 5 gal. |

---

### Rope Preservative Fungicide

| | Formula No. 1 | No. 2 |
|---|---|---|
| Copper | | |
| Naphthenate | 35 | — |
| 80% Hard Asphalt in Mineral | | |
| Spirits | 12 | — |
| Mineral Oil | | |
| S.A.E. 10 | 17 | 19 |
| Scale Wax | 33 | 33 |
| Mineral Spirits | 3 | 3 |
| Zinc Naphthenate | — | 45 |

Warm and mix. Apply at 130°F.

---

### Agricultural Fungicide
### Formula No. 1

| | |
|---|---|
| Copper Resinate | 20.0 |
| Mercury Phenyl | |
| Salicylate | 0.2 |
| Mineral Diluent | 79.8 |

### No. 2
*British Patent 775,981*

| | |
|---|---|
| "Cetrimide" | 2 g. |
| Beta-Naphthoxyacetic | |
| Acid | 1 g. |
| Carboxymethyl- | |
| cellulose Sodium | 10 g. |
| Water | To make 2 gal. |

---

### Citrus Fruit Fungicide
*Spanish Patent 235,781*

Immerse fruit in 0.1 to 5.0% chloroacetic acid for 1 to 10 minutes, drain and wash.

---

### Oak Wilt Fungus Spray

| | |
|---|---|
| Tartaric Acid | 0.6 |
| Water | 99.0 |
| Wetting Agent | 0.4 |

Cotton Insecticides and Acaricides (pounds per acre)

| Pesticides | Boll Weevil | Boll Worms | Cotton Aphid | Cotton Flea-hopper | Cut-Cotton | Fall Army-worm | Grass-hoppers | Lygus & Other Mirids | Cotton Leaf-worm | Spider Mites | Stink Bugs | Pink Boll-worms | Thrips |
|---|---|---|---|---|---|---|---|---|---|---|---|---|---|
| Aldrin | 0.25-0.75 | — | — | 0.20-0.25 | — | 0.25-0.50 | 0.10-0.25 | 0.25-0.75 | — | — | — | — | 0.10-0.15 |
| Aramite | — | — | — | — | — | — | — | — | — | 0.33-1.0 | — | — | — |
| BHC (Gamma) | 0.30-0.75 | — | 0.3-0.6 | 0.10 | — | 0.3-0.6 | 0.30-0.50 | 0.30-0.45 | — | — | 0.5 | — | 0.1-0.2 |
| Calcium Arsenate[2] | 7-10 | 10-15 | — | — | — | — | 0.5-1.5 | 1.0-1.5 | 7-10 | — | — | — | — |
| Chlordane | — | — | — | 0.20 | — | 1.5-2.0 | — | — | — | — | — | — | — |
| Demeton | — | — | 0.125-0.4 | — | — | — | 0.07-0.125 | 1.0-1.5 | — | 0.125-0.40 | — | — | — |
| DDT | — | 1-2 | — | 0.5 | 1-2.5 | 0.5-1.5 | 0.25-0.50 | 0.15-0.50 | — | — | — | 2-3 | 0.5-1.0 |
| Dieldrin | 0.15-0.50 | — | — | 0.10-0.15 | 0.37-0.50 | 0.15-0.30 | — | — | 0.2-0.5 | — | 0.5 | — | 0.25-1.5 |
| Endrin[3] | 0.20-0.50 | — | — | 0.10 | — | — | — | 0.20-0.50 | — | — | — | — | 0.10-0.15 |
| Heptachlor | 0.25-0.75 | — | — | 0.20-0.25 | — | — | — | — | — | — | — | — | 0.10-0.15 |
| Nicotine | — | — | 0.3-0.45 | — | — | — | — | — | — | — | — | — | 0.10-0.15 |
| Parathion | — | — | 0.10-0.25 | — | — | — | — | 0.25-0.75 | 0.125 | 0.1-0.4[5] | — | — | — |
| Poison Baits | — | — | — | — | 6 | — | 6 | — | — | — | — | — | — |
| Sulfur[2] | — | — | — | — | — | — | — | — | — | 20-60[5] | — | — | — |
| TEPP3,4 | — | — | 0.5 Pt. | — | — | — | — | — | — | — | — | — | — |
| Toxaphene | 2-3 | 2-4 | — | 1.0 | 2-5 | 2-3 | 1.0-2.5 | 2-3 | 1.5-2.0 | — | 6.0 | — | 0.75-1.0 |

[1] All compounds may be applied either as dusts or as emulsion sprays unless otherwise indicated.

[2] Dust only.

[3] Spray only.

[4] 40% concentrate.

[5] Does not control all species of spider mites on cotton.

[6] As recommended by state or federal agencies.

Insect Repellent Stick for the
Skin

*U. S. Patent 2,465,470*

| | |
|---|---|
| o-Dimethyl Phthalate | 12.2 · |
| 2-Ethyl 1,3-Hexanediol | 18.0 |
| 91% Isopropyl Alcohol | 31.5 |
| Powdered Sodium | |
| Stearate | 20.3 |
| Glycerin | 12.1 |
| Distilled Water | 3.5 |
| Color Solution | 1.2 |
| Perfume | 1.2 |

The first three ingredients are mixed together and the soap, glycerin, water, and coloring are added. The combination is heated at 81 to 82°C., with occasional stirring. When a clear solution is obtained (usually after 7 to 10 minutes), the solution is removed from the heat and allowed to cool to 60°C. At this point, the perfume is added, the mass is poured into molds, and allowed to cool at room temperature.

---

Insect Repellent

Formula No. 1

| | |
|---|---|
| Dimethyl Phthalate | 210 |
| Magnesium Stearate | 30 |
| Zinc Stearate | 70 |
| Diglycol Stearate | 10 |

Grind together until uniform.

No. 2

| | |
|---|---|
| Benzyl Benzoate | 45 |
| Dibutyl Phthalate | 45 |
| Nonionic Emulsifier | 10 |

No. 3

*U. S. Patent 2,496,270*

| | |
|---|---|
| 2-Ethyl 1,3-Hexanediol | 10 |
| Phenolphthalein (10% | |
| Alcoholic Solution) | 1 |
| Glycerin | 2 |
| Alcohol | 2 |

These ingredients are blended and then agitated with an excess of sodium carbonate. The undissolved solids are removed by centrifuging. The resultant product is a somewhat viscous liquid with an intense red color. All of the color disappears within a few minutes after the dressing is applied to the skin. The glycerin is beneficial to the skin and at the same time serves to intensify the red color of the solution.

No. 4

| | |
|---|---|
| Dimethyl Phthalate | 90 |
| "Antarox" A–400 | 10 |

No. 5

| | |
|---|---|
| p-Tolyl Benzoate | 20 |
| "Antarox" B–201 | 10 |
| Xylene | 70 |

## Self-Propagating Thermal Insecticide
### U. S. Patent 2,590,529

| | |
|---|---|
| Ammonium Nitrate | 30 |
| Potassium Nitrate | 2 |
| Charcoal (40 Mesh) | 13 |
| Water | 1½ |

Mill together for 20 minutes and add:

| | |
|---|---|
| China Clay | 10 |

Dry and mix with:

| | |
|---|---|
| Benzene Hexa - chloride or | 45 |
| DDT | 50 |

---

## Self-Vaporizing Insecticide
### U. S. Patent 2,633,444

| | |
|---|---|
| Sodium Nitrite | 55 |
| Ammonium Chloride | 43 |
| Magnesium Oxide | 2 |
| Insecticide* | 82-90 |

On ignition, this will continue burning of itself and give off insecticide fumes.

---

## Insecticide Aerosols

### Formula No. 1

| | |
|---|---|
| "Thanite" | 1.00 |
| Pyrethrins | 0.10 |
| DDT | 2.00 |
| Piperonyl Butoxide | 0.80 |
| Deodorized Kerosene | 1.40 |

---

*DDT, lindane, malathon or others.

| | |
|---|---|
| Methylated Naphthalenes | 9.70 |
| "Freon" 11 and 12 (1:1) | 85.00 |

### No. 2

| | |
|---|---|
| "Thanite" | 2.00 |
| Pyrethrins | 0.10 |
| DDT | 2.00 |
| Piperonyl Butoxide | 0.25 |
| Deodorized Kerosene | 2.40 |
| Methylated Naphthalenes | 8.25 |
| "Freon" 11 and 12 (1:1) | 85.00 |

### No. 3

| | |
|---|---|
| "Thanite" | 2.00 |
| Allethrin | 0.10 |
| DDT | 2.00 |
| Piperonyl Butoxide | 0.25 |
| Deodorized Kerosene | 8.65 |
| Methylated Naphthalenes | 7.00 |
| "Freon" 11 and 12 (1:1) | 80.00 |

---

## Herbarium Aerosol Insecticide

| | |
|---|---|
| Pyrethrum Extract | 40 |
| Cyclohexane | 60 |
| DDT | 100 |
| "Freon" | 1800 |

Aerosol-Generating and
Propelling Composition
*U. S. Patent 2,637,678*

| | |
|---|---|
| Potassium Nitrate | 2 |
| Sulfamic Acid | 1-1½ |
| Thiourea | ½-1 |

---

Protecting Stored Tobacco Leaf
*U. S. Patent 2,577,453*
Tobacco is protected from insect infestation by spraying all layers with 3½ oz. triglycol chloride per 500 lb. of tobacco.

---

Grape-Vine Fungicide
*Italian Patent 459,302*

| | |
|---|---|
| Ammonium Biphosphate | 5 |
| Ammonium Sulfate | 5 |
| Aluminum Sulfate | 26 |
| Copper Sulfate | 64 |

## Chapter X

# LEATHER, SKINS, AND FURS

Removing Hair &
Wool from Hides
*British Patent 735,783*

| | |
|---|---|
| Calcium Hydrosulfide | 7.5 |
| Calcium Oxide | 15.0 |
| Barium Hydroxide | 11.0 |
| Water | 67.0 |

Washed and cleaned hides are painted with above and hung over poles for 24 hours.

---

Non-ionic Fat Liquoring
Stock Emulsion

| | |
|---|---|
| "Ethofat" O/15 or 0/20 | 5 |
| Oil | 45 |
| Water | 50 |

Dissolve the "Ethofat" and oil while agitating. Add water to produce an emulsion.

---

Non-ionic Fat Liquoring
Emulsifiable Concentrate

| | |
|---|---|
| 35° Neatsfoot Oil | 70 |
| Mineral Oil | 20 |
| "Ethofat" 0/20 | 10 |

The ingredients are mixed to form a solution. This mixture can be added directly to the fat liquoring drum containing the wet leather to form the emulsion.

---

Liquoring Emulsifiable
Concentrate

| | |
|---|---|
| Prime Neatsfoot Oil | 49 |
| Sulfonated Oil | 43 |
| "Ethofat" 0/15 | 8 |

---

Cationic Fat Liquoring
Concentrate

| | |
|---|---|
| Sheep Oil or Extra Neatsfot Oil | 94.3 |
| "Ethomeen" 18/15 | 5.0 |
| Acetic Acid | 0.7 |

The "Ethomeen" 18/15 is dissolved in the oil and the acetic acid added to the mixture.

## Cationic Fat Liquoring Emulsifiable Concentrate

| | |
|---|---|
| "Ethomeen" T/15 | 3 |
| "Arquad" 2C | 1 |
| Mineral Oil | 48 |
| Neatsfoot Oil | 48 |

## Fat Liquoring Leather Treatment

| | |
|---|---|
| Neatsfoot Oil | 10 |
| "Span" 20 | 3 |
| Acetylated Lanolin Alcohols | 2 |
| Water | To make 100 |

## Non-Mercury Rabbit Hide Cure

| | | |
|---|---|---|
| a | Potassium Dichromate | 25 g. |
| | Hot Water | 750 cc. |
| b | Potassium Chlorate | 100 g. |
| | Hot Water | 100 cc. |
| c | Hydrochloric Acid (Sp.Gr.1.15) | 175 cc. |

Hydrogen Peroxide (30%) 500 cc.

Add a to b and then mix in c. Smear skins with this; dry; crop and blow. The down is left for 2 to 3 months.

## Leather Pasting
### U. S. Patent 2,805,952

| | |
|---|---|
| Methylcellulose | 89.54 |
| Propyleneglycol | 8.97 |
| Casein | 1.35 |
| Morpholine | 0.14 |
| Water | To suit |

## Fur Skin Softener
### British Patent 798,464

| | |
|---|---|
| Sulfated Sperm Oil (Neutralized with ammonia) | 90 |
| "Carbowax" 9000 Distearate | 5 |
| Sodium Lauryl Sulfate | 2½ |
| Water | 2½ |
| Woolfat | 20 |

## Fur Cleaning & Glazing

| | Formula No. 1 | No. 2 | No. 3 |
|---|---|---|---|
| Polyethyleneglycol (400) Monolaurate | 5 | 5 | 5 |
| Polyethyleneglycol (400) Monooleate | 3 | — | — |
| Silicone (D.C. 200-350) | — | 3 | — |
| "Acetulan" | — | — | 3 |
| Perchlorethylene | 120 | 120 | 120 |

Dilute 1 part of above with 7 parts of water; mix to emulsify and then mix into sawdust in tumbling barrel.

No. 4

*German Patent 937,434*

| | |
|---|---|
| Hydrogen Peroxide | 1 cc. |
| Acetic Acid | 3 cc. |
| Glycerin | 1 cc. |
| Formaldehyde | 80 cc. |
| Methanol | 15 cc. |
| Water | 1 l. |

Spray, brush or dip method may be used.

No. 5

*U. S. Patent 2,861,949*

| | |
|---|---|
| Soybeandimethyl Ammonium Chloride | 3 fl. oz. |
| Isopropanol | 1 fl. oz. |
| Poly (Dimethyl-siloxane) | 1 fl. oz. |
| Nonylphenol poly-oxyethylene Ether | 1 fl. oz. |
| Water | To make 1 gal. |

The above is diluted with water (4 vol.), sprayed on fur and ironed at 325°F.

## CHAPTER XI

# LUBRICANTS

## Metal Cutting Soluble Oil

### Formula No. 1

| | |
|---|---|
| Mineral Oil | 94 |
| "Ethomeen" S/15 | 4 |
| "Ethomeen" S/12 | 2 |

The emulsifiers are dissolved in the oil. The concentrate disperses readily in water to form a stable, white emulsion. With different grades of mineral oil the ratio of "Ethomeen" S/15 to "Ethomeen" S/12 may require changing to produce the most satisfactory emulsion.

### No. 2

| | |
|---|---|
| Mineral Oil | 95.0 |
| "Ethomeen" S/12 | 2.5 |
| "Arquad" 2C | 2.5 |

The emulsifiers dissolve easily in the mineral oil. The emulsifiable concentrate disperses easily in water forming a stable emulsion. Metals dipped into an emulsion of this type will pick up an oil coating which acts as a protective layer against corrosion during storage and shipment.

---

## Heavy Duty Soluble Oil
### U. S. Patent 2,846,393

| | |
|---|---|
| Sodium Mahogany Sulfonate | 15-30 |
| Potassium Rosinate | 15-0 |
| Mineral Lubricating Oil | 20-55 |
| Potassium Hydroxide | 0.01-0.5 |
| Hexyleneglycol | 2.5-3.5 |
| Sulfurized Alkyl Esters of Animal Fatty Acids | 20-30 |
| Water | 0.01-0.5 |

## Rustproofing Cutting Oil
### U. S. Patent 2,780,598

| | |
|---|---|
| Sodium Petroleum | |
| Sulfonate | 9.00 |
| Sodium Rosinate | 1.00 |
| Triethanolamine | 0.20 |
| Oleic Acid | 0.20 |
| Tetrachlorophenol | 0.25 |
| Alcohol | 1.40 |
| Water | 1.45 |
| Mineral Oil (80 s.s.u. | |
| at 100°F.) | 85.50 |
| Tetrasodium EDTA | 0.50 |

## Water Dispersible Cutting Oil Concentrate

| | |
|---|---|
| Mineral Oil | 93 |
| "Barolan" L (Acetylated | |
| Lanolin Alcohols) | 7 |

## Drawing Compound for Stainless Steel

### Formula No. 1
| | |
|---|---|
| Chlorinated Paraffin Wax | 33 |
| Petrolatum | 67 |

### No. 2
| | |
|---|---|
| Sodium Chloride | 36 |
| Mineral Oil | 53 |
| Paraffin Wax | 11 |

### No. 3
| | |
|---|---|
| Cotton Seed Oil | 50 |
| Talcum | 15 |
| Ammonium Chloride | 35 |

## Metal Drawing Compound
### U. S. Patent 2,760,931

| | |
|---|---|
| Soap | 1-5 |
| Tallow | 5-12 |
| Starch, Expanded | ½-2 |
| Water | 80-92 |

## Drawing Lubricant for Galvanized and Other Coated Wires

| | |
|---|---|
| Barium Lanolate | 30 |
| Calcium Stearate | 70 |

## Lead Extrusion Lubricant
### U. S. Patent 2,815,560

| | Formula No. 1 | No. 2 |
|---|---|---|
| Glycol | 45 | — |
| Glycerol | 35 | 85 |
| Water | 20 | 5 |
| Castor Oil | — | 5 |
| Soluble Oil | — | 5 |

## Titanium Extrusion Lubricant

| | |
|---|---|
| Flake Graphite | 25 |
| Molybdenum Disulfide | 15 |
| Mica, Powdered | 5 |
| Bentonite Grease | 55 |

## Wire Drawing Lubricant
### U. S. Patent 2,736,699

| | |
|---|---|
| Calcium Stearate | 8-35 |
| Iron Sulfate | 20-80 |
| Calcium Oxide | 10-45 |

## Flange Lubricant
### U. S. Patent 2,863,833

| | |
|---|---|
| Sulfur | 54-70 |
| Hydrated Magnesium Silicate | 10-22 |
| Graphite | 10-22 |
| Chlorinated Diphenyl | 2-10 |

## Electrical Apparatus Lubricant
### German Patent 872,242

| | |
|---|---|
| Petrolatum | 45 |
| Polyisobutylene | 5 |
| Rosin | 20 |
| Graphite | 30 |

## Valve Spindle Lubricant

| | |
|---|---|
| Glyceryl Monoricin-oleate (S-1153) | 77 |
| Acrawax | 23 |

Heat and mix until uniform.

## Knitting-Needle Oil
### U. S. Patent 2,882,331

| | |
|---|---|
| Neutral Mineral Oil | 98.45 |
| Tritolyl Phosphate | 1.25 |
| Polyoxyethylene Dodecyl Alcohol | 0.25 |
| Zinc Naphthenate | 0.05 |

## Textile Spindle Lubricant
### U. S. Patent 2,830,951

| | |
|---|---|
| Methyl Ricinoleate | 0.50 |
| Tetramethyl-p, p'-diaminodiphenyl-methane | 0.10 |

| | |
|---|---|
| Zinc Naphthenate | 0.05 |
| Mineral Oil | 99.35 |

## Synthetic Lubricant
### U. S. Patent 2,760,934

| | |
|---|---|
| Tricresyl Phosphate | 3 |
| Petroleum Sulfonate, Sodium | ¾ |
| Mixed Alkyl Hexyl and Heptyl Sebacates | To make 100 |

## Plug Valve Lubricant
### U. S. Patent 2,878,184

| | |
|---|---|
| Glycerin | 78-79 |
| Sodium Carboxy-methylcellulose | 7-3 |
| Water Soluble Vegetable Gum Powder | 2-1 |
| Sodium Stearate | 2-1 |
| Detergent Wetting Agent | 4-2 |
| Starch | 2-1 |
| Animal Glue | 3-1 |
| Sodium Glycer-oxide | 0.75-3.75 |

## Extreme Pressure Soluble Oil
### U. S. Patent 2,848,416

| | |
|---|---|
| Chlorinated Dibenzyl Disulfide | 4-6 |
| Chlorinated Paraffin Wax | 3.5-5 |
| Sulfurized Sperm Oil | 3.5-5 |
| Mineral Oil | To make 100 |

Extreme Pressure Lubricant
*German Patent 872,623*

| Chloranil | 4 |
|---|---|
| Octadecylamine | 1 |
| Mineral Oil | 95 |

Heat until dissolved.

---

Aerosol Lubricant for
Plastic Molding

| Acrawax C Atomized | 5-10 |
|---|---|
| "Genetron" 12 | 30 |
| "Genetron" 11 | 70 |

---

Latex Mold Release Lubricant

| "Carbowax" 1000 | 50 |
|---|---|
| "Glucarine" B | 50 |
| Water | To suit |

---

Dry Film Lubricant

| Molybdenum Disulfide | 55 |
|---|---|

| Graphite | 6 |
|---|---|
| Sod. Silicate | 39 |

This film is stable from 300 to 750°F.

---

Washable Dry Graphite
Lubricant
*British Patent 817,003*

| Graphite | 10.0 |
|---|---|
| Sod. Dodecyl Alcohol Sulfate | 2.5 |

---

Colloidal Graphite
*Japanese Patent 3767 (1959)*

| Graphite (200 mesh) | 500 |
|---|---|
| Phosphoric Acid | 2500 |

Heat at 100°C. for 4 hours and powder. Mix in ball mill with:

| Tannic Acid | 15 |
|---|---|
| Water | 2000 |

---

Stable Lubricating Grease

| | Formula No. 1 | No. 2 | No. 3 | No. 4 | No. 5 |
|---|---|---|---|---|---|
| Caprylic Acid | 5.82 | — | — | 16.0 | 9.14 |
| Capric Acid | — | 8.7 | 12.4 | — | — |
| Stearic Acid | 3.42 | 5.4 | — | — | — |
| Acetic Acid | 6.00 | 1.6 | 2.2 | 10.5 | 6.03 |
| Lithium Hydroxide Monohydrate | — | 4.2 | 4.7 | 12.0 | — |
| Sodium Hydroxide | 6.85 | — | — | — | 6.52 |
| Glycerin | 5.14 | 2.2 | 3.3 | 4.3 | 4.89 |
| 100 Second Paraffin Oil | 72.77 | 77.9 | — | — | — |
| Naphthenic Oil (750 sec. @ 100°C.) | — | — | 77.4 | 57.2 | 73.42 |

*Penetration*

| | | | | | |
|---|---|---|---|---|---|
| Unworked at 77°F. | 405 | 358 | 368 | — | 365 |
| Worked at 77°F. | 409 | 339 | 349 | — | 360 |
| *Dropping Point* °F. | 461 | 455 | 494 | 420 | 378 |
| *Description of Product* | — | — | Plastic | Semi-Fluid | Plastic |

Lubricating Grease, Thick
*U. S. Patent 2,900,338*

| | |
|---|---|
| Glass Fibers ( < 3 microns) | 5-15 |
| Lubricating Oil | To make 100 |

Ship Launching Grease
*British Patent 804,252*

| | |
|---|---|
| Paraffin Wax (m. p. 135-140°F.) | 50 |
| Microcrystalline Wax | 20 |
| Petrolatum | 20 |
| Rosin | 10 |

Heavy Duty Hydraulic
Brake Fluid

Formula No. 1

| | |
|---|---|
| "Hydricin" 4 | 19. 00 gal. |
| 3-Methoxy Butanol | 21.00 gal. |
| "Celanese" Solvent 901-H | 23.00 gal. |
| Hexylene Glycol | 37.00 gal. |
| Bisphenol A | 0.75 lb. |
| Dibutylamine | 1.50 lb. |

The bisphenol A should be dissolved in the 3-methoxy butanol by heating to 125°F. accompanied by vigorous stirring. It is unnecessary to add any corrosion inhibitor since the "Hydricin" 4 has already been processed to inhibit any corrosion tendency of the finished fluid. For pH adjustor, add Dibutylamine preferably to "Celanese" Solvent 901-H with stirring but order of addition is not critical.

No. 2

| | % Volume |
|---|---|
| "Polyglycol" 15-200 | 20 |
| "Dowanol" EB | 20 |
| "Dowanol" DE | 45 |
| Ethylene Glycol | 15 |

| | % Weight of Total Fluid |
|---|---|
| *Inhibitor System* | |
| Bisphenol A (Oxidation) | 0.2 |
| Borax (Corrosion) | 0.5 |

Moderate Duty Brake Fluid

Formula No. 1

| | |
|---|---|
| Hydroquinone | 0.1 |
| Castor Oil | 26.7 |
| Butanol | 66.7 |
| Ethyleneglycol | 6.6 |

## No. 2

| | |
|---|---|
| Butanol | 46 |
| Castor Oil AA | 54 |
| Hydroquinone | 0.02-0.06 |

---

## Hydraulic Shock Absorber Fluid
*French Patent 1,112,026*

| | |
|---|---|
| Castor Oil | 59 |
| Butanol | 25 |
| Ethanol | 10 |
| Cyclohexanal | 6 |

Works well from −20 to + 50°C.

---

## Hydraulic Fluid
*German Patent 853,487*

| | | |
|---|---|---|
| *a* | Ethylene Oxide | 50 |
| | Castor Oil | 500 |
| *b* | "Carbitol" | 40 |
| | Triethyleneglycol Monomethyl Ether | 20 |
| | Butyl "Cellosolve" | 20 |

Mix *a*. Add 20 parts to *b*.

---

## Automatic Transmission Hydraulic Fluid
*U. S. Patent 2,751,355*

| | |
|---|---|
| Ethylene Glycol (50% Solution) | 1 gal. |
| Sodium Oleate | ¼ oz. |
| Sodium Nitrate | ¼ oz. |
| Borax | 1 oz. |
| Triethanolamine | 1/200 oz. |
| Tributyl Phosphate | 1/100 oz. |

## Hydraulic Fluid

| | |
|---|---|
| Chlorinated Biphenyl (54% Cl) | 62 |
| o-Terphenyl | 25 |
| Biphenyl | 13 |

---

## Hydraulic Fluid With Lubricating Properties
*German Patent 939,045*

| | |
|---|---|
| Triethyleneglycol | 55 |
| Butylpolyglycol | 10 |
| Ethylpolyglycol | 10 |
| Methylpolyglycol | 10 |
| Ethyleneglycol Borate | 15 |

---

## Heat Transfer & Hydraulic Fluid
*U. S. Patent 2,741,598*

| | |
|---|---|
| Chlorinated Biphenyl (54% Cl) | 62 |
| o-Terphenyl | 25 |
| Biphenyl | 13 |

---

## Nonskid for Oil Rails

| | |
|---|---|
| "Ludox" | 33.3 |
| Isopropanol | 66.7 |

---

## Oil Well Drilling Mud Additive

| | |
|---|---|
| Caustic Soda | 53 kg. |
| Quebracho | 21 kg. |
| Slaked Lime | 60 kg. |
| Potato Starch | 100 kg. |

The above is added to 7000 liters mud.

## Chapter XII

# MATERIALS OF CONSTRUCTION

High Temperature Furnace
Lining

| | |
|---|---|
| Fireclay ( < 5 mm., 50-60% 0.088 mm.) | 90 |
| Aluminum Hydroxide | 10 |
| Sodium Silicate | 15 |
| Water | 25 |

---

Heat Insulation

| | |
|---|---|
| Sawdust | 1440-1610 |
| Infusorial Earth | 180 |
| Lime | 90 |
| Cement | 30 |
| Sodium Silicate | 11-13 |
| Bleaching Powder | 16-18 |
| Water | To suit |

---

Low Shrinking Light Thermal
Insulation
U. S. Patent 2,875,075

| | |
|---|---|
| Quicklime | 120.0 |
| Tripoli | 34.6 |
| Diatomite | 129.0 |

| | |
|---|---|
| Chrysotile Asbestos | 75.0 |
| Kaolin | 16.4 |
| Calcium Chloride | 23.2 |
| Water | 2250.0 |

Autoclave at 401°F. and 250
lb. per sq. in. for 3 hours.

---

Light Weight Building Block
U. S. Patent 2,899,325

| | |
|---|---|
| Sawdust | 5.00 |
| Sand | 1.00 |
| Cement | 1.00 |
| Burned Gypsum Powder | 0.85-1.25 |
| Rosin Powder | 0.15-0.25 |
| Water | To suit |

Mix well and cure.

---

Hydraulic Cement
U. S. Patent 2,752,261

| | |
|---|---|
| Granulated Blast Furnace Slag | 56 |
| Fly Ash | 30 |

Portland Cement 7
Limestone 3
Gypsum 3
Barium Chloride 1

Dry, crush and grind. This has a strength and setting speed equal to or higher than Portland Cement.

---

Slow-Setting Cement
U. S. Patent 2,848,340

Add 0.1-2.0% ethyl acid phosphate to Portland cement.

---

High Temperature Cement for Oil Wells
U. S. Patent 2,875,835

Slaked Lime 40-60
Quartz Sand 15-25
Kaolinate 15-25
Urea-Resin (Preform) 8-12

Make into a slurry of 47 to 53% water. This is pumpable for 134 to 272 minutes.

---

Quick-Setting Cement Mortar
U. S. Patent 2,763,561

Silica (200 mesh) 15
Portland Cement 94
Asbestos Fiber 1-5

Add water as required.

---

Quick-Setting Underwater Cement
U. S. Patent 2,918,385

Portland Cement 30-70
Anhydrous Calcium
  Sulfate 10-36
Kaolin 20-45

---

Waterproofing for Sorel Cement
German Patent 905,834

Montan Wax 10
Olein 10
Butyl Acetate 150
Sodium Hydroxide 10
Magnesium Flurosili-
  cate 180

---

Decreasing Shrinkage of Concrete

Bentonite 0.25-2
Sodium Silicate 1-5

Based on weight of cement.

---

Prepared Sand for Mortar
U. S. Patent 2,757,096

Sand 59.735
Diethylhexanol 0.005
Sodium Mononaptha-
  lene Sulfonate 0.250
Sodium Dodecylben-
  zene Sulfonate 0.010
Infusorial Earth 20.000
Active Volcanic Puz-
  zuolana 20.000

## Asbestos Plaster
### Italian Patent 484,910

| | |
|---|---|
| Asbestos Fiber | 50 |
| Activated Bentonite | 50 |
| Colloidal Cellulose | 5 |
| Colloidal Kaolin | 15 |
| Flaked Talc | 10 |

Apply in a paste containing 40% water.

---

## High Expansion Plaster
### U. S. Patent 2,741,562

| | |
|---|---|
| High-Strength Gypsum Plaster | 96.40 |
| Powdered Bone Glue | 0.82 |
| Sodium Fluoride | 0.28 |

## Finely Ground Set
### Gypsum Powder

Gypsum Powder 2.50

---

## Colored Interior Plaster
### U. S. Patent 2,868,660

| | |
|---|---|
| Powdered Marble | 1000 |
| Hydrated Lime | 450 |
| Keene Cement | 200 |
| Zinc Stearate | 5 |
| Pigment | 3 |

---

## Cold Paving Composition
### U. S. Patent 2,750,297

| | |
|---|---|
| Mineral Aggregate | 89.50 |
| Asphalt Emulsion | 9.00 |
| Water | 1.50 |
| Methylcellulose | 0.01 |

## CHAPTER XIII

# METALS AND THEIR TREATMENT

### Dry Tinning Composition
#### U. S. Patent 2,822,325

| | |
|---|---|
| Potassium Hydroxide | 3-8 |
| Potassium Stannate | 60-120 |
| Potassium Carbonate | 0-10 |
| "Versene" | 50-100 |
| Trisodium Phosphate | 30-60 |
| Sodium Lauryl Sulfate | ½-1 |

---

### Lead Coating Metals
#### German Patent 958,796

| | |
|---|---|
| Lead Chloride | 25 |
| Tin Chloride | 25 |
| Zinc Chloride | 25 |
| Potassium Chloride | 10 |
| Sodium Chloride | 15 |

Dip metal into above at 280 to 500°C. for 1 to 60 minutes. An alloy of lead (60) and tin (40) is deposited in thickness of 2 to 40 millimicrons.

### Phosphating Electrical Steel
#### U. S. Patent 2,743,203

| | |
|---|---|
| Phosphoric Acid (85%) | 31 |
| Aluminum Hydrate | 1 |
| Wetting Agent | 1 |
| Water | 62 |

Dip silicon-iron sheets in mixture, then drain and heat at 500°C.

---

### Hot Zinc Coating for Iron

| | |
|---|---|
| Aluminum | 0.30 |
| Silicon | 0.05 |
| Zinc | 99.65 |

---

### Metal Cleaner

#### Formula No. 1

| | |
|---|---|
| "Ethofat" 0/20 | 5 |
| Kerosene | 95 |

The "Ethofat" is dissolved in the kerosene. Heating is not necessary. A small amount of water is necessary to clarify.

### No. 2

| | |
|---|---|
| "Ethofat" 60/15 | 2.5 |
| "Ethomeen" S/15 | 2.5 |
| Kerosene | 95.0 |

### No. 3

| | |
|---|---|
| "Ethomeen" S/12 | 1.5 |
| "Ethomeen" S/15 | 3.0 |
| Kerosene | 95.5 |

The emulsifiers are soluble in the kerosene at room temperature.

### No. 4
### (Acid)

| | |
|---|---|
| Kerosene | 37 |
| "Arquad" 2C | 2 |
| "Ethomeen" S/12 | 1 |
| 20% Phosphoric Acid | 60 |

Dissolve the emulsifiers in the kerosene and add to the phosphoric acid solution with vigorous agitation. "Arquad" 2C in addition to its emulsifying power is also a good acid corrosion inhibitor.

### No. 5
(Metal Cleaning Emulsions

| | |
|---|---|
| Kerosene | 13 |
| Acetylated Lanolin Alcohols | 3 |
| "Span" 20 | 2 |
| "Tween" 20 | 2 |
| Water | 80 |

### Alkali Metal Cleaner
### (Hot Soak)

| | |
|---|---|
| Sodium Hydroxide (10% Solution) | 99.5 |
| "Ethomeen" C/25 | 0.5 |

### Rust Removing Gel
*British Patent 812,745*

| | |
|---|---|
| Sulfamic Acid | 11.1 |
| Glacial Acetic Acid | 10.5 |
| Silica Aerosol | 6.0 |
| Polyvinyl Alcohol | 0.2 |
| Water | 72.2 |

### Iron Scale Remover

### Formula No. 1
*French Patent 1,107,489*

| | | |
|---|---|---|
| a | Stannous Chloride | 2 |
| | Polyvinyl Alcohol | 2 |
| | Inhibitor | ½ |
| | Hydrochloric Acid (22 Bé) | 53 |
| b | Bentonite | 42½ |

Mix a until dissolved, then mix in b to get a dry paste.

### No. 2
*U. S. Patent 2,847,384*

| | |
|---|---|
| Sodium Xylene Sulfonate | 60 |
| Sodium Bisulfate | 30 |
| Tartaric Acid | 6 |
| Polyoxypropylene Polyethylene Ether | 1 |

Make a 28% water solution of above. Allow steel to soak in it. It is less corrosive than inhibited hydrochloric acid and does not require neutralization.

### No. 3
### U. S. Patent 2,846,341

| | |
|---|---|
| Sulfuric Acid | 10 |
| Oxalic Acid | 10 |
| Ferrous Sulfate | 1 |
| Water | 79 |

### Recovering Steel Pickling Liquor
### British Patent 794,386

Air is blown into the liquor heated to 160°F. An aqueous paste of magnesium hydroxide is added to pH of 7.2 to 7.3. Ferric hydroxide is precipitated, filtered and washed. The magnesium sulfate solution is treated with calcium chloride to precipitate calcium sulfate. The magnesium chloride solution is drawn off and treated with slaked lime to precipitate magnesium hydroxide for reuse.

### Pickling Inhibitor
### German Patent 864,185

| | |
|---|---|
| Xylenesulfonic Acid | 41.5 |
| Thiocarbanilide | 8.5 |
| Paraformaldehyde | 25.0 |
| Sugar | 25.0 |

Melt together and mix. Use at 0.5 to 2% in 10% Hcl at 20°C.

### Pickling Titanium - Zirconium
### U. S. Patent 2,876,144

| | |
|---|---|
| Nitric Acid | 20 |
| Hydrofluoric Acid | 3 |
| Phosphoric Acid | 2 |
| Water | To make 100 |

Use at 100 to 190°F.

### Deoxidizing Iron & Steel

### Formula No. 1
### British Patent 791,887

| | |
|---|---|
| Ammonium Chloride | 20 g. |
| Hydrochloric Acid (35%) | 100 cc. |

Heat to 70 to 75°C. and immerse metal for 30 seconds; rinse and dry.

### No. 2
### U. S. Patent 2,808,327

| | |
|---|---|
| Aluminum | 80 |
| Zinc | 3-5 |
| Magnesium | 1.75-3 |

The above is mixed into molten steel.

### Removing Poisonous Metals from Surfaces
### U. S. Patent 2,774,736

| | |
|---|---|
| EDTA | 1 |
| Sodium Thiosulfate | 9 |
| Water (warm) | 415 |

Mix in slowly and apply with brush, mop or by spray.

---

### Removing Iron Stains from Silver

| | |
|---|---|
| Sulfuric Acid | 95 |
| Quinoline | 5 |

Treat silver with above and wash thoroughly.

---

### Metal Etching
*U. S. Patent 2,846,294*

| | |
|---|---|
| Nitric Acid | 0.2-2.5 |
| Sodium Tetradecyl Sulfate | 0.2-2.5 |
| Petroleum Distillate (B.P. 70-390°C.) | 1-75 |
| Nitric Acid, Dilute | To make 100 |

---

### Cerium Etchant

| | |
|---|---|
| Nitric Acid | 6 cc. |
| Glycerin | 5 cc. |

---

### Etching Germanium
*U. S. Patent 2,849,296*

| | |
|---|---|
| Glacial Acetic Acid | 20 |
| Nitric Acid (70%) | 15 |
| Hydrofluoric Acid (48%) | 10 |
| Water | 3 |

Use at 45 to 65°C. in a rotating container at 100 r.p.m. at an angle of 30-45° to vertical.

---

### Magnesium Etch
*U. S. Patent 2,849,297*

| | |
|---|---|
| Alcohol | 65 |
| Nitric Acid | 6⅔-20 |
| Hydrochloric Acid | 3⅓-10 |
| Acetic Acid | 5 |

The magnesium must be degreased and cleaned before etching.

---

### Slow-Acting Steel Etch

| | |
|---|---|
| Chloroacetic Acid | 70 g. |
| Alcohol (75%) | 100 cc. |

---

### Etchant for Zinc Alloys

| | |
|---|---|
| Sulfuric Acid | 15 |
| Hydrofluoric Acid (48%) | 1 |
| Water | To make 100 |

Use at 100°F. for 10 seconds.

---

### Brown Patina on Copper or Brass
*U.S.S.R. Patent 107,179*

| | |
|---|---|
| Minium | 59.9 |
| Chalk, Ground | 12.0 |
| Glacial Acetic Acid | 3.0 |
| Alcohol | 6.0 |
| Water | 19.1 |

---

### Blackening Stainless Steel
*Japanese Patent 3355 (1957)*

| | |
|---|---|
| Sodium Sulfide | 55 g. |
| Nickel Chloride | 30 g. |
| Sod. Hydroxide (50% Solution) | 1 l. |

Immerse object in above and heat at 130°C. for 2 to 5 hours.

---

Blackening Titanium

*Japanese Patent 6716 (1956)*

| | |
|---|---|
| Sulfuric Acid | 50 cc. |
| Manganese Dioxide | 10 g. |
| Potassium | |
| Permanganate | 20 g. |
| Copper Sulfate | 10 g. |
| Water | 1 l. |

Treat at boiling point of solution for 20 to 30 minutes.

---

Volatile Corrosion Inhibitor

Formula No. 1

4-Biphenylamine impregnated paper gives good protection for polished steel in neutral, acid and salty soils.

No. 2

Dicyclohexylamine *o* - nitrophenolate is an effective inhibitor.

No. 3

*U. S. Patent 2,837,432*

| | |
|---|---|
| Polyglycol (M.W. 6000) | 80 |
| Ethyleneglycol | 10 |
| Dicyclohexylamine | 10 |

This is used to impregnate wrapping paper.

No. 4

*U. S. Patent 2,829,945*

Morpholine Caprylate 8-15

Dicyclohexylamine

| | |
|---|---|
| Caprylate | 1-2 |
| Paraffin Oil | 83-91 |

Paper is coated with the above to give a moisture proof and vapor corrosion proofing.

| | No. 5 | No. 6 |
|---|---|---|
| *U. S. Patent 2,739,870* | | |
| Sodium Nitrite | 30 | 30 |
| Urea | 30 | 30 |
| Sodium Benzoate | 20 | — |
| Water | 60 | 50 |
| 2-Ethyl-1, | | |
| 3-Hexanediol | — | 30 |
| Ethanol | — | 10 |

---

Corrosion Inhibitor

Formula No. 1

*U. S. Patent 2,877,188*

| | |
|---|---|
| Sodium Nitrite | 75 |
| Borax | 25 |
| Benzotriazole | 1 |

Use 1000 to 2000 p.p.m.

No. 2

*Japanese Patent 860 (1956)*

| | |
|---|---|
| *a* Hexadecylamine | 100 |
| Linolenic Acid | 100 |
| *b* Linseed Oil | 600-1000 |

Heat *a* until uniform, before adding *b*.

No. 3

| | |
|---|---|
| *a* Octadecylamine | 100 |
| Lauric Acid | 230 |
| *b* Kerosene | 1650 |

Heat *a* until uniform, before adding to *b*. Use 0.005 to 0.1% to oil used for lubrication.

## No. 4
*Japanese Patent 262 (1958)*

| | | |
|---|---|---|
| *a* | Itaconic Acid | 130 |
| | Dodecylamine | 185 |
| *b* | Castor Oil | 99 |

Heat and mix *a* at 180 to 200°C. Add 1 part *a* to *b*, then warm and mix.

---

## Corrosion Inhibitor for Steel
*U. S. Patent 2,878,191*

| | |
|---|---|
| Triethanolamine | 1.1 |
| Coal Acids | 0.1 |
| Water | 175.0 |

---

## Corrosion Resistant Coating for Aluminum
*U. S. Patent 2,843,513*

| | |
|---|---|
| Chromium Oxide | 23.8 |
| Sodium Silicofluoride | 38.8 |
| Boric Acid | 38.5 |

Dissolve 10 to 80 g. of above in 1 liter of water. This coating absorbs various dyes.

---

## Corrosion Proofing Magnesium
*U. S. Patent 2,864,730*

| | |
|---|---|
| Chromic Acid | 19.2 |
| Nitric Acid | 24.0 |
| Hydrochloric Acid | 46.0 |
| Water | 1000.0 |

Treat at 75°F. at pH 0.2 for 30 seconds.

## Corrosion Proofing Tin
*Japanese Patent 261 (1958)*

| | |
|---|---|
| Sodium Phenol Sulfonate or Sodium Naphthol Sulfonate | 2-20 g. |
| Sodium Chromate | 3-30 g. |
| Water | 1 l. |

Immerse tin for a short time; rinse and dry.

---

## Anticorrosive Dip for Zinc Alloy Castings
*Japanese Patent 259 (1956)*

| | |
|---|---|
| Chromic Acid | 100 |
| Sulfuric Acid | 5 |
| Hydrochloric Acid | 3 |
| Hydrofluoric Acid | ½ |
| Water | 1000 |

Immerse for 20 seconds at 60°C. in this solution.

---

## Corrosion Inhibitor for Antifreeze

### Formula No. 1

| | |
|---|---|
| Sodium Benzoate | 1.5 |
| Sodium Nitrate | 0.1 |

### No. 2
*U. S. Patent 2,832,742*

| | |
|---|---|
| Oxidized Petroleum Fatty Acids | 27.1 |
| Diethanolamine | 27.7 |
| Water | 18.1 |
| *p*-tert-Butylbenzoic Acid | 27.1 |

0.1 to 1% of above is used. It does not affect copper, brass, aluminum, lead or solder.

### No. 3
U. S. Patent 2,764,553

Lithium Chromate .2 $H_2O$ is added to methanol at rate of 3 oz. per gal.

---

### Corrosion Proofing for Razor Blades
U. S. Patent 2,776,917

| | |
|---|---|
| Zinc Naphthenate | |
| (14.5% Zinc) | 16.0 |
| Benzoic Acid | 3.2 |
| Mineral Oil | 63.0 |
| Mineral Spirits | 17.8 |

---

### Corrosion Inhibitor for Fuel Gas Tanks

80 p.p.m. of sodium silicate sprayed on inside of tanks shows an increase of life of 30 to 60%.

---

### Corrosion Inhibitor for Well Pipes

#### Formula No. 1
U. S. Patent 2,885,359

| | |
|---|---|
| Arsenous Oxide | 40 |
| Sodium Hydroxide | 8-12 |
| Boric Acid | 8-12 |
| Water | 60-350 |

### No. 2
U. S. Patent 2,882,227

| | |
|---|---|
| Formaldehyde | 0.1-10 |
| Neutralized Mahogany Sulfonate | 0.1-10 |

Use 1 to 50 pt. of above per 1000 bbl. of well fluid.

---

### Anticorrosive for Rubber-Metal Assemblies

| | |
|---|---|
| Castor Oil | 85 |
| Barium Ricinoleate | 15 |
| Chloroxylenol | 1/5 |

---

### Slushing Oil
For Metal Protection
U. S. Patent 2,852,396

| | |
|---|---|
| Unrefined Petroleum | |
| (110 SUS at 100°F.) | 86 |
| Microcrystalline Wax | 10 |
| Silica Aerogel | 1 |
| Powdered Copper | 1 |
| Stearic Acid | 2 |

---

### Emulsifiable Rustproofing
U. S. Patent 2,862,823

| | |
|---|---|
| Mineral Oil | 20 |
| Sod. Petroleum Sulfonate | 20 |
| Neutral Degras | 40 |
| Butyl "Carbitol" | 50 |
| Triethanolamine | 1 |
| Water | 14 |

Emulsify with 1-10 water before use.

## Rustproofing Coating

### Formula No. 1
*U. S. Patent 2,861,892*

| | |
|---|---|
| Petroleum Tar | 28 |
| Rosin | 1 |
| Gilsonite | 9 |
| Poly Dodecylbenzene Sulfonate | 0.002-0.1 |
| Solvent Naphtha | 62 |

### No. 2
*German Patent 922,430*

| | |
|---|---|
| Stearic Acid | 150 |
| Talc | 25 |
| Bentonite | 50 |

## Boiler Water Corrosion Inhibitor
*U. S. Patent 2,882,171*

| | |
|---|---|
| Octadecylamine | 90 |
| "Tetronic" 90 | 9 |
| Cyclohexylamine | 1 |

Use about 2 p.p.m.

## Steam-Condensate Line Corrosion Inhibitor
*U. S. Patent 2,771,417*

| | |
|---|---|
| Oleic Acid | 45.75 |
| Tall Oil | 45.75 |
| Diethylenetriamine | 8.50 |

Δ to 115-116°C. in 24 hours.
Dissolve in methanol and use at rate of 10 p.p.m.

## Gun Barrel Corrosion Inhibitor
*Italian Patent 487,073*

| | |
|---|---|
| Mineral Oil | 0.30 |
| Kerosene | 0.15 |
| Colloidal Graphite | 1.00 |

Coat shot lightly.

## Aluminum Solder
*Japanese Patent 9214 (1957)*

| | |
|---|---|
| Tin | 70-85 |
| Zinc | 8-15 |
| Lead | 5-20 |

## Corrosion Resistant Aluminum Solder
*Japanese Patent 7359 (1958)*

| | |
|---|---|
| Tin | 70-6 |
| Zinc | 15-21 |
| Cadmium | 8-14.7 |
| Tellurium | 0.3-1 |

## Hard Solder for Aluminum, Steel-Non-Ferrous Alloys
*German Patent 946,405*

| | |
|---|---|
| Zinc | 70.0 |
| Aluminum | 29.5 |
| Manganese | 0.5 |

A *flux for this formula is*

| | |
|---|---|
| Zinc Chloride | 20-30 |
| Calcium Chloride | 15-20 |
| Potassium Chloride | 30-40 |
| Sodium Chloride | 15-20 |
| Sodium Fluoride | 1½ |
| Potassium Fluoride | 3½ |

## Solder for Graphite
### U.S.S.R. Patent 104,434

| | Formula No. 1 | No. 2 |
|---|---|---|
| Silver | 90 | 85 |
| Zirconium | 10 | — |
| Titanium | — | 15 |

## Low Melting Silver Solder
### U.S.S.R. Patent 103,701

| | |
|---|---|
| Silver | 39.1-41.1 |
| Copper | 15.5-16.5 |
| Cadmium | 25.1-26.5 |
| Zinc | 17.3-18.5 |
| Nickel | 0.1- 0.3 |

## Magnesium Alloy Solder

### Formula No. 1
### U.S.S.R. Patent 102,676

| | |
|---|---|
| Aluminum | 21-22 |
| Cadmium | 25-26 |
| Zinc | 0.2-0.5 |
| Manganese | 0.1-0.3 |
| Magnesium | To make 100 |

### No. 2
### U.S.S.R. Patent 111,928

| | |
|---|---|
| Aluminum | 0.75-2.5 |
| Zinc | 13-25 |
| Magnesium | To make 100 |

### No. 3
### U.S.S.R. Patent 103,370

| | |
|---|---|
| Aluminum | 30-1 |
| Manganese | 0.1-0.3 |

| | |
|---|---|
| Zinc | 1-1.5 |
| Magnesium | To make 100 |

## Solder Stable to 300°C.
### U.S.S.R. Patent 112,948

| | |
|---|---|
| Silver | 2.5-3.5 |
| Magnesium | 0.03-0.08 |
| Cadmium | To make 100 |

## All-Metals Solder
### U.S.S.R. Patent 113,111

| | |
|---|---|
| Silver | 23-27 |
| Zinc | 21-27 |
| Cadmium | 12-18 |
| Copper | To make 100 |

## Soldering Semiconductors to Leads
### German Patent 872,602

Silver oxide is placed between ends and heated to form a silver joint.

## Soldering Stainless Steel

Addition of a teaspoonful of glycerin per cup of stainless steel soldering flux is recommended to help flux wet steel surfaces. A 50-50 tin-lead mixture is the most popular solder for working stainless, but a 60-70 combination can also be used. A clean surface is essential and roughening of smooth surfaces greatly increases adhesion.

## Alumina-Zirconia to Metal Solder
### *Japanese Patent 3514 (1958)*

|  | Formula No. 1 | No. 2 | No. 3 | No. 4 |
|---|---|---|---|---|
| Silver | 60 | 60 | 60 | 60 |
| Gold | 20 | — | 20 | 20 |
| Titanium | 5 | 5 | 10 | 5 |
| Tin | 15 | 35 | — | 9 |
| Lead | — | — | 10 | 6 |
| Working Temperature | 830°C. | 700°C. | 900°C. | 840°C. |

Use in a non-oxidizing atmosphere.

---

## Quartz to Metal Solder
### *Japanese Patent 3515 (1958)*

|  | Formula No. 1 | No. 2 | No. 3 |
|---|---|---|---|
| Tin | 92 | — | 96 |
| Silver | 5 | 2.5 | 2 |
| Titanium | 3 | 3.0 | 2 |
| Lead | — | 94.5 | — |
| Working Temperature | 250°C. | 300°C. | 230°C. |

---

## Soldering Flux
### *U.S.S.R. Patent 114,400*

| Rosin | 30.0 |
|---|---|
| Salicylic Acid | 2.0 |
| Triethanolamine | 1.5 |
| Alcohol | 55.3 |
| Ammonium Chloride | 4 |

---

## Aluminum Soldering Flux
### *U. S. Patent 2,756,497*

| Zinc Chloride | 14 |
|---|---|
| Stannous Chloride | 185 |
| Hydrazine Hydrochloride | 16 |

---

## Steel Soldering Paste
### *French Patent 1,098,244*

| Silver | 19.98 |
|---|---|
| Copper | 6.66 |
| Zinc | 3.35 |
| Cadmium | 1.68 |
| Gold | 1.66 |
| Boric Acid | 10.00 |
| Pot. Bifluoride | 23.31 |
| Petrolatum | 33.36 |

## Residueless Soldering Flux
### U. S. Patent 2,803,572

| | |
|---|---|
| Hexamethylene-tetramine | 89 g. |
| Hydrobromic Acid (40%) | 1 cc. |
| Wetting Agent | ½ g. |
| Water | To make 1000 cc. |

---

## Welding & Hard Soldering Flux
### German Patent 932,828

| | |
|---|---|
| Alum | 8.2 |
| Ammonium Chloride | 6.2 |
| Copper Sulfate | 0.6 |
| Boric Acid | 0.3 |
| Borax | 84.7 |

---

## Improved Soldering Flux
### U. S. Patent 2,827,408

| | |
|---|---|
| Stannous Chloride | 50 |
| Cadmium Chloride | 20 |
| Lead Chloride | 25 |
| Ammonium Chloride | 8 |
| Perchloroethylene | 30 |
| Silicone Oil (400 centistokes) | 2 |

Use at 675°C. This has an improved shelf life and volatilizes at 600°C. without residue.

---

## Welding Flux
### U. S. Patent 2,820,725

| | |
|---|---|
| Titanium Dioxide | 25 |
| Calcium Fluoride | 6 |
| Calcium Carbonate | 18 |
| Ferrous Oxide | 10 |
| Manganese Oxide | 8 |
| Calcium Silicate | 8 |
| Alumina | 2 |
| Potassium Titanate | 20 |
| Bentonite | 3 |

---

## Arc-Welding Flux
### U. S. Patent 2,814,579

| | |
|---|---|
| Manganese Ore (75% Manganese Oxide) | 49.5 |
| Zirconium Oxide | 25.0 |
| Calcium Fluoride | 4.0 |
| Ferrosilicon | 4.5 |
| Sodium Silicate (43% Solids) | 17.0 |

Mix and dry at 1400 to 1600°F.

---

## Universal Flux for Aluminum-Silicon Alloys

| | Formula No. 1 (M.P. 750°C.) | No. 2 (M.P. 600°C.) |
|---|---|---|
| Sodium Fluoride | 30 | 30 |
| Sodium Chloride | 25 | 50 |
| Cryolite | 15 | 10 |
| Potassium Chloride | — | 10 |

## Aluminum Welding Flux
### U.S.S.R. Patent 102,672

| | |
|---|---|
| Potassium Chloride | 40 |
| Sodium Chloride | 30 |
| Cryolite | 30 |

---

## Brass Scrap Refining Flux
### Spanish Patent 228,351

| | |
|---|---|
| Silica | 40 |
| Potassium Nitrate | 16 |
| Salt | 40 |
| Vegetable Carbon | 4 |

Mix until uniform. Use 2% on weight of scrap.

---

## Galvanizing Flux

| | |
|---|---|
| Ammonium Chloride | 91.0 |
| Sodium Chloride | 4.0 |
| Sodium Fluoride | 2.5 |
| Glycerin | 2.5 |

---

## Liquid Flux for Hard Soldering

| | Formula No. 1 | No. 2 |
|---|---|---|
| Ethyleneglycol | 38 | — |
| Anhydrous Borax | 17½ | 9.1 |
| Methanol | 160 | — |
| Glycerin | — | 36.4 |
| Alcohol | — | 54.5 |

*Formula No. 2 is diluted with:*

| | |
|---|---|
| Alcohol | 135 |

## Boiler Plate Welding Flux
### U. S .Patent 2,748,040

| | |
|---|---|
| Silica | 23.0-34.2 |
| Titanium Dioxide | 3.0-3.6 |
| Aluminum Oxide | 42.0-64.3 |
| Calcium Oxide | 4.7-5.1 |
| Manganese Dioxide | 5.0-8.5 |
| Soda Ash | 0.0-7.9 |
| Potassium Carbonate | 0.0-7.9 |

---

## Cast Iron Welding Flux
### U. S. Patent 2,799,607

| | | |
|---|---|---|
| Potassium Pentaborate | 40 | -50 |
| Borax | 20 | -30 |
| Boric Acid | 7½-10½ | |
| Potassium Nitrate | 8 | -10 |
| Iron Powder | 2 | -6 |
| Ilmenite | 4 | -6 |

---

## Ceramic Flux for Welding Low Carbon Steel

| | |
|---|---|
| Manganese Ore | 54 |
| Fluorite | 7 |
| Silica Sand | 30 |
| Ferrosilicon (Silicon 75) | 7 |
| Aluminum Powder | 2 |
| Sodium Silicate (Sp.Gr. 1.35) | 15 |

## Iron Soldering Flux
### U. S. Patent 2,772,192

| | |
|---|---|
| Ammonium | |
| Chloride | 50-200 g. |
| Urea | 200-2500 g. |
| Mono- | |
| ethanolamine | 100-2500 g. |
| Hydrochloric | |
| Acid | 500-800 g. |
| Water | 0-150 cc. |
| Alcohol | To make 1 gal. |

This is not readily charged by high-melting solder.

## Light Metal Soldering Flux
### U. S. Patent 2,845,700

Light metal soldering with lead-tin solder is improved if the former is first coated thinly with tritolyl phosphate.

## Magnesium Alloy Soldering Flux
### U.S.S.R. Patent 103,450

| | |
|---|---|
| Cryolite | 8 |
| Zinc Oxide | 3 |
| Carnalite | 89 |

## Cover Flux for Melting Magnesium

| | Formula No. 1 | No. 2 | No. 3 |
|---|---|---|---|
| Boron Oxide | 30-60 | 40-70 | 45-55 |
| Cryolite | 25-40 | 25-40 | — |
| Borax | 7-11 | 7-11 | — |
| Calcium Fluoride | 0-5 | 3-6 | 15-20 |
| Aluminum Fluoride | 20-27 | — | 12-18 |
| Magnesium Fluoride | — | 3-6 | 15-20 |

## Flux for Magnesium Zirconium Alloys
### British Patent 739,660

| | |
|---|---|
| Barium Chloride | 45-55 |
| Calcium Chloride | 55-45 |

*To above add:*

| | |
|---|---|
| Magnesium Fluoride | 15-35 |

## High-Nickel Welding Alloy Flux
### U. S. Patent 2,820,732

| | |
|---|---|
| Calcium Carbonate | 43.5 |
| Cryolite | 29.0 |
| Ferrocolumbion | 18.2 |
| Titanium Dioxide | 1.5 |
| Ferromolybdenum | 2.9 |
| Ferroaluminum | 2.0 |
| Bentonite | 2.9 |

## Plastic Welding Flux
### U. S. Patent 2,841,513

| | |
|---|---|
| Asbestos | 16 |
| Titanium Dioxide | 12 |
| Ferromanganese | 8 |

| | |
|---|---|
| Sodium Silicate | 34 |
| Glycerin | 30 |

---

Dental Silver Solder Flux

*Japanese Patent 3617 (1958)*

| | |
|---|---|
| Borax | 70-80 |
| Potassium Chloride | 10-15 |
| Sodium Fluoride | 5-8 |
| Silicon | 3-6 |
| Lithium Chloride | 3-5 |
| Zirconium | 2-5 |

---

Titanium Brazing Flux

*U. S. Patent 2,882,593*

| | |
|---|---|
| Potassium Bifluoride | 35 |
| Potassium Chloride | 50 |
| Barium Chloride | 10 |
| Barium Fluoride | 5 |

---

Uranium Melting Flux

*U. S. Patent 2,849,307*

| | |
|---|---|
| Calcium Fluoride | 35-55 |
| Magnesium Fluoride | 35-55 |
| Uranium Tetrafluoride | 5-15 |

---

Welding Wire for
Aluminum Bronze

*German Patent 840,943*

| | |
|---|---|
| Aluminum | 2-8 |
| Manganese | 1-10 |
| Silicon | < 0.2 |
| Copper | To make 100 |

Exothermic Aluminum
Welding Mix

*U. S. Patent 2,831,760*

| | |
|---|---|
| Red Phosphorus | 8.33 |
| Aluminum Powder | 8.33 |
| Copper-Aluminum Alloy (60-40) | 16.67 |
| Black Copper Oxide | 30.56 |
| Red Copper Oxide | 36.11 |

---

Self-Fluxing Brazing Alloy

*U. S. Patent 2,903,353*

| | |
|---|---|
| Lithium | 0.25-8 |
| Boron | 0.25-2.5 |
| Nickel | 5-50 |
| Copper | 1-80 |
| Palladium | 14.5 |

---

Copper Brazing Alloy

*U. S. Patent 2,891,860*

| | |
|---|---|
| Copper | 85-95 |
| Nickel | 3.5-10.5 |
| Chromium | 0.825-2.44 |
| Boron | 0.175-0.56 |

---

Chromium-Manganese
Welding Alloy

*U. S. Patent 2,820,725*

| | |
|---|---|
| Carbon | 0.11 |
| Manganese | 23.80 |
| Nickel | 0.90 |
| Chromium | 14.50 |
| Silicon | 1.20 |

| | |
|---|---|
| Molybdenum | 2.10 |
| Nitrogen | ¼ |
| Iron | To make 100 |

---

### Self-Fluxing Silver Brazing Alloy
*U. S. Patent 2,777,767*

| | |
|---|---|
| Copper | 53-56 |
| Zinc | 9-11 |
| Manganese | 0.05-0.5 |
| Iron | 0.10-0.5 |
| Nickel | 6-9 |
| Silver | 25-28 |

---

### Waterproof Welding Electrode Coating
*U.S.S.R. Patent 102,508*

| | |
|---|---|
| Titanium Dioxide | 3-4 |
| Hematite | 25-30 |
| Feldspar | 25-30 |
| Ferrotitanium | 4-6 |
| Ferromanganese | 28-35 |
| Starch | 3-7 |
| Sodium Silicate | 20-5 |

---

### Electrode Coating for Welding Copper to Carbon Steel
*U.S.S.R. Patent 104,208*

| | |
|---|---|
| Ferromanganese | 26 |
| Ferrosilicon | 18 |
| Copper Silicon | 12 |
| Feldspar | 14 |
| Fluorspar | 10 |
| Sodium Silicate | 20 |

### Welding Point Alloy
*Italian Patent 494,819*

| | |
|---|---|
| Copper | 73.0 |
| Zinc | 25.2 |
| Aluminum | 1.5 |
| Nickel | 0.3 |

This is resistant to acids and active gases.

---

### Lighter Flint Alloy (Pyrophoric)

#### Formula No. 1

Mischmetal with 12 to 17% zinc is extruded at 380 to 400°C. at 4100 to 4800 kg. per sq.in. The pyrophoric properties are improved by the addition of 2 to 4% cadmium and 1.5% copper.

The alloy is chill cast at 100 to 500°C.

#### No. 2
*Austrian Patent 203,223*

| | |
|---|---|
| Thorium | 80 |
| Iron | 20 |

Addition of cerium increases pyrophority; addition of zirconium (0.1 to 10%) increases wear resistance.

#### No. 3
*Austrian Patent 200,348*

| | |
|---|---|
| Cerium | 20-80 |
| Titanium | 5-55 |
| Antimony | 10-60 |

## No. 4
### Austrian Patent 194,620

| | |
|---|---|
| Antimony | 25-80 |
| Bismuth | 2-50 |
| Zirconium | 5-38 |

## Noncorrosive Pyrophoric Alloy
### U. S. Patent 2,815,281

| | |
|---|---|
| Titanium | 45-54 |
| Lead | 30-52 |
| Antimony | < 25 |

## Aluminum Bearing Alloy
### Formula No. 1
### U. S. Patent 2,760,860

| | |
|---|---|
| Copper | 0.2-4 |
| Nickel | 0.2-4 |
| Cadmium | 0.5-5 |
| Aluminum | To make 100 |

## No. 2
### U. S. Patent 2,852,365

| | |
|---|---|
| Indium | 0.05-5 |
| Silicon | 0.5-10 |
| Silver | 0.05-5 |
| Aluminum | To make 100 |

## Aluminum Base Bearing
### Formula No. 1
### U. S. Patent 2,752,239

| | |
|---|---|
| Silicon | 0.2-10 |
| Lead | 0.1-3 |
| Chromium | 0.1-0.8 |
| Aluminum | To make 100 |

## No. 2
### U. S. Patent 2,752,240

| | |
|---|---|
| Silicon | 0.2-10 |
| Bismuth | 0.1-5 |
| Chromium | 0.1-8 |
| Aluminum | To make 100 |

## No. 3
### U. S. Patent 2,763,546

| | |
|---|---|
| Magnesium | 0.2-0.5 |
| Cadmium | 0.2-2.5 |
| Silicon | 2.0-5.0 |
| Nickel | 0.3-1.3 |
| Iron | < 0.5 |
| Aluminum | To make 100 |

## Low Shrinkage Aluminum Alloy

| | |
|---|---|
| Silicon | 3.5 |
| Copper | 2.0 |
| Magnesium | 0.4 |
| Manganese | 0.5-2 |
| Titanium | 0.4 |
| Aluminum | To make 100 |

## Copper Casting Alloy

| | |
|---|---|
| Nickel | 10 |
| Aluminum | 10 |
| Manganese | 2 |
| Iron | 5 |
| Copper | To make 100 |

## Sliding Bearing Bronze Alloy
### Swiss Patent 309,977

| | |
|---|---|
| Copper | 90 |
| Tin | 5 |

| | |
|---|---|
| Zinc | 2 |
| Lead | 2 |
| Graphite | 1 |

### Dental Alloy
*Japanese Patent 3611 (1958)*

| | |
|---|---|
| Cobalt | 20-30 |
| Chromium | 20-25 |
| Nickel | 20-35 |
| Iron | 10-20 |
| Tungsten | 1-6 |
| Molybdenum | 1-5 |
| Titanium | ½-1 |

### Electrical Resistance Alloy
*U. S. Patent 2,850,383*

| | |
|---|---|
| Aluminum | 2-4.5 |
| Vanadium | 1-7 |
| Molybdenum | 0.1-10 |
| Chromium | 15-25 |
| Nickel | To make 100 |

### Forming Die Alloy
*U. S. Patent 2,857,267*

| | |
|---|---|
| Carbon | 0.5-6.0 |
| Manganese | 1.5-2.5 |
| Silicon | 1.5-2.5 |
| Chromium | 3.0-4.0 |
| Tungsten | 0.75-1.5 |
| Nickel | 1.0-2.0 |
| Vanadium | 0.75-1.5 |
| Iron | To make 100 |

### Hard Facing Alloy
*U. S. Patent 2,757,084*

| | |
|---|---|
| Cobalt | 35.0 |

| | |
|---|---|
| Boron | 3.5 |
| Nickel | 36.5 |
| Silicon | 5.0 |
| Iron | 20.0 |

### Ink-Resistant Pen Points
*German Patent 829,503*

| | Formula No. 1 | No. 2 |
|---|---|---|
| Tungsten | 65 | 65 |
| Ruthenium | 27 | 25 |
| Nickel | 5 | — |
| Platinum | 3 | — |
| Iridium | — | 5 |
| Cobalt | — | 5 |

### Low Expansion Alloy
*U. S. Patent 2,767,086*

| | |
|---|---|
| Manganese | 31.1 |
| Antimony | 63.8 |
| Arsenic | 5.1 |

### Long Wearing Type Alloy

| | |
|---|---|
| Zinc | 94 |
| Aluminum | 4 |
| Magnesium | 2 |

M.P. 340 to 365°C. It outwears lead-type metal 100 times.

### Leadless Typographic Alloy

| | |
|---|---|
| Zinc | 94 |
| Aluminum | 4 |
| Magnesium | 2 |

This is cheaper, lighter and more durable than lead alloys.

## Powdered Metal Friction Element
### U. S. Patent 2,783,529

Add 2 to 8% powdered molybdenum oxide to the following:

|                | Formula No. 1 | No. 2 |
|----------------|---------------|-------|
| Copper         | 78.9          | 78.9  |
| Tin            | 8.9           | 8.9   |
| Lead           | 6.7           | 3.3   |
| Flake Graphite | 5.5           | 8.9   |

These are used for brake linings and clutch facings to increase wear resistance.

---

## Nontarnishing Silver Alloy

### Formula No. 1
*Japanese Patent 6108-11 (1954)*

| Palladium | 5-30 |
|-----------|------|
| Indium    | 2-20 |
| Cadmium   | 0.5-5 |
| Silver    | To make 100.0 |

### No. 2

| Indium   | 5-30 |
|----------|------|
| Aluminum | 1-15 |
| Copper   | 0.5-10 |
| Zinc     | 0.5-10 |
| Silver   | To make 100.0 |

### No. 3

| Indium   | 10-40 |
|----------|-------|
| Zinc     | 2-10 |
| Cadmium  | 2-10 |
| Silver   | To make 100 |

## Ferromagnetic Composition
### U. S. Patent 2,773,039

| Magnesium Oxide | 8-25 |
|-----------------|------|
| Zinc Oxide      | 25-14 |
| Copper Oxide    | 2-11 |
| Nickel Oxide    | 3-12 |
| Lithium Oxide   | 0-3 |
| Ferric Oxide    | 47.4-52.4 |

This is fired in the usual way.

---

## Permanent Magnet Alloy
### U. S. Patent 2,768,427

| Nickel   | 13.50 |
|----------|-------|
| Aluminum | 8.00 |
| Cobalt   | 24.00 |
| Copper   | 3.00 |
| Titanium | 0.30 |
| Silicon  | 0.05 |
| Iron     | 50.00 |

---

## Sintered Permanent Magnet

| Iron-Aluminum Alloy | 60 |
|---------------------|----|
| Cobalt              | 23 |
| Nickel              | 14 |
| Copper              | 3 |

Powdered metals are pressed at 5 tons per sq. cm. and sintered at 1375°C.

---

## Magnetic Compensating Alloy
### *Japanese Patent 2763 (1958)*

| Nickel    | 29-50 |
|-----------|-------|
| Manganese | 1-20 |
| Carbon    | $<0.1$ |
| Iron      | To make 100 |

## High Specific Gravity Metal Sinter
### Swiss Patent 283,493

| | |
|---|---|
| Powdered Tungsten | 91.5 |
| Powdered Nickel | 5.0 |
| Powder Silver | 3.5 |

Sintered under protective atmosphere at 1300 to 1500°C. This alloy has a Sp.Gr. of 17 to 18.

---

## Electric Fuse Alloy
### Austrian Patent 199,724

| | |
|---|---|
| Zinc | 4-12 |
| Tin | 60-75 |
| Cadmium | 15-25 |

---

## High-Temperature Electric Resistance Alloy
### Swedish Patent 155,511

| | |
|---|---|
| Chromium | 10-40 |
| Aluminum | 1-9 |
| Iron | 51-89 |
| Tantalum and/or Titanium | 0-5 |
| Cerium and/or Calcium | 0.01-1 |
| Cobalt | >5 |
| Zirconium | 0.01-1 |
| Manganese and Silicon | 2 |

This will withstand temperatures up to 1350°C.

---

## Electrical Contact Amalgam
### U. S. Patent 2,850,382

| | |
|---|---|
| Silver | 7.20-14.33 |
| Palladium | 0.1 |
| Mercury | To make 100 |

---

## Metal to Ceramics Bonding Alloy
### U. S. Patent 2,805,944

| | |
|---|---|
| Titanium | 3 |
| Copper | 3 |
| Lead | 94 |

---

## Electrolytic Galvanizing
### German Patent 942,366

| | |
|---|---|
| Zinc Oxide | 45.6 |
| Sodium Cyanide | 78.0 |
| Sodium Hydroxide | 31.1 |
| Water | 1000.0 |
| Zinc Carbonate | 50.0 |

*Anode surface* = 7 sq. dm.
*Amperage* = 75-100

---

## Bright Nickel Plating
### East German Patent 10,268

| | |
|---|---|
| Nickel Sulfate | 150 g. |
| Ammonium Sulfate | 15 g. |
| Sodium Chloride | 10 g. |
| Sodium Sulfate | 100 g. |
| Tartaric Acid | 10 g. |
| Additive | 0.1-2 g. |
| Water | 1 l. |

Temp. 55°C.; c.d. 2 amp. per sq.dm.

## Continuous Chemical Nickel Plating
### U. S. Patent 872,353

|  | Mols |
| --- | --- |
| Nickel Chloride | 0.6750 |
| Sodium Hypophosphite | 0.2250 |
| Maleic Acid | 0.0600 |
| Lactic Acid | 0.2025 |
| Sodium Succinate | 0.1200 |
| Water | 1000.0000 |

## Black Nickel Plating Solution
### U. S. Patent 2,844,530

| | |
| --- | --- |
| Nickel Chloride 6H$_2$O | 60-90 g. |
| Ammonium Chloride | 5-30 g. |
| Sodium Sulfocyanide | 10-20 g. |
| Zinc Chloride | 20-50 g. |
| Water | 1 l. |

## Silver Plating by Brushing
### Italian Patent 491,099

| | |
| --- | --- |
| Silver Nitrate | 200 |
| Potassium Binoxalate | 300 |
| Cream of Tartar | 300 |
| Sodium Chloride | 420 |
| Ammonium Chloride | 80 |

The above is applied with a wet brush to brass to give a silver deposit.

## Silver Plating Without Cyanide

| | |
| --- | --- |
| Silver Sulfate | 20-25 g. |
| Potassium Iodide | 300-350 g. |
| Water | To make 1000 cc. |

0.3 amp. per sq. dm. at 18 to 25°C. Dispersion is increased by adding sodium pyrophosphate, 30 to 50 g. per liter.

## Silver Dip Brightener

| | |
| --- | --- |
| Thiourea | 3-5 |
| Sulfuric Acid | 1-2 |
| "Triton" X-100 | ½ |
| Water | To make 100 |

## Electropolishing Titanium
### Japanese Patent 9060 (1957)

| | |
| --- | --- |
| Lithium Chloride | 26.0 |
| Potassium Chloride | 30.6 |
| Magnesium Chloride | 22.0 |
| Titanium Chloride | 2.0 |

Use titanium piece as anode in fused bath at 550°C. with direct current of 1080 milliamperes, c.d. of 2.17 amp per sq.cm. A tungsten rod is used as cathode.

## Electropolishing Tantalum Electrolyte
### Japanese Patent 10,562 (1957)

| | |
| --- | --- |
| Ammonium Acid Fluoride | 100 g. |
| Phosphoric Acid | 300 cc. |
| Ethyleneglycol | 200 cc. |
| Water | 300 cc. |

## Bright Dip for Zinc Alloys
### U. S. Patent 2,904,413

| | |
|---|---|
| Chromic Anhydride | 6.5-20 |
| Nitric Acid | 2.7-15 |
| Hydrofluoric Acid | 1.0-8 |
| Water | To make 100 |

---

## Electropolishing Zirconium
### Japanese Patent 10,561 (1957)

| | |
|---|---|
| Water | 50-60 |
| Ethyleneglycol | 34-40 |
| Phosphoric Acid | 5-10 |
| Hydrofluoric Acid | 1-3 |

Immerse for 20 to 60 seconds at 35 to 50°C.

---

## Porous Gypsum Mold
### Japanese Patent 706 (1956)

| | Formula No. 1 | No. 2 |
|---|---|---|
| Calcined Gypsum | 70.0 | 70.0 |
| Quartzite (100 mesh) | 28.0 | 28.3 |
| Aluminum Powder | 0.4 | — |
| Caustic Soda | 0.6 | — |
| Borax | 0.9 | — |
| Water | 80.0 | — |
| Calcium Carbide | — | 1.4 |
| Talc Powdered | — | 0.1 |

Mold without removing hydrogen; heat at 80 to 100°C. for one hour and at 220°C. for one hour.

## Casting Shell Mold
### U. S. Patent 2,881,082

| | |
|---|---|
| Split Hard Stearine Pitch | 10.0 |
| Sulfur | 2.5 |
| Talc | 5.0 |
| Grind to powder and mix with Quartz Sand (Dried & Cooled) | 100.0 |

---

## Foundry Molding Sand
### U. S. Patent 2,822,278

| | |
|---|---|
| Sand | 85.5 |
| Western Bentonite | 4.0 |
| Olivine Flour | 10.0 |
| Starch (Partially Dextrinized) | 0.5 |

---

## Foundry Molding Mix
### U.S.S.R. 106,866

| | |
|---|---|
| Bitumen or Peat Wax | 1.5-4.5 |
| Bentonite | 2-5 |
| Graphite or Alum | 1-2 |
| Water | 0.5 |
| Quartz Sand | To make 100 |

---

## Jewelers' Casting Sand Binder

| | |
|---|---|
| Glycerin | 3 oz. |
| Saturated Solution of Boric Acid (At Room Temperature) | ½ gal. |

Liquid Wetting
  Agent          ¼ oz.
Molasses     To make 1 gal.

This binder is especially suitable for sands containing equal parts of fine sand, 2F pumice and zircon sand, and can be used indefinitely without losing its effectiveness.

---

Foundry-Core Binder

Formula No. 1
U. S. Patent 2,718,681

| | |
|---|---|
| Kerosene | 30.0 |
| "Belloid" F.R. | 4.0 |
| "Nervan" C.S. | 0.2 |
| Water | 150.0 |
| Urea-Formaldehyde Resin (45% Water Solution) | 150.0 |

No. 2
U. S. Patent 2,861,893

| | |
|---|---|
| Sand (60-120 Mesh) | 95.0 |
| Sodium Silicate | 2.7 |
| Water | 1.5 |
| Sugar | 0.8 |

The cores are exposed to carbon dioxide gas for 4 to 5 minutes for hardening. This gives increased compressive strength.

---

Foundry Sand Additive
U. S. Patent 2,848,338

| | |
|---|---|
| Powdered Charcoal (80-200 Mesh) | 70-75 |

Fuel Oil
  (Fl. Pt. 120-180°F.) 25-30
0.5-1.5% of above is added to coat particles of foundry core sand.

---

Foundry Mold Facing
U. S. Patent 2,798,817

| | |
|---|---|
| Sand (140 Mesh) | 75 |
| Bentonite | 5½ |
| Corn Flour | 1 |
| Graphite | 1 |
| Wood Flour | 13½ |
| Water | 4 |

---

Foundry Sand Core and Mold
U. S. Patent 2,765,507

| | |
|---|---|
| Water-Soluble Poly-acrylic Acid | 0.3-1 |
| Moisture | 2.5-5 |
| Kerosene | 0.25-0.75 |
| High Grade Foundry Sand | To make 100 |

Mold and air-oven dry at 200 to 350°F. for 10 to 30 minutes. This gives a high hot strength core or mold.

---

Foundry Core Wash
U. S. Patent 2,786,771

| | |
|---|---|
| Calcium Oxide | 21-30 |
| Magnesium Oxide | 2.6-4.0 |
| Flake Graphite | 46.5-65.0 |
| Soda Ash | 1-5 |
| Water | To make 100 |

Foundry Sand Core Coating
U. S. Patent 2,809,117

Magnesium
Chloride                9
Magnesium Oxide         6-6¼
Calcium Oxide           35½-36½
Graphite                107-109
Water         To make 30-35°Bé.
                  Suspension

---

Foundry Mold Coating
U. S. Patent 2,755,192

Cork Dust          8 lb. 2 oz.
Casein Glue        1 lb. 14 oz.
Bentonite          3 lb. 6 oz.
Sodium Silicate
  (40°Bé.)         3½ qt.
Water              To suit

---

Foundry Mold Filler
U.S.S.R. 102,545

Potassium Nitrate      5.5
Sodium Nitrate         4.5
Powdered Charcoal      5-10
Quartz
  Sand        To make 100

---

Solid Carburizer
U.S.S.R. 111,009

Charcoal           10-15
Semi-Coke          75-80
Barium Acetate     5-10

---

Revivifying Carburizing Agents
East German Patent 11,840
Carburizing Salts        50

Ferric Hydroxide        5
Heat to 900°C. for 30 minutes.

---

Iron & Steel Hardening Salts
French Patent 1,113,571

Soda Ash               2.5
Salt                   5.0
Sodium Sulfate         5.0
Iron Cyanide           30.0
Potassium Cyanide      30.0
Borax                  27.5

---

Steel Rimming Agent
U. S. Patent 2,854,329

Cryolite               17-50
Sodium Nitrate         10-15
Aluminum Fines         2-7
Globular Ferric Oxide
              To make 100

---

Steel Quenching Bath
U. S. Patent 2,817,612

Sodium Chloride     24 oz.
Sodium Hydroxide    19 oz.
Sodium Sulfate      2 oz.
Alum                1 oz.
Water               1 gal.

---

Metal Defect Detector
U.S.S.R. 115,104,

"Sudan" IV (Dye)       10
Mineral Oil            50
Benzol                 940

## Cast Steel Mold Releasing Agent

*Japanese Patent 7563 (1954)*

| | |
|---|---|
| Calcined Steatite | 97 |
| Iron Oxide | 32 |
| Bentonite | 5 |

## Ore Pellet Binder

*U. S. Patent 2,833,642*

| | |
|---|---|
| Alkaline Soybean Meal | 0.03-0.1 |
| Bentonite | 0.2 -0.5 |

## Plastic Metal Alloy

*U. S. Patent 2,585,393*

| | |
|---|---|
| Nickel | 49-74 |
| Gallium | 25-45 |
| Silicon | ½-7½ |

This alloy sets at room temperature.

## Neutron-Absorbing Low-Melting Alloy

*U. S. Patent 2,680,071*

| | |
|---|---|
| Indium | 54-62 |
| Cadmium | 8-18 |
| Bismuth | To make 100 |

## Low-Melting Gallium Alloy

| | |
|---|---|
| Gallium | 82 |
| Tin | 12 |
| Zinc | 6 |

This forms a ternary eutectic melting at 17°C.

## Wood's Metal Substitute for Pipe Bending

| | |
|---|---|
| Beeswax | 45 |
| Ethyl Cellulose | 15 |
| Triethanolamine | 1 |
| Calcium Sulfate or Other Inert Filler | 5 |

Melt the beeswax, stir in the triethanolamine and calcium sulfate, and heat to 350°F., with stirring. Then add the ethyl cellulose slowly to avoid lumps and stir till smooth. The mass may be poured at about 300°F. By increasing the amount of ethyl cellulose, it is possible to obtain stronger waxlike substitutes, at the expense of pourability. Other waxes which may be used include candelilla, carnauba, and montan. The formula will not work with mycrocrystalline, ozokerite, and paraffin waxes. If the compound is too sticky, add more filler.

For production bending of small tubes, the casting formula may be cast into slightly tapered rods, which can be driven into the tubing to be bent. After bending, heat the parts in an oven at 350°F. and blow out the wax blown in by air pressure, or allow the parts to soak overnight in a lacquer solvent to remove the wax.

Electrical Resistance Wire
U. S. Patent 2,506,420

| Cobalt | 54.5 |
| Manganese | 39.6 |
| Copper | 4.5 |
| Aluminum | 1.4 |

Low-Resistance Electrical
Fuse Alloy
U. S. Patent 2,486,202

| Bismuth | 14.5 |
| Cadmium | 2.5 |
| Antimony | 1.0 |
| Lead | 82.0 |

Resistance Wire for Explosives
U. S. Patent 2,514,765

| Platinum | 93-98 |
| Chromium | 1-2 |
| Nickel | 1-6 |

Corrosion-Resistant
Thermostatic Alloy

| a Manganese | 0.2-0.6 |
| Carbon | 0.01-0.2 |
| Chromium | 12-20 |
| Silicon | 0.2-1.5 |
| Copper | 0.5-1.5 |
| Iron | To make 100 |

| b Copper | 96-98 |
| Silicon | 0.5-1.5 |
| Tin | 0.75-2.5 |

The surfaces of a and b are fused together at the junction.

Silver Alloy
(Resistant to Oxygen and
Sulfur)
French Patent 936,021

| Silver | 800 |
| Zinc | 50-100 |
| Nickel | 150-100 |

Nontarnishing Silver Alloy
French Patent 972,734

Formula No. 1

| Silver | 78.8 |
| Cadmium | 20.0 |
| Aluminum | 1.2 |

No. 2

| Silver | 78.6 |
| Cadmium | 13.0 |
| Zinc | 7.0 |
| Calcium | 1.4 |

No. 3

| Silver | 75 |
| Cadmium | 20 |
| Tin | 5 |

No.4

| Silver | 72.4 |
| Cadmium | 20.0 |
| Tin | 5.0 |
| Magnesium | 2.6 |

### Rhodium

Because of its whiteness, hardness, high reflectivity and corrosion resistance, rhodium makes an ideal electroplate for many special applications. Its most important uses are in the optical, jewelry, and electrical fields, where its high reflectivity, hardness, and chemical stability make it particularly desirable. Table 1 compares the reflectivity of rhodium with that of certain other metals. The only metals used for reflecting surfaces which have a reflectivity higher than that of rhodium are freshly polished silver and aluminum which, however lose their high reflectivity after exposure to the atmosphere and drop to values lower than shown in the table.

Rhodium is reported to be harder than any of the other metals shown in Table 1 with the exception of chromium, the reflectivity of which is much lower than that of rhodium. This makes rhodium very desirable for a reflecting surface since it combines hardness with high reflectivity and can be frequently cleaned by wiping, if necessary, without danger of scratching.

Rhodium electrodeposits up to 2 or 3 mg. per sq. ft. are bright and are suitable for searchlight reflectors, jewelry and ornamental work, wave guides, and other applications where the desired characteristics are corrosion resistance, color, good electrical conductivity and high reflectivity. Deposits in excess of this amount become frosty in appearance and although their corrosion resistance, as reported, is not much higher than that of the thinner deposits, they are very useful for sliding or wiping electrical contacts, where light contact pressures and low contact resistance are required.

TABLE 1

Approximate Reflectivity of Certain Metals

| | % |
|---|---|
| Freshly Polished Silver | 94* |
| Aluminum | 90 |
| Rhodium | 76° |
| Iridium | 70 |
| Nickel | 67* |
| Chromium | 67* |
| Platinum | 65 |
| Palladium | 63 |

*Determined by the National Bureau of Standards.

TABLE 2

Rhodium-plating Baths

*Concentration*

| | | Concentration |
|---|---|---|
| *a* | Concentrated C.P. Sulfuric Acid | 20 cc. per liter (2.523 liquid oz. per gal.) |
| | Rhodium Metal in Prepared Concentrated Solution | 2 g. per liter (5 dwt. per gal.) |
| *b* | 85% Orthophosphoric Acid | 40 cc. per liter (5 to 6 liquid oz. per gal.) |
| | Rhodium Metal in Prepared Concentrated Solution | 2 g. per liter (5 dwt. per gal.) |

(Based on data from Baker and Co., Inc.)

## Rhodium-plating Baths

Two general types of rhodium-plating baths are used in which rhodium is present either as phosphate or as sulfate; a special bath is sometimes used in which the metal is present as a combination of the two salts. The metal content of the rhodium sulfate or rhodium phosphate bath is usually around 2g. per liter (5 dwt. per gal.). The acid concentration is not very critical and may range from 10 cc. per liter upward for either sulfuric acid or phosphoric acid. Higher acid concentration tends to lower the cathode efficiency slightly.

Representative compositions for rhodium sulfate and phosphate plating baths are given in Table 2. It is important in making up either bath to add the acid to the water before adding the rhodium. If this is not done, the rhodium compound may be partially precipitated by hydrolysis. The best working temperature is 110 to 125°F. The current density may be 10 to 100 amp. per sq. ft. Insoluble platinum anodes are used and the metal content is replenished by the addition of a concentrated rhodium solution. A satisfactory rhodium solution of this type should produce a bright deposit of approximately 0.00006 in. in 30 minutes at 200 amp. per sq. ft. Deposits so produced do not require buffing and the throwing power approaches that of a gold or silver cyanide plating bath.

Heavy rhodium deposits (up to 0.0005 in. thickness) found limited application during World War II where highest wear and corrosion resistance were required. A bath consisting of 10 to 20 g. per liter of rhodium metal as rhodium phosphate and 20 cc. per liter of concentrated sulfuric acid has been recommended for producing heavy deposits. A current density of 5 amp. per sq. ft. or less should be used at a bath temperature of around 120°F .

It has become common practice to use a nickel or bright nickel undercoat for rhodium, except in the case of gold alloys and platinum bases. Zinc, tin, and lead alloys must receive a heavy nickel undercoat to protect the metal from the action of the acid rhodium plating solution completely. All basic metals, as well as silver, should receive an underplate of nickel. Soft metals, such as the tin alloys, should be copper flashed prior to nickel plating. After nickel plating and a very thorough rinsing, the work should be rhodium plated without delay.

One point to be kept in mind is that imperfections and flaws in the basic metals are not covered by plating. Instead, they are emphasized and will show up more conspicuously after rhodium plating. For this reason, all work to be rhodium plated should be given a "perfect" finish. After polishing, it is desirable to remove the bulk of the polishing compound by vapor degreasing or treatment in an emulsion cleaner. The articles are then rinsed, preferably in hot water, and placed on racks for cleaning and plating.

A commercial electrocleaner, maintained at 200°F., may be used for further cleaning prior to plating. A solution of 4 to 6 oz. per gal. trisodium phosphate is also a satisfactory cleaner. Cathodic cleaning at 6 to 8 volts for about 1 minute is generally satisfactory. The electrocleaning should be followed by several rinses in running water. This should be followed by a dip in 10% potassium or sodium cyanide solution and then by a very thorough water rinse. A dip in 10% hydrochloric or sulfuric acid and a final rinse conclude the preparatory cycle before the actual plating.

For general work with conspicuous polished areas, the

lower current density ranges are recommended for rhodium plating in order to avoid burning or cloudiness. For fine chains or similar work, a high current density will generally give better results. Considering the slow rate of deposition of rhodium, one minute is considered a minimum for bringing out the true rhodium color.

___

## Platinum-plating Baths

Since platinum deposits are used mostly for ornamental purposes and as flash plates, for reasons of economy, it has become common practice to use a nickel undercoat. This is necessary whenever lead, tin, or zinc alloy articles are to be plated. When heavier than flash deposits of platinum are applied, they must be scratch brushed intermittently.

The oldest and most widely used platinum-plating solution is based on a chloroplatinic acid solution. A solution of this type recommended for flash plating has the following[*] composition:

### Platinum-plating Bath

| | Concentration |
|---|---|
| Diamonnium Phosphate | 20 g. per liter (2.2 oz. per gal.) |
| Disodium Phosphate | 100 g. per liter (13.4 oz. per gal.) |
| Choloroplatinic Acid | 4 g. per liter (0.535 oz. per gal.) |

This solution is operated close to the boiling temperature at a current density of 5 to 10 amp. per sq. ft. Bright flash deposits can be obtained in 30 to 120 seconds; later the deposit turns cloudy and finaly dull gray.

Another type of plating bath, based on the same formula, uses platinum diaminonitrite instead of the chloride. The bath has a longer useful life since no chlorides are introduced on replenishing the platinum content. An improvement[†] on this bath is said to consist of 100 g. per liter ammonium nitrate, 10 g. per liter sodium nitrite, 16.5 g. per liter of platinum diaminonitrite, and 50 g. per liter of 28% ammonium hydroxide solution. A more recently proposed platinum-plat-

___

[*]*Trans. Electrochemical Soc.*, 80, 489 (1941).

[†]*Trans. Electrochem. Soc.*, 59, 273 (1931).

ing bath‡ represents a combination of the phosphate and aminonitrite. It contains 100 g. per liter disodium phosphate, 20 g. per liter diammonium phosphate, and 6.5 g. per liter diaminoplatinous nitrite.

### Palladium-Plating Baths

As a precious metal, palladium is tarnish resistant and also is more than nickel or chromium to most acids except nitric acid. In color it closely resembles nickel and its hardness equals or surpasses a nickel deposit from a Watts solution. Real progress was made in palladium plating with the introduction of the amino compounds and alkali double nitrites of palladium.

A typical bath consists of 8 g. per liter palladium aminonitrite dissolved in a small quantity of ammonia, 100 g. per liter ammonium nitrate, and 10 g. per liter sodium nitrite. If this solution is operated with a pH above 7, by the addition of ammonia, the deposits obtained become smoky rather quickly, but heavier deposits can be built up by

‡Blum and Hogaboom, *Principles of Electroplating and Electroforming*, (3rd Edition) p. 385, McGraw-Hill Book Co., New York (1949).

means of intermittent scratch brushing. In all cases, it is necessary to keep the voltage below the gassing point. If the pH is below 7, the deposits have a tendency to be brighter, but at the same time the danger of peeling increases.

Palladium - plating solutions made up with the alkali double nitrites of palladium are interesting in that under certain conditions they can be made self-sustaining. While at a $pH$ above 7, palladium anodes are hardly attacked, they become soluble at a $p$H below 7. To prepare a bath of this type, 10 g. of sodium palladium chloride (about 4 g. of metal) are dissolved in 500 cc. distilled water and 10 g. sodium nitrite are added.

This bath is operated at about 120°F. with palladium anodes and at a current density of about 10 amp. per sq. ft. Both the anode and cathode efficiencies are close to 100%. In making up the bath, as described, the $pH$ should be somewhere around 4. If it is not, it can be easily adjusted by the addition of dilute hydrochloric acid or sodium carbonate. For best results, the $pH$ should be held between 4 and 5. The deposits from this bath stay bright for a long time and if a slight cloudiness appears, it can be easily buffed away. Palladium can be plated on most basic metals, on sterling or fine silver, and on gold.

---

### Stripping Chrome Plate from Steel Molds

| | |
|---|---|
| "Anodex" 61–X | 8 oz. |
| Water | 1 gal. |
| Temperature | 180-190°F. |
| Work as Anode | 6 volts |

Use a steel tank with the tank as the cathode.

---

### Metal Stop-Off

In tinning steel, it is often desirable to prevent the tin from wetting certain parts of the steel. This can be done with a stop-off.

#### Formula No. 1

| | |
|---|---|
| Water Glass | 2 |
| Water | 2 |
| Clay | 1 |

#### No. 2

| | |
|---|---|
| Xylene | 1 |
| "Oildag" | 1 |

This stop-off is easily removable by brushing.

#### No. 3

This is a satisfactory stop-off coating for plating chromium and cadmium and for anodizing. It can be used in acid baths up to 200°F. Its adhesion to the metal is quite good and it can be stripped readily.

| | |
|---|---|
| "Glyco" Wax S–1167 | 75 |
| "Acrawax" B | 16 |
| "Flexo Wax" C | 9 |

Melt together and mix until uniform. Apply hot.

---

### Brightening Nickel
#### U. S. Patent 2,638,410

| | |
|---|---|
| Phosphoric Acid | 45-60 |
| Sulfuric Acid | 15-25 |
| Nitric Acid | 8-15 |
| Water | 10-20 |

Use at above 65°C.

---

### Stainless-Steel Electropolishing

| | |
|---|---|
| Phosphoric Acid | 47 |
| Glycerin | 49 |
| Water | 4 |

Use at 110 to 120°F., 4.5 amp. per sq. in., and 14 volts.

---

### Metal-Casting Mold
#### U. S. Patent 2,681,860

Use silica powder of at least 99% silicon oxide content and grind it to such a fineness that 99% will pass through a 250-mesh sieve, the average particle size being 3 to 5 microns. Mix this powder with an aqueous solution of sodium hexametaphosphate by weight, in the proportion of 100 g. of powder to 34 ml. of solution. Thus, the sodium hexametaphosphate amounts to 0.085% of the silica. Mixing mechanically for at least 10 minutes, the fluidity increases as the mixing continues. Pour the resultant slurry around one or more wax patterns mounted on absorbent asbestos board within a ring of heat-resisting metal. This ring is prolonged upward by a paper extension and the ring and extension are filled with the slurry. Liquid will separate from the slurry as a supernatant layer within the paper extension. After about 2 hours, decant this liquid and, 30 minutes later, cut off the extension level with the top of the metal ring. Dry the investment mold so preduced in the usual way, e.g., by heating for ½ hour periods at successively increasing temperatures, such as 100, 200, and 300°C., and then raising the temperature to 800° C. in 1 hour. Hold the investment mold at 800°C. for 30 minutes or more and then allow to cool. Evacuate the melted wax from the mold while drying.

## Foundry-Core Mix

### Formula No. 1

(For Gray and Malleable Iron)

| | |
|---|---|
| Sand (60-90 A.F.S.) | 1000 |
| Cereal Flour | 7-10 |
| "Resimene"–970 | 10 |
| Release Agent* | 4 |
| Water | 20-40 |

Where high green strength is a requisite, 5 lb. of southern bentonite can be added to the mix. Improvement in green strength may also be obtained by substituting some fine sand (130 to 150 A.F.S.) for the regular sand. For example, substituting 100 lb. of fine sand in the mix will improve both green strength and flowability. Sticking in the core box will also be minimized. The addition of bentonite in "Resimene" 970 core mixes also decreases collapsability of the cores on shake out.

### No. 2

(For Brass and Bronze)

| | |
|---|---|
| Sand (90-130 A.F.S.) | 1000 |
| Cereal Flour | 7-10 |
| "Resimene" 970 | 10-12 |
| Release Agent* | 4 |
| Water | 20-40 |

Almost any standard brass-foundry core wash will work well on this.

## No. 3
### (For Aluminum)

| | |
|---|---|
| Sand (90-130 A.F.S.) | 1000 |
| Cereal Flour | 6 |
| "Resimene" 970 | 7 |
| Release Agent* | 4 |
| Water | 20-30 |

In aluminum core work, the weakest possible core, consistent with core-handling requirements, is desirable. Often a much "weaker" mix than this is used and the cores are sprayed with water before being placed in the baking oven. This produces a core with a strong skin and maximum collapsability.

## No. 4
### (For Magnesium)

| | |
|---|---|
| Sand (90-130 A.F.S.) | 1000 |
| Sulfur, Potassium, | |
| Fluoborate, Etc. | 7 |
| Cereal Flour | 4 |
| "Resimene" 970 | 7 |
| Release Agent* | 4 |
| Water | 20-30 |

### Precision-Casting Refractory Mixture

To function as an investment binder, ethyl silicate is first hydrolyzed. To prepare a completely hydrolyzed ethyl silicate solution, stir 15 volumes of condensed ethyl silicate or tetraethyl orthosilicate into a mixture of 8 volumes of "Synasol" solvent (denatured ethyl alcohol) and 2 volumes of 1% hydrochloric acid. Hydrochloric acid serves as the hydrolysis catalyst. Allow the mixture to stand a minimum of 2 hours before use to insure complete hydrolysis. This solution will contain 17.5% silica by weight or 16 g. of silica per 100 cc. of solution and is called "anhydrous," since it contains just enough water to give complete hydrolysis. It is referred to as *Solution 1* in the rest of this article. A similar solution of the same silica content can be made from ethyl silicate 40 by using a mixture of 37.6% of this product with 59.8% "Synasol" solvent and 2.6% of 3% hydrochloric acid.

The anhydrous hydrolyzed ethyl silicate solution is made ready for use by adding more water to give shorter setting times. Under most conditions, 50 to 60 minutes setting time is suitable. At 68°F., about 6.5 volumes of water added to 100 volumes of ethyl silicate Solution 1 will give a setting time of 50 minutes. This water should be

---

*Distilled oleic acid (red oil).

added just prior to mixing the binder solution with the refractory.

---

Refractory Mixture

| | |
|---|---|
| Magnesium Oxide | 0.4 |
| Silica Flour | 34.6 |
| Sand (140 Mesh) | 45.0 |
| Sand (35-60 Mesh) | 20.0 |

To insure even distribution of the magnesium oxide, mix it with the silica flour and then add this mixture to the sand. Magnesium oxide is incorporated as a neutralizing agent for the hydrochloric acid.

Satisfactory results are obtained by mixing 100 cc. (3.4 oz.) of the binder solution with 1 lb. of the refractory mixture. This blend is poured around the wax pattern, settled on a vibrating table, and allowed to gel.

Where a smooth casting and reproduction of fine detail are necessary, a fine-grained refractory composition should be in contact with the fusible pattern. This can be accomplished most economically by using a primary dip coat followed by backing up with the regular investment.

---

Dip Coat

| | |
|---|---|
| Ethyl Silicate | |
| (Solution 1) | 33 |
| Silica (120 Mesh) | 40 |
| China Clay | 27 |

Immediately after application, dust with 35- to 80-mesh Ottawa sand or similar.

If erratic results are obtained, they are usually caused by improper control of the setting time. To correct the trouble, check the factors summarized here.

DELAYED SETTING TIME IS CAUSED BY:

Insufficient water for complete hydrolysis; or drop in the temperature of the sand-binder mixture; or insufficient neutralizing agent (magnesium oxide) in the investment mixture; or too low a silicate content—appreciable effect only below 10%; or too low a ratio of dry investment to ethyl silicate solution.

ACCELERATED SETTING TIME IS CAUSED BY:

Excess water over the amount required to complete hydrolysis; or increase in temperature of the sand-binding mixture; or increase in silica content—appreciable effect only above about 17% silica; or prolonged aging of solution containing excess water.

## Fusible Alloys and Eutectics

| Trade Name or Type of Alloy | Composition, per cent | | | | | Melting Range | | | |
| | | | | | | Solidus* | | Liquidus** | |
| | Sn | Bi | Pb | Cd | Others | °C. | °F. | °C. | °F. |
| --- | --- | --- | --- | --- | --- | --- | --- | --- | --- |
| Binary Eutectic .... | 99.25 | ... | ... | ... | 0.75 Cu | 227 | 441 | 227 | 441 |
| Binary Eutectic .... | 96.5 | ... | ... | ... | 3.5 Ag | 221 | 430 | 221 | 430 |
| Binary Eutectic .... | ... | ... | ... | 17.0 | 83.0 Tl | 203 | 397 | 203 | 397 |
| Binary Eutectic .... | 92.0 | ... | ... | ... | 8.0 Zn | 199 | 390 | 199 | 390 |
| Binary Eutectic .... | ... | 47.5 | ... | ... | 52.5 Tl | 188 | 370 | 188 | 370 |
| Binary Eutectic .... | 62.0 | ... | 38.0 | ... | ... | 183 | 361 | 183 | 361 |
| Binary Eutectic .... | 67.0 | ... | ... | 33.0 | ... | 176 | 349 | 176 | 349 |
| Binary Eutectic .... | 56.5 | ... | ... | ... | 43.5 Tl | 170 | 338 | 170 | 338 |
| .................... | 40.0 | ... | 42.0 | 18.0 | ... | 145 | 293 | 160 | 320 |
| Ternary Eutectic ... | 51.2 | ... | 30.6 | 18.2 | ... | 145 | 293 | 145 | 293 |
| Binary Eutectic .... | ... | 60.0 | ... | 40.0 | ... | 144 | 291 | 144 | 291 |
| .................... | 48.8 | 10.2 | 41.0 | ... | ... | 142 | 288 | 166 | 331 |
| Binary Eutectic— "Cerrotru" ....... | 43.0 | 57.0 | ... | ... | ... | 138 | 281 | 138 | 281 |
| .................... | 41.6 | 57.4 | 1.0 | ... | ... | 134 | 273 | 135 | 275 |
| Ternary Eutectic .... | 40.0 | 56.0 | ... | ... | 4.0 Zn | 130 | 266 | 130 | 266 |
| Ternary Eutectic .... | 46.0 | ... | ... | 17.0 | 37.0 Tl | 128 | 262 | 128 | 262 |
| Binary Eutectic— "Cerrobase" ...... | ... | 55.5 | 44.5 | ... | ... | 124 | 255 | 124 | 255 |
| Binary Eutectic .... | 48.0 | ... | ... | ... | 52.0 In | 117 | 243 | 117 | 243 |
| "Cerroseal"—35 ..... | 50.0 | ... | ... | ... | 50.0 In | 117 | 243 | 127 | 260 |
| .................... | 1.0 | 55.0 | 44.0 | ... | ... | 117 | 243 | 120 | 248 |
| "Cerromatrix" ...... | 14.5 | 48.0 | 28.5 | ... | 9.0 Sb | 103 | 217 | 227 | 440 |
| .................... | 34.5 | 44.5 | ... | 21.0 | ... | 103 | 217 | 120 | 248 |
| .................... | 25.0 | 50.0 | ... | 25.0 | ... | 103 | 217 | 113 | 235 |
| Ternary Eutectic .... | 25.9 | 53.9 | ... | 20.2 | ... | 103 | 217 | 103 | 217 |
| .................... | 33.0 | 34.0 | 33.0 | ... | ... | 96 | 205 | 143 | 289 |
| Malotte's .......... | 34.2 | 46.1 | 19.7 | ... | ... | 96 | 205 | 123 | 253 |
| Rose's ............. | 22.0 | 50.0 | 28.0 | ... | ... | 96 | 205 | 110 | 230 |
| D'Arcet's ......... | 25.0 | 50.0 | 25.0 | ... | ... | 96 | 205 | 98 | 208 |
| Newton's .......... | 18.8 | 50.0 | 31.2 | ... | ... | 96 | 205 | 97 | 207 |
| Onion's or Lichtenberg's ...... | 20.0 | 50.0 | 30.0 | ... | ... | 96 | 205 | 100 | 212 |
| Ternary Eutectic .... | 15.5 | 52.5 | 32.0 | ... | ... | 96 | 205 | 96 | 205 |
| Ternary Eutectic .... | ... | 51.7 | 40.2 | 8.1 | ... | 92 | 198 | 92 | 198 |
| .................... | 15.4 | 38.4 | 30.8 | 15.4 | ... | 70 | 158 | 97 | 207 |
| "Cerrosafe" ........ | 11.3 | 42.5 | 37.7 | 8.5 | ... | 70 | 158 | 90 | 194 |

| | | | | | | | | | |
|---|---|---|---|---|---|---|---|---|---|
| .................. | 24.5 | 45.3 | 17.9 | 12.3 | ... | | 70 | 158 | 88 | 190 |
| .................. | 13.0 | 40.0 | 37.0 | 10.0 | ... | | 70 | 158 | 85 | 185 |
| .................. | 13.0 | 42.0 | 35.0 | 10.0 | ... | | 70 | 158 | 80 | 176 |
| Lipowitz's or "Cerrobend" | 13.3 | 50.0 | 26.7 | 10.0 | ... | | 70 | 158 | 73 | 163 |
| Wood's | 12.5 | 50.0 | 25.0 | 12.5 | ... | | 70 | 158 | 72 | 162 |
| Quaternary Eutectic | 13.1 | 49.5 | 27.3 | 10.1 | ... | | 70 | 158 | 70 | 158 |
| .................. | 13.2 | 49.3 | 26.3 | 9.8 | 1.4 | Ga | 65 | 149 | 66 | 151 |
| "Cerrolow" 147 | 12.77 | 48.0 | 25.63 | 9.6 | 4.0 | In | 61 | 142 | 65 | 149 |
| "Cerrolow" 136B | 15.0 | 49.0 | 18.0 | ... | 18.0 | In | 57 | 136 | 69 | 156 |
| "Cerrolow" 136 | 12.0 | 49.0 | 18.0 | ... | 21.0 | In | 57 | 136 | 57 | 136 |
| "Cerrolow" 140 | 12.6 | 47.5 | 25.4 | 9.5 | 5.0 | In | 56 | 134 | 65 | 149 |
| "Cerrolow" 117 | 8.3 | 44.7 | 22.6 | 5.3 | 19.1 | In | 47 | 117 | 47 | 117 |
| "Cerrolow" 117B | 11.3 | 44.7 | 22.6 | 5.3 | 16.1 | In | 47 | 117 | 52 | 126 |
| "Cerrolow" 105 | 7.97 | 42.91 | 21.7 | 5.09 | 18.33 In $\rbrace$ 4.0 Hg | | 38 | 100 | 43 | 110 |
| Binary Eutectic | 8.0 | ... | ... | ... | 92.0 | Ga | 20 | 68 | 20 | 68 |

\* Temperature at which melting begins
\*\* Temperature at which melting is complete

## CHAPTER XIV

# PAINT, VARNISH, AND LACQUER

### Alkyd Primer-Surfacer

a "Titanox" RCHT-X    268.0
"Ti-Pure" R-610        91.0
"Celite" 266           21.0
SP-323-50-El
    "Beckosol"        224.0
Mineral Spirits        72.0

b SP-323-50-El
    "Beckosol"        127.0
Mineral Spirits        73.0
6% Cobalt
    Naphthenate         1.5
24% Lead
    Naphthenate         3.0
Anti-Skinning Agent     1.0

Grind a 16 to 18 hours in a pebble mill and add b.

---

### Primer for Concrete Floor Paint

R-16-A "Neville" Resin,
    70% Solution in
    Mineral Spirits   75.000

### Fed. Spec. TT-R-266, Type I, Alkyd

Resin                 25.000
6% Cobalt
    Naphthenate        0.126
24% Lead
    Naphthenate        0.310

Blend together at room temperature.

---

### Maritime Primer

"Neville" LX-685,135    75.0
Modified Phenolic
    Resin              25.0
Bodied Linsed Oil,
    Z-8 Viscosity     200.0
Mineral Spirits       245.0
24% Lead Naphthenate
    or Tallate          4.2
6% Manganese
    Naphthenate or
    Tallate             1.7

Place both resins and 20 gal. of bodied linseed oil in the kettle

217

and heat to 540 to 545°F. Hold at this temperature for about 30 minutes. Check with the remaining 5 gal. of bodied linseed oil, reduce, and add driers.

---

Industrial Anti-Corrosive Primer

| | |
|---|---|
| 97% Grade Red Lead | 970 |
| "Celite" 110 | 91 |
| Magnesium Silicate | 76 |
| "Bentone" 18-C | 15 |
| "Epon" Resin 1001 (Shell) | 253 |
| "Beetle" Resin 216-8 (Cyanamid) | 15 |
| Butyl "Cellosolve" | 21 |
| Methyl Isobutyl Ketone | 153 |
| Toluene | 175 |

*For Baking:*

The above composition is used as shown.

*For Force Dry:*

Add to the above volume of paint the folowing "catalyst" solution:

| | |
|---|---|
| Diethylene Triamine (90% grade) | 17 |
| Butyl "Cellosolve" | 25 |

After mixing, the paint should be used within 24 hours.

*For Air Dry:*

Substitute ethylene diamine (87% grade) for the diethylene triamine in the above catalyst mixture. The mixed paint should be used within a day.

---

Etching Primer (Wash Primer)

*Base*

| | |
|---|---|
| Polyvinyl butyral Resin | 13 |
| Zinc Tetroxychromate | 12 |
| "Asbestine" | 2 |
| Denatured Alcohol | 36 |
| Methylethylketone | 37 |

*Acid Diluent*

| | |
|---|---|
| Phosphoric Acid | 6½ |
| n-Butanol | 36½ |
| Denatured Alcohol | 7 |

---

Wash Primer

| | MIL-C-15328A Lead Chromate type | |
|---|---|---|
| | Formula No. 1 | No. 2 |
| | (WP-1) | (Single Package) |
| **Base** | | |
| Polyvinyl Butyral Resin | 7.2 | 9.0 |
| Basic Zinc Chromate | 6.9 | — |
| Lead Chromate (Low Soluble Salts) | — | 8.6 |

| | | |
|---|---|---|
| Talc | 1.1 | 1.4 |
| Lampblack | Trace | Trace |
| Isopropyl Alcohol (99%) | 48.7 | 53.0 |
| Butyl Alcohol | 16.1 | — |
| Methyl Isobutyl Ketone | — | 13.0 |

*Acid Component*

| | | |
|---|---|---|
| 85% Phosphoric Acid | 3.6 | 2.9 |
| Water | 3.2 | 2.9 |
| Isopropyl Alcohol | 13.2 | 9.2 |

All of the ingredients of the base are ground together in the pebble mill until the requisite grind is obtained. Good practice is to cut the resin in the mill first with the solvent, then add the pigments. A portion of the reducer can be held out and used for washing the mill. In the case of Formula No. 1 (WP-1) the acid component is added only just before use. Formula No. 2 is a one-package material, and the acid component may be added immediately after manufacturing. It is stable for at least one year.

| | No. 3 (MIL-C-14504) | No. 4 (Chromic Phosphate) | No. 5 (Chromic Phosphate & Zinc Chromate) |
|---|---|---|---|
| *Base Grind* | | | |
| Basic Zinc Chromate | 4.3 | — | 4.5 |
| Strontium Chromate | 4.3 | — | — |
| Chromic Phosphate | — | 9.0 | 4.5 |
| Talc | 2.0 | 1.4 | 1.4 |
| Polyvinyl Butyral Resin | 9.4 | 9.0 | 9.0 |
| Isopropyl Alcohol (99%) | 58.9 | — | — |
| Ethyl Alcohol | — | 54.5 | 54.5 |
| Methyl Isobutyl Ketone | 16.5 | 16.1 | 16.1 |
| *Acid Component* | | | |
| 85% Phosphoric Acid | 2.8 | 1.8 | 1.8 |
| Water | 1.8 | 1.8 | 1.8 |
| Ethyl Alcohol | — | 6.4 | 6.4 |

Manufacturing instructions are similar to those given for Formulas No. 1 and No. 2. However, in the case of Formula No. 3, the laboratory of the Corps of Engineers at Fort Belvoir Va., prefers an alternate manufacturing procedure. The dispersion should be satisfactory in the pebble mill by the end of 24 hours, if the mill is not overloaded. At the end of 12 hours' grinding, the mill is stopped and the phosphoric acid added while the contents are still warm. The mill is then allowed to run for the remaining 12 hours, when the balance of the acid component is added. In this way an increase in stability and performance is obtained.

The three formulas shown in this table may all be pre-mixed with a stability of at least 6 months.

------

### Two-Package Wash Primer

| Base Grinds | Formula No. 1 | No. 2 |
| --- | --- | --- |
| Ethyl Alcohol 2B | 378.0 | 284.0 |
| n-Butyl Alcohol | – | 97.0 |
| Methyl Ethyl Ketone | 400.0 | 277.0 |
| "Butvar" B-76 | 100.0 | 100.0 |
| Basic Zinc Chromate Y563D | 133.3 | 97.6 |
| "Resimene" 881 | 77.7 | – |
| "Celite" 266 | 55.6 | 38.5 |
| *Reducer and Catalyst* | | |
| Phosphoric Acid, 85% | 75.6 | 75.5 |
| n-Butyl Alcohol | 933.4 | 738.0 |

The recommended reduction with the reducer-catalyst mixture is 1:1 by volume for spraying. If it is desired to add more reducer, only the solvent portion, not the acid catalyst, should be used.

Formula No. 1 is recommended for aluminum surfaces, Formula No. 2 for steel. However, Formula No. 2 may be used on aluminum if the amount of phosphoric acid is reduced to 39.2 lb.

## Emulsion Paint

### Formula No. 1
### (Interior White)

*Paint Base*
(Part I)

| | |
|---|---|
| Dibutyl Phthalate | 22.0 |
| Hexylene Glycol | 16.5 |
| Ethylene Glycol | 27.5 |
| "Tergitol" NPX | 2.2 |
| "Aerosol" OT, 75% | 3.0 |
| Water | 16.5 |

(Part II)

| | |
|---|---|
| "Gelva" Emulsion TS-30 | 267.0 |

*Pigment Slurry*

| | |
|---|---|
| Water | 131.0 |
| "Tamol" 731, 10% | 3.0 |
| "Methocel" HG 4000 cps, 2% | 200.0 |
| "Titanox" RA50 | 220.0 |
| Wollastonite P1 | 41.0 |
| "Gold Bond" "R" Silica | 39.0 |

### Formula No. 2

*Paint Base*
(Part I)

| | |
|---|---|
| Hexylene Glycol | 15.0 |
| Ethylene Glycol | 25.0 |
| "Tergitol" NPX | 2.0 |
| "Aerosol" OT, 75% | 2.7 |
| Water | 15.0 |

(Part II)

| | |
|---|---|
| "Gelva" Emulsion TS-70 | 285.0 |

*Pigment Slurry*

| | |
|---|---|
| Water | 189.0 |
| "Tamol" 731, 10% | 10.0 |
| "Methocel" 4000 cps, 2% | 180.0 |
| "Titanox" RA 50 | 197.0 |
| "ASP" 400 | 88.0 |
| "Gold Bond" "R" Silica | 88.0 |

---

### Alkyd Semi-Gloss Paint

| *Premix:* | | *lb.* |
|---|---|---|
| *a* | "Amsco" 140L | 100 |
| | "Thixcin" | 3½ |
| | "D-65A" | 137 |
| | "Dustex" Micro-Silica | 200 |
| | "RCHTX" | 200 |
| | "R-110" | 125 |
| *b* | Dipentene | 21 |
| | "D-65A" | 15 |

Mix *a* 15 minutes. Mix *b* 5 minutes

*Mill:*

Use 80 Stator, 46 Rotor. One pass through Morehouse Mill. Set stones 2-5/1000 off positive tight, to control temperature to best gelling of "Thixcin."*

---

* Can be controlled by change in stone clearance (with water on mill) to a range of 100 to 150°F. When 80 over 46 suggested stones are employed they should be sealed (used) stones. To get same results with new stones use 80 over 60 grit.

| Let Down: | lb. |
|---|---|
| "D-65A" | 260 |
| "Castung" 103 GH | 32 |
| 24% Lead Naphthenate | 6 |
| 6% Cobalt Naphthenate | 3 |
| 4% Calcium Naphthenate | 4 |
| "Amsco" 140L | 40 |
| (to satisfactory viscosity) | |

*Production Rate:*
8-12 gal.
  per hour –Model –M
Over 100 gal.
  per hour –Model B–1400

*Grind:*
3-4 N.S.

---

### Alkyd Eggshell Paint

| Premix: | lb. |
|---|---|
| a "Amsco" 140L | 100 |
| "Thixcin" | 3½ |
| "D-65A" | 137 |
| "Dustex" Micro-Silica | 200 |
| "RCHTXX" | 250 |
| "Ti-pure" R-110 | 100 |
| b "D-65A" | 13 |
| Dipentene | 21 |

Mix *a* 15 minutes. Mix *b* 5 minutes.

*Mill:*

Use 80 Stator, 46 Rotor. One pass through Morehouse Mill. Set stones 2/1000 off positive tight, at a cold start. Reset 4-5/1000 off positive tight, after maximum warm up of mill.

| Let Down: | lb. |
|---|---|
| "D-65A" | 260 |
| "Castung" 102 G-H | 24 |
| 24% Lead Naphthenate | 6 |
| 6% Cobalt Naphthenate | 2 |
| 4% Calcium Naphthenate | 3 |
| "Amsco" 140L | 40 |
| (to satisfactory viscosity) | |

*Production Rate:*
8-12 gal.
  per hour –Model –M
Over 100 gal.
  per hour –Model B–1400

*Grind:*
4-5 N.S.

Temperature can be controlled to range of 110 to 150°F. by stone clearance. Appearance (24 hours) very good in all respects. When 80 over 46 suggested stones are employed they should be sealed (used) stones. To get same results with new stones use 80 over 60 grit.

## Alkyd Trim Paint

| Premix: | lb. |
|---|---|
| a "Amsco" 140L | 73 |
| "Thixcin" | 3 |
| White Refined Linseed Oil | 78 |
| Medium Oil Soya Alkyd (50% solids) | 77 |
| "Kadox" Zinc Oxide | 10 |
| "Titanox" | 225 |
| "Dustex" Micro-Silica | 200 |
| b "D-65A" | 40 |
| Dipentene | 21 |

Mix *a* 15 minutes. Mix *b* 5 minutes.

### Mill:

Use 80 Stator, 46 Rotor. One pass through Morehouse Mill. Set stones 2/1000 off positive tight, at a cold start. Reset 4-5/1000 off positive tight, after maximum warm up of mill.

| Let Down: | lb. |
|---|---|
| "D-65A" | 280 |
| 24% Lead Naphthenate | 7 |
| 6% Cobalt Naphthenate | 2½ |
| 4% Calcium Naphthenate | 4 |
| "Amsco" 140L | 47 |

### Production Rate:

7-10 gal.
  per hour –Model –M
Over 100 gal.
  per hour –Model B–1400

### Grind:

6-7 N.S.

Quality of grind will always be dependent upon inert portion of formulation. When 80 over 46 suggested stones are employed they should be sealed (used) stones. To get the same results with new stones use 80 over 60 grit.

---

## Alkyd Gloss Paint

| Premix: | lb. |
|---|---|
| a "Amsco" 140L | 100 |
| "D-65A" | 90 |
| "Thixcin" | 3 |
| "Dustex" Micro-Silica | 100 |
| "RCHTX" | 100 |
| "Ti-pure" R-110 | 200 |
| b "D-65A" | 48 |
| "Dipentene | 7 |

Mix *a* 15 minutes. Mix *b* 5 minutes.

### Mill:

Use 80 Stator, 46 Roter. One pass through Morehouse Mill. Set stones 2/1000 off positive tight, at a cold start. Reset 4-5/1000 off positive tight, after maximum warm up of mill.

*Let Down:*      *lb.*
"D-65A"      300
"Castung" 103 G-H    40
24% Lead
   Naphthenate      4
6% Cobalt
   Naphthenate      2
4% Calcium
   Naphthenate      6
"Amsco" 140L      33
( to satisfactory
   viscosity)
Dipentene      24

*Production Rate:*
8-12 gal.
   per hour –Model –M
Over 100 gal.
   per hour –Model –B1400

*Grind:*
6-7 N.S.

When 80 over 46 suggested stones are employed they should be sealed (used) stones. To get the same results with new stones use 80 over 60 grit.

---

### Alkyd Flat Paint

*Premix:*      *lb.*
*b* "Amsco" 140L      175
   "Thixcin"      4
   "D-65A"      120
   "Dustex" Micro- Silica 225
   "RCHTX"      450
   "Ti-Pure" R-110      50
*b* Dipentene      21

Mix *a* 15 minutes. Mix *b* 5 minutes.

*Mill:*
Use 60 Stator, 36 Rotor. One pass through Morehouse Mill. Set stones 2/1000 off positive tight, at a cold start. Reset 4-5/1000 off positive tight, after maximum warm up of mill.

*Let Down:*      *lb.*
"Sealkyd" 19-147A (at
   60% solids)      240
"Castung" 103 GH      16
24% Lead Naphthenate   5
6% Cobalt Naph-
   thenate      1½
"Amsco" 140L      27

*Production Rate:*
10-15 gal. per hour—Model –M
Over 125 gal. per hour—Model B-1400

*Grind:*
3½-4½ N.S.

---

### Lead Free House Paint

*Premix:*      *lb.*
*a* Mineral Spirits      126
   Refined Linseed Oil
   (Acid #2 to 4)      180
   "Dustex" Micro-Silica 169
   Acicular Zinc Oxide    292
   "Titanox" RCHT      296
   "Titanox" AMO      84

*b.* Bodied Linseed Z-Z₁ 54

Mix *a* 15 minutes. Mix *b* 5 minutes.

*Mill:*

Use 60 Stator, 36 Rotor. One pass through Morehouse Mill. Set stones 2/1000 off positive tight, at a cold start. Reset 4-5/1000 off positive tight, after maximum warm up of mill.

| *Let Down:* | *lb.* |
|---|---|
| Refined Linseed Oil | 137.0 |
| Bodied Linseed Z-Z₁ | 50.0 |
| 6% Manganese Naph-thenate | 8.0 |
| 6% Cobalt Naphthen-ate | 1.6 |

*Production Rate:*

8 gal. per hour—Model —M

Over 100 gal. per hour—Model B-1400

*Grind:*

4-5 N.S.

The use of X-Y bodied Linseed Oil raises gloss appreciably.

---

### Limed Linseed Flat Wall Paint

| *Premix:* | *lb.* |
|---|---|
| "Amsco" 140L | 160 |
| "Bentone" 34 | 4 |
| "Ardanco" V-160 | 400 |
| Litharge | 3 |
| "Silver Bond" "B" Silica | 100 |

| "Lorite" | 100 |
|---|---|
| Titanium Calcium | 600 |
| 24% Lead Naphthenate | 2 |
| 6% Cobalt Naphthen-ate | 1 |

Mix 15 minutes.

*Mill:*

Use 60 Stator, 36 Rotor. One pass through Morehouse Mill. Set stones 2/1000 off positive tight, at cold start. Reset 4-5/1000 off positive tight, after maximum warm up of mill. Fill direct from cooling chamber.

*Production Rate:*

15 gal. per hour—Model —M

Over 125 gal. per hour—Model B-1400

*Grind:*

2½-3½ N.S.°

For production beyond 250 gal. per hour on B-1400 use 36 Stator and 36 Rotor.

---

### Polyvinyl Acetate Interior Flat Tint Base

| Water | 150 |
|---|---|
| Potassium Tripolyphos-phate | 1 |
| "Advawet" #33 | 4 |
| "Nopco" JMK | 2 |
| 2% 4000 cp. "Methocel" | 25 |

---

* The 60/36 combination will give a 3-4 N.S. grind, the 36/36 combination gives a 2½-3½ N.S. grind.

| | |
|---|---|
| Titanium Dioxide | 150 |
| "Gold Bond" "R" Silica | 165 |
| Wollastonite P-1 | 45 |
| "Micromite" | 42 |

*Grind:*

| | |
|---|---|
| Water | 23 |
| 2% 4000 cp. "Methocel" | 195 |
| Polyvinyl Acetate Co-polymer Emulsion (55% solids) | 277 |
| Hexylene Glycol | 16 |
| "Cellosolve" | 24 |
| "Dowicide" A (20%) | 10 |

### White Wall Paint

| | |
|---|---|
| a Pigment Dispersion | |
| "Titanox" RA | 267.0 |
| "Albalith" 11 litho-pone | 76.3 |
| "Mineralite" 3X Mica | 51.7 |
| "Silver Bond" "B" Silica | 81.0 |
| "Tamol" N | 6.8 |
| Diethylene Glycol | 7.2 |
| Water | 190.0 |
| b. "Rhoplex" AC-33 (46% solids) | 516.0 |

Mix *a.* Then mix with *b.*

### Flat Wall Finish, White or Light Tint Base

| | |
|---|---|
| a Water | 153 |
| Potassium Tripoly-phosphate | 1 |

| | |
|---|---|
| "Advawet" #33 | 4 |
| "Ti-Pure" R-610 | 150 |
| "Alsilate" W | 100 |
| "Gold Bond" "R" Silica | 100 |
| "Micromite" | 50 |
| "Methocel" 400 cps. (2% sol.) | 100 |
| "Foamicide" 581-B | 1 |
| b "Resyn" 12K51 | 295 |
| "Methocel" 400 cps. (2% sol.) | 40 |
| Water | 90 |
| Hexylene Glycol | 15 |
| Ethylene Glycol | 30 |

Charge *a* into mixer with agitation, disperse in Morehouse or other high speed mill, then slowly add *b.*

### High Lead Zinc Type LTZ Outside White Paint

| *Premix:* | lb. |
|---|---|
| a. Mineral Spirits | 78 |
| Refined Linseed Oil (Acid #2 to 4) | 200 |
| "Dustex" Micro-Silica | 125 |
| Anatase Titanium Dioxide | 135 |
| 35% Leaded Zinc | 500 |
| Lead Carbonate | 300 |
| b Y Bodied Linseed Oil | 32 |
| Mineral Spirits | 27 |

Mix *a* 15 minutes. Mix *b* 5 minutes.

*Mill:*

Use 60 Stator, 36 Rotor. One pass through Morehouse Mill. Set stones 2/1000 off positive tight, at a cold start. Reset 4-5/1000 off positive tight, after maximum warm up of mill.

| *Let Down:* | *lb.* |
|---|---|
| Refined Linseed Oil | 100 |
| Y Bodied Linseed Oil | 94 |
| Mineral Spirits | 27 |
| 24% Lead Naphthenate | 8 |
| 6% Cobalt Naphthen-ate | 2 |

*Production Rate:*

15 gal. per hour—Model —M
Over 150 gal. per hour—Model B-1400

*Grind:*

3½-4½ N.S.

---

## Interior & Exterior Polyvinyl Acetate Paint

| Potassium Tripoly-phosphate | 2.0 |
|---|---|
| Titanium Dioxide | 200.0 |
| Velveteen 'R' Silica | 100.0 |
| "Mistron" HGO-55 | 100.0 |
| Water | 300.0 |
| "Resyn" 12K51 | 360.0 |
| Hexylene Glycol | 20.0 |
| "Methocel" 4000 cps. | 2.5 |
| "Triton' X-100 | 1.0 |
| "Balab" 259 | 0.5 |

Dissolve Potassium Tripolyphosphate in 250 parts of water. Disperse Titanium Dioxide, "Mistron" and Silica in this solution. Use remaining 50 parts of water for "Triton" X-100. "Methocel" solution, add the other ingredients in indicated order. Disperse entire formula by means of a single pass through a Morehouse Mill.

### Suggested Formulation

This paint gives exceptionally good application properties and shows good hiding power and scrub resistance. It may be used for both interior and exterior finishes applied over wood or masonry surfaces.

---

## Alkyd Modified White Latex Paint

### Formula No. 1

*Pigment Dispersion* (in pebble mill)

| "Titanox" RA-50 | 150.0 |
|---|---|
| "Celite" 281 | 141.0 |
| "Dowicide" G | 1.0 |
| "Dowicide" A | 1.0 |
| "Tamol" 731—25% Solution | 3.0 |
| "Tergitol" NPX | 2.0 |
| "Foamicide" 581-B | 4.0 |

CMC 120 High, 5.5%
Solution*                  47.0
Water                     303.9

*Add after 1½ hours:*
"Penglo" 65 ⎫           45.5
"Tergitol"  ⎬ Blended
NPX         ⎭            3.0

*Rinse:*
Water                      50.0

*Latex thindown:*
"Glen-Flo" 67 (48%
Solids)                   201.0

*Thickener:*
CMC 120 High, 5.5%
Solution*                  78.0

### No. 2

*a* Rutile Titanium
Dioxide                  252.8
Lithopone                 36.0
Magnesium Silicate        29.0
"Celite" 281              35.1
Potassium Tripoly-
phosphate                  1.0
Soya Lecithin, Water
Dispersible                5.3
"Dowicides" 1 and
7, 1:1, 20% in
Carbitol                  28.1

---

* To prepare the CMC solution, start
with 94.45 parts of water at 180°F in a
suitable vessel; add 0.05 parts "Dowicide" G and then sift in 5.50 parts
CMC. Mix with high speed stirrer until
a smooth gel is obtained.

15% Alpha Protein
Solution                   72.0
Morpholine: Soya
Acids, 1:1                  3.5
Pine Oil                    2.4
Water (for Good
paste Consistency)
Up to 235.5

Mix the above with portion
of the indicated water, and add
the following, premixed:

*b.* Long Oil Soya Alkyd,
100% Solids               43.0
Mineral Spirits, odor-
less                      11.0
"Triton" X-100             3.1
24% Lead Drier,
odorless                   3.1
6% Cobalt Drier,
odorless                   2.2

Combine the above with the
pigment paste, using enough
water to adjust consistency to
the preferred milling procedure.
Then add:

*c.* "Sindar" No. 19         1.5
"Pliolite" Latex 165
48% N.V.                 309.0
"Nalco" 211, 50%           4.0
Potassium Polyacryl-
ate, 15%                  17.0

Mix *a* with portion of water,
then add *b* premixed. Combine
with the pigment paste, using
enough water to adjust con-

sistency to the preferred milling procedure. Add the balance of the water, then add c.

---

### Interior Latex Paint
### Formula No. 1
### (Deep Green)

a Pigment Green "B"          20.0
  "ASP" 400 Clay            100.0
  "Celite" 281              40.0
  "Titanox" RA–50           75.0
  Water                     250.0
  "R & R" 551 Lecithin       5.0
  "Blancol" Solution,
    20% in water            12.5

b. "EVT" 50 ("Cargill"
    Modified Latex)         434.0
  24% Lead Naph-
    thenate                  4.0
  6% Cobalt Naph-
    thenate                  2.0
  Water                     33.0
  "Polyco" 530              11.0

Grind a 2 to 4 hours in pebble mill and add b.

### No. 2
### (Deep Gray)

Rutile Titanium Di-
  oxide                    214.2
Lithopone                   65.3
"Superjet" Lampblack         1.4
Magnesium Silicate          36.1
"Celite" 281                42.0
Potassium Tripoly-
  phosphate                  1.0

Soya Lecithin, Water
  Dispersible                5.3
Water                      233.0
15% Alpha Protein
  Solution                  73.5
"Dowicide" 1/7 (1:1),
  20% in "Carbitol"         30.0
Pine Oil                     2.5
"Sindar" No. 19              0.5
"Pliolite" Latex 165
  (48%)                    374.0
"Nalco" 211 (50%)            3.3
Potassium Polyacrylate
  (15%)                     15.0

### No. 3
### (Tint Base)

Rutile Titanium Di-
  oxide                    200.0
Lithopone                   70.0
Magnesium Silicate          27.0
"Celite" 281                33.0
Potassium Tripoly-
  phosphate                  1.0
Soya Lecithin, Water
  Dispersible                3.4
"Dowicides" 1 and 7,
  1:1, 20% in Carbitol      29.0
Water                      181.0
15% Alpha Protein
  Solution                  88.0
Pine Oil                     2.1
"Sindar" No. 19              0.5
Pliolite" Latex 165,
  48% N.V.                 425.0
"Nalco" 211 (50%)            4.0

Potassium Polyacryl-
ate, 15%     12.5

### No. 4
### (Interior Flat Wall)

| | |
|---|---|
| "Titanox," RA-50 | 240.0 |
| "Atomite" | 71.0 |
| "Celite" 281 | 52.0 |
| Tetrapotassium pryo-phosphate | 1.2 |
| 15% Casein, KOH cut | 118.5 |
| "Dylex" K-34 Latex (48% solids) | 372.0 |
| Water | 252.0 |
| "Butrol" | 3.3 |
| "Defoamer ED" | 1.3 |

---

### Polyvinyl Acetate Emulsion Paint

### Formula No. 1
### (White Flat Ceiling)

| | |
|---|---|
| "R & R" 551 Lecithin | 5 |
| "Butrol" | 1 |
| "Tergitol" NP-14 | 4 |
| Ethylene Glycol | 18 |
| "Carbitol" | 17 |
| "Celluflex" 179C | 20 |
| "Titanox" RA-50 | 200 |
| "Celite" 110 | 200 |
| "Nytal" 300 | 150 |
| Water | 133 |

Add in the order listed and mix as a heavy paste for a minimum of 20 minutes, then add:

| | |
|---|---|
| "Foamicide" 581B | 2 |
| Water Balance | |

Disperse on a three-roll mill and let down with:

| | |
|---|---|
| 2% "Methocel" 4000 | 230 |
| "Celanese" CL-102 | 234 |
| Water and/or 2% Methocel" 4000 (for viscosity adjustment) | 85 |

### No. 2
### (Interior White)

| | |
|---|---|
| "Elvacet" 81-900 Poly-vinyl Acetate | 287 |
| Dibutyl Phthalate | 32 |

*Pigment Grind:*

| | |
|---|---|
| Water | 140 |
| "Tamol" 731, 10% Solution | 18 |
| "Emulphor" El-719 | 2 |
| "Polyglycol p"-1200 | 2 |
| "Carbitol" | 35 |
| 2% "Methocel" 4000 | 50 |
| "Ti-Pure" R-510 Titanium Dioxide | 200 |
| "HGO"-55 Talc | 75 |
| "Celite" 281 | 75 |

*Reduction:*

| | |
|---|---|
| Water | 56 |
| 2% "Methocel" 4000 | 130 |
| "Dowicide" A, 20% in Water | 5 |

### No. 3
### (White Interior Wall)

| | |
|---|---|
| 3% "Methocel" 4000 | 80.0 |
| "Polyglycol p"-1200 | 2.0 |

| | |
|---|---|
| Water | 110.0 |
| "Titanox" RA-50 | 225.0 |
| "Mica C"-3000 | 50.0 |
| "Celite" 281 | 25.0 |
| Water | 125.0 |
| Water (Mill wash) | 31.0 |
| "Emulphor" EL-719 | 2.2 |
| "Aerosol" OT–B | 2.6 |
| "Foamicide" 581B | 0.8 |
| "Carbitol" Acetate | 20.0 |
| 20% "Dowicide" A | 11.5 |
| "Everflex" G (51%) | 353.0 |

*Viscosity Adjustment:*

| | |
|---|---|
| Water and/or 3% "Methocel" 4000 | 40.0 |

### No. 4
(Dark Green Interior Wall)

| | |
|---|---|
| 3% "Methocel" 4000 | 60.0 |
| "Polyglycol p"-1200 | 2.0 |
| Water | 208.0 |
| "Igepal" CTA-639 | 2.0 |
| "Aerosol" OT-B | 2.6 |
| "Butrol" | 2.5 |
| "Titanox" RA-50 | 50.0 |
| Phthalocyanine Green | 20.0 |
| "Mapico" Yellow Dark Orange | 35.0 |
| Mica, 325 Mesh | 50.0 |
| "Celite" 281 | 75.0 |
| "Carbitol" Acetate | 20.0 |
| "Everflex" G (51%) | 440.0 |

*Viscosity Adjustment:*

| | |
|---|---|
| Water and/or 3% "Methocel" 4000 | 40.0 |

### No. 5
(Universal Tint Base)

| | | |
|---|---|---|
| *a* | 2% "Methocel" 4000 | 67.0 |
| | "Polyglycol p"-1200 | 2.0 |
| | 25% "Tamol" 731 | 5.0 |
| | Water | 25.0 |
| | "Titanox" RA-50 | 200.0 |
| | Wollastonite P-1 | 100.0 |
| | "Nytal" 300 | 20.4 |
| | "Celite" 281 | 25.0 |
| | Phenyl Mercuric Borate | 0.2 |
| | Water | 47.6 |
| *b* | Water (Mill Wash) | 10.0 |
| | 2% "Methocel" 4000 | 151.5 |
| | Water | 15.0 |
| | "Carbitol" Acetate | 15.0 |
| | "Nonic" 260 | 8.0 |
| | "Hallco" C-451 | 4.0 |
| *c* | "Everflex" G (51% N.V.) | 410.0 |

Combine *a* ingredients in a change can mixer. Mix to a smooth homogeneous paste, then add remaining water and give one pass through a high speed stone mill. Premix *b*. Then add *b* to *a*. Add *c* with good agitation.

### No. 6
(Light Tint Base)

| | | |
|---|---|---|
| *a* | Water | 232.0 |
| | 2% "Methocel" 4000 | 125.0 |

| Potassium | |
|---|---|
| Tripolyphosphate | 1.0 |
| "Nopco" JMK-1 | 4.0 |
| "Dowicide" A | 2.0 |
| "Brij" 30 | 5.0 |
| Rutile Titanium | |
| Dioxide, | |
| Non-chalking | 200.0 |
| "Atomite" | 70.0 |
| b "Celite" 281 | 17.5 |
| "Poly-Tex" 600 | 362.0 |
| Ethylene Glycol | 20.0 |
| "Brij" 30 | 2.0 |
| "Atlas" G-3300 | 13.0 |
| Water | 6.0 |
| "Carbitol" Acetate | 15.0 |

Grind *a* 4 to 6 hours. Rinse mill and add *b*.

## No. 7
### (Deep Tint Base)

| a Water | 203.0 |
|---|---|
| 2% "Methocel" | |
| 4000 | 83.5 |
| Potassium | |
| Tripolyphosphate | 1.0 |
| "Nopco" JMK-1 | 4.0 |
| "Dowicide" A | 2.0 |
| Rutile Titanium | |
| Dioxide, | |
| Non-chalking | 100.0 |
| "Atomite" | 100.0 |
| "Celite" 281 | 100.0 |
| "Brij" 30 | 5.0 |
| b "Poly-Tex" 600 | 362.0 |

| 2% "Methocel" | |
|---|---|
| 4000 | 74.2 |
| Ethylene Glycol | 20.0 |
| "Carbitol" Acetate | 15.0 |
| "Brij" 30 | 2.0 |
| "G-3300" | 12.5 |

Grind *a* 4 to 6 hours in pebble mill. Rinse mill and add *b*.

## No. 8
### (Interior-Exterior White)

| 3% "Methocel" 4000 | 60.0 |
|---|---|
| Polyglycol P-1200 | 2.0 |
| 25% "Tamol" 731 | 8.0 |
| Phenyl Mercuric Borate | 0.2 |
| Water | 150.0 |
| "Titanox" RA-50 | 200.0 |
| "Mica C-3000" | 25.0 |
| "Celite" 281 | 25.0 |
| Talc | 135.0 |
| "Carbitol" Acetate | 15.0 |
| Ethylene Glycol | 40.0 |
| Water | 35.6 |
| "Emulphor" EL-719 | 2.0 |
| 3% "Methocel" 4000 | 80.0 |
| "Everflex" G (51%) | 350.0 |

## No. 9
### (Texture Paint)

| a Water | 80.0 |
|---|---|
| Potassium | |
| Tripolyphosphate | 1.0 |
| "Advawet" No. 33 | 2.0 |
| "Titanox," RANC | 87.0 |
| "Hydrite" Flat | 93.0 |
| Ethylene Glycol | 25.0 |
| "Camel Carb" | 60.0 |

2% "Methocel"

| | |
|---|---|
| 4000° | 75.0 |
| "Foamicide" 581-B | 1.5 |

b No. 873 Quartz

| | |
|---|---|
| (stir in) | 245.0 |
| "Resyn" 12K-51 | 221.0 |
| "Celite" 281 (stir in) | 42.0 |
| Water | 147.0 |

2% "Methocel"

| | |
|---|---|
| 4000° | 85.0 |
| 'Carbitol" Acetate | 5.5 |

Perlite No. 600

| | |
|---|---|
| (stir in) | 53.5 |

Charge *a* into mixer with agitation. Disperse in Morehouse or other high speed mill, then slowly add *b*.

### No. 10
### (Stipple Paint)

| | |
|---|---|
| a Water | 175.0 |
| Potassium | |
| Tripolyphosphate | 1.0 |
| "Advawet" No. 33 | 4.0 |
| "Titanox" RANC | 125.0 |
| "Asbestine" 3X | 115.0 |
| Bentonite | 10.0 |

2% "Methocel"

| | |
|---|---|
| 4000° | 100.0 |
| b "Resyn" 12K-51 | |
| (55% N. V.) | 290.0 |
| Water | 25.0 |
| c "Celite" 281 | 180.0 |

---

2% "Methocel"

| | |
|---|---|
| 4000° | 95.0 |
| Hexylene Glycol | 15.0 |

Grind *a*, then add *b* and stir in *c*.

---

### Chlorinated Rubber
### Exterior Paint

Chromium Oxide

| | |
|---|---|
| Green | 107.5 |
| Wet Ground Mica | 107.5 |
| "Celite" 110 | 70.7 |
| "PE-16" Clear | 681.0 |
| Water | 56.2 |

"PE-16" Clear is an emulsion of chlorinated natural rubber. In the preparation of paints, it should not be milled or overheated. The pigments should be dispersed in water, and the dispersion then let down with the emulsion vehicle.

---

### Alkyd Paints

### Formula No. 1
### (Interior White Flat)

| | |
|---|---|
| "Ti-Pure" R-110 | 200.0 |
| Whiting | 450.0 |
| Talc | 100.0 |
| "Celite" 281 | 50.0 |
| Aluminum Stearate "R" | 4.0 |
| Soya Lecithin | 5.0 |

---

* Protect "Methocel" solution with ¼% "Dowicide" A based on total weight of solution.

* Protect "Methocel" solution with ¼% "Dowicide" A based on total weight of solution.

Alkyd Resin Solution,
50% Solids ("Syntex"
2964)     223.0
24% Lead Drier
(Odorless)     2.3
6% Cobalt Drier
(Odorless)     0.9
Anti-Skinning Agent     1.0
Mineral Spirits
(Odorless)     223.0

### No. 2

"Ti-Pure" R-610     275
"Snowflake" (Whiting)     432
"Celite" 281     45
"Bentone" 34     6
Soya Lecithin     3
1% Soap Solution*     12
FAFL Alkyd Solution
(30% N. V.)     453
Mineral Spirits     50
6% Cobalt Naphthenate     3
24% Lead Naphthenate     8
Anti-skinning Agent     1

### No. 3

(Odorless White Flat)
Rutile Titanium
Dioxide     185.0
"Kalmac"     490.0
"Celite" 281     50.0
Aluminum Stearate     3.6
Soya Lecithin     3.6
"Syntex" 179 Alkyd,
35% N. V.     301.0

---

* Add soap solution to pigment paste just before grinding.

"Syntex" 60 Odorless
Alkyd, 70% N.V.     16.2
24% Lead Drier
(Odorless)     1.9
6% Cobalt Drier
(Odorless)     1.6
Anti-oxidant B     1.0
Odorless Mineral
Spirits     149.3
"RL-80" (added to
finished paint)     1.0
Grind overnight in pebble mill.

### No. 4

(Thixotropic Flat White Wall)
Rutile Titanium
Dioxide     204
"Cal-White" Whiting     440
"Celite" 281     55
"Burnok" 3540
(40% N. V.)     350.00
Mineral Spirits     150.00
24% Lead
Naphthenate     1.75
6% Cobalt
Naphthenate     1.39
Anti-skinning Agent     1.67

---

### Interior Wall Paint

### Formula No. 1
(Low Solids White)

"Titanox" RA     161.2
"Celite" 281     77.5
"Bentone" 34*     13.2

---

* Include with pigment grind.

"Admerol" 400 RS
(32% N. V.) 460.0
Light Mineral Spirits 158.0
24% Lead
Naphthenate 2.4
6% Cobalt
Naphthenate 1.2
Soya Lecithin 1.1
Anti-skinning agent 0.6
Polar Solvent**

No. 2
(Deep Blue)
a "Norlin" L-70
(70% N.V.) 255.0
Rutile Titanium
Dioxide 87.0
Zinc Oxide 25.0
Ultramarine Blue 149.0

** Use methanol or denatured ethanol,
20 to 40 per cent on weight of "Bentone."

Bone Black 37.0
Medium Chrome
Yellow 9.5
Aluminum
Stearate TD 4.2
Mix and add:
b Mineral Spirits 126.0
"Celite" 281 173.0

Grind on roller mill and reduce with:

c "Apco" 467 Thinner 42.0
"Norlin" L-70
(70% N.V.) 63.0
Mineral Spirits 67.0
24% Lead
Naphthenate 4.0
6% Cobalt
Naphthenate 1.6

Mix a and add b. Grind on roller mill and reduce with c.

## Enamel

### Formula No. 1
(Interior Low Odor)

|  | White | Green |
| --- | --- | --- |
| GE. 7332 "Glyptal" Alkyd Resin | 432.5 | 433.1 |
| Titanium Calcium Pigment | 403.0 | 403.0 |
| Titanium Dioxide Rutile | 40.3 | — |
| "Celite" 281 | 40.3 | 40.3 |
| Calcium Carbonate | 161.5 | 184.0 |
| Light Chrome Green | — | 21.0 |
| Lampblack | — | 0.8 |
| Mineral Spirits, low odor | 85.7 | 86.0 |

| | | |
|---|---|---|
| 24% Lead drier, low odor | 3.0 | 3.1 |
| 6% Cobalt drier | 1.1 | 1.1 |
| 4% Calcium drier | 1.8 | 1.8 |
| Anti-skinning Agent | 0.8 | 0.8 |

### No. 2
### (Eggshell)

| | |
|---|---|
| "Titanox" RCHT-X | 150.0 |
| "Titanox" RA | 300.0 |
| "Kadox" 15 | 10.0 |
| "Atomite" | 157.5 |
| "Celite" 165-S | 48.0 |
| Litharge | 3.2 |
| Limed Safflower Oil, 40% N. V. | 135.3 |
| Blown Safflower Oil, Z3 Viscosity | 16.0 |
| Heat Bodied Safflower Oil, Z7½ | 10.1 |
| Safflower 22, Z3 | 96.9 |
| Cold Cut Penta Ester Gum, 60% N. V. | 75.0 |
| Cold Cut Penta Ester Gum, 60% N. V. | 30.2 |
| Mineral Spirits | 203.9 |
| 24% Lead Naphthenate | 5.5 |
| 6% Cobalt Naphthenate | 1.5 |

| | |
|---|---|
| Anti-skinning Agent | 1.5 |

### No. 3
### (Flat Stipple)

| | |
|---|---|
| "Titanox" RCHT-X | 486.0 |
| Whiting, Natural, Low Oil Absorption | 155.0 |
| Magnesium Silicate | 65.8 |
| "Celite" 281 | 64.1 |
| "Bentone" 34 | 4.5 |
| Limed Safflower Oil, 40% N. V. | 209.8 |
| Reinforced Drying Oil | 47.9 |
| Cold-cut Ester Gum Solution, 60 % N. V. | 19.1 |
| Heavy Mineral Spirits | 39.8 |
| Mineral Spirits | 87.4 |
| Mineral Spirits | 92.9 |
| Pine Oil | 2.0 |
| 24% Lead Naphthenate | 4.7 |
| 6% Cobalt Naphthenate | 1.9 |

## Chapter XV

# PAPER

Grease Resistant Paper Coating

Formula No. 1

| | |
|---|---|
| Zein G200 | 100 |
| Oleic Acid | 15 |
| Sodium Resinate | 20 |
| Water | 400 |
| Aqueous Sodium Hydroxide (5%) | < 20 |
| Alkali Stable Wax Emulsion (50%) | 10 |
| Water | 250 |

Slurry the Zein in cold water below 30°C., then add the sodium resinate and sodium hydroxide. After heating to produce the alkaline dispersion of resin and Zein in water, the oleic acid, wax emulsion and dilution water are added.

No. 2

| | |
|---|---|
| Zein G200 | 100 |
| Hydrogenated Rosin | 40 |
| Oleic Acid | 20 |
| Alcohol, Denatured | 210 |
| Water | 10 |

A similar coating can also be obtained from a water system. This is made at about 18% solids by using water instead of alcohol and by adding 20 parts of aqueous ammonia (28%). In preparing such a solution, the oleic acid is added after the Zein and resin have been thoroughly dispersed.

---

Improved Paper Coating Wax

U. S. Patent 2,783,161

| | |
|---|---|
| Microcrystalline Wax | 99-99.5 |
| Nonaethyleneglycol Distearate | 0.5-1 |

This gives improved slip and higher speed of packing of waxed paper.

237

Telegraphic Facsimile
Transmission Paper

*U. S. Patent 2,751,310*

| | |
|---|---|
| Stearic Acid | 10 |
| Carnauba Wax | 5 |
| Mineral Oil | 6 |
| Zinc Sulfide | 7½ |

---

Vegetable Parchment Paper
*German Patent 957,181*

Pulp (65% solids) is treated with 80% sulfuric acid at 17°C. for 2 seconds. The acid is recovered by countercurrent washing.

---

Translucent Duplicating Paper
*U. S. Patent 2,851,378*

| | |
|---|---|
| "Carbowax" 4000 Monostearate | 37.83 |
| Cold Pressed Castor Oil | 18.91 |
| Denatured Alcohol | 43.22 |
| Antioxidant | 0.03 |
| Perfume | 0.02 |

Bond paper is coated with above.

---

Transparentizing Bond Paper
*U. S. Patent 2,851,378*

| | |
|---|---|
| Volatile Solvent | 35-55 |
| Castor Oil | 10-25 |
| Polyglycol 4000 Monostearate | 20-50 |

Bond paper is treated with above to give a nonbleeding and ink-receptive surface.

De–inking Printed Waste Paper
*U. S. Patent 2,743,178*

| | |
|---|---|
| Waste Paper | 200 |
| Water | 12000 |
| Sulfated Fatty Alcohol | 2 |
| Tetrasodium Pyrophosphate | 8 |

Cook for one hour, with agitation, at 75 to 90°C. Strain and wash on wire screen.

---

Wallpaper Remover

Formula No. 1

| | |
|---|---|
| "Tergitol" Anionic 08 | 12.5 |
| Water | 75.0 |
| Butyl Carbitol | 12.5 |

No. 2

| | |
|---|---|
| "Tergitol" Nonionic NPX | 12.5 |
| Water | 87.5 |

To use these formulations, dilute them with 6 or 7 parts of water, apply them with a sponge or brush, and allow them to remain on the paper for several minutes. If they do not immediately penetrate and loosen the paper, application of more solution and the allowance of adequate time will produce the desired ease of removal.

## Metallic Decorative Paper Coating

| | |
|---|---|
| Zein | 100 |
| "Santolite" MHP | 35 |
| "Santicizer" 3 | 50 |
| Metallic Powder | As desired |
| Denatured Alcohol | 300 |

For aluminum powder ½ to 1 part of powder per 1 part of total binder solids (Zein, "Santolite" MHP and "Santicizer") is suggested. With bronze powder, 1 to 3 parts per part of binder solids is recommended.

---

## Black Cover-Paper Coating

| | |
|---|---|
| Black Pulp (34% Solids) | 63.0 lb. |
| Violet Lake | 5.0 lb. |
| Steel-Blue Pulp (29% Solids) | 4.5 lb. |
| "Calgon" | 0.5 lb. |
| Casein Solution (18% Solids) | 68.0 lb. |
| Evener | 1.0 qt. |
| Plasticizer | 1.0 gal. |
| Water | 12.0 gal. |

---

## Greaseproofing Paper Containers
### U. S. Patent 2,544,509

| | |
|---|---|
| Low-Viscosity Glue | 10.00 |
| Glycerin | 18.00 |
| Sorbitol | 12.00 |
| Hexamethylene-tetramine | 0.03 |
| Water | 20.00 |

## Water-Resistant Coating for Paper

### Formula No. 1

| | |
|---|---|
| Zein | 75 |
| Ammonium Resinate | 15 |
| Oleic Acid | 10 |
| Water | 400 |
| 28% Aqueous Ammonia | 15 |

### No. 2

| | |
|---|---|
| Zein | 50 |
| Rosin | 40 |
| Oleic Acid | 10 |
| Water | 300 |
| 28% Aqueous Ammonia | 15 |
| Solvent | 20-50 |

In making formula 2, the rosin and oleic acid are first dissolved in the solvent. Zein is slurred in the water with stirring; ammonia is added to the slurry, followed by the cool solvent–oleic acid–rosin solution. Stirring is continued until the solution is complete. The most satisfactory solvents are n-butyl alcohol, ethylene glycol monobutyl ether, or methyl isobutyl ketone.

---

## Waterproof and Vaporproof Paper Coating
### U. S. Patent 2,538,397

| | |
|---|---|
| Cumarone Resin | 55-65 |
| Ethyl Cellulose | 8-12 |
| Methyl Abietate | 10-20 |

| Polyisobutylene | 3- 6 |
|---|---|
| Paraffin Wax | 8-14 |
| Zinc Stearate | 1- 3 |

---

## Transparent Waterproofing for Paper

| "Ethocel" (10 cp.) | 8 |
|---|---|
| Denatured Alcohol | 18 |
| Toluol | 74 |

Wet the "Ethocel" with the alcohol and then add the toluol. Use agitation until thoroughly cut in. This can be applied by brush or spray. It stiffens the paper or fabric on which it is applied.

---

## Corrosion-Inhibiting Paper Wrapping
### U. S. Patent 2,653,854

| a | Decyclohexylamine | 169 |
|---|---|---|
| | Caprylic Acid | 144 |
| b | Paraffin Wax | 1 |
| | Microcrystalline Wax | 2 |

Mix a and stir for about 10 minutes.

Add a to an equal weight of b in a melted state.

Place in the squeeze-roll impregnator of the paper-coating machine at 205°F. and apply to the paper at the rate of 30 lb. per 3000 sq. ft.

## Antiseptic Paper
### British Patent 663,575

| Phenyl Mercury | |
|---|---|
| Chloride | 5-20 |
| Glycerin | 200-20000 |
| Water | 1,000,000 |

Impregnate the paper with this solution.

---

## Facsimile-Recording Paper
### U. S. Patent 2,555,321

| a | Gas Black | 20.0 |
|---|---|---|
| | "Methocel" (400 cp.) | 6.0 |
| | Water | 200.0 |
| b | Zinc Sulfide | 40.0 |
| | Mercuric Sulfide | 0.3 |
| | "Ethocel" (10 cp) | 10.0 |
| | "Dow Plasticizer" | |
| | No. 6 | 1.0 |
| | Toluene | 144.5 |
| | Alcohol | 25.5 |

Coat white writing paper with a; then dry and coat with b.

---

## Reconditioning Documents
### U. S. Patent 2,576,320

When documents, letters, drawings, or other papers are photostated or blueprinted, the copies often contain "ghosts," i.e., exaggerated reproductions of folds, erasures, or other flaws in the original. A solution consisting of 35 parts glycerin and 65 parts ethyl alcohol is an excellent means for removing all traces of such flaws on the original paper. This solution is applied by brush or glass rod directly to the paper at the point of the flaw or fold. After blotting and drying, the paper will regain its original over-all uniform appearance.

## Deinking Paper

| | |
|---|---|
| Hydrogen Peroxide | 0.53-0.61 |
| Caustic Soda | 3.00 |
| Sodium Silicate (42°Bé.) | 5.00 |
| Temperature | 180-190°F. |

The alkali is added to the stock during the defibration process and the peroxide is added after the stock has been heated to bleaching temperature and at such time as to insure retention to the end of the process. The "Hydrapulper" would be preferred equipment because it requires less total processing time than chests. If chests are used, adequate agitation should be insured.

## Deinking Waste Paper
*U. S. Patent 2,525,594*

| | |
|---|---|
| Waste Paper | 1000 |
| Caustic Soda | 15 |

| | |
|---|---|
| Sodium Peroxide | 15 |
| Water | 20,000 |

Pulp at 70°C. and, after cooking at 70°C. for 2 hours, screen, wash, and optionally bleach.

## Paper Deinking Detergent

| | |
|---|---|
| Sodium Silicate B.W. | 5 |
| Caustic Soda | 1½ |
| Hydrogen Peroxide (130 Vol.) | 2 |
| "Ethofat" 242/25 | 1 |

## Bleaching Recovered Waste Paper
(Magazine and Newspaper)

To prepare 2500 gal. of a typical peroxide solution for the pulping operation, the following procedure is recommended:

Run about 2000 gal. of fresh, cold water into the dissolving tank and start the agitator.

Add and dissolve completely 5 to 10 lb. of Epsom salt; add 60 lb. of sodium silicate. If additional silicate is desired in the pulping operation, it can be added at this time.

With vigorous agitation, add 412 lb. of caustic soda and 336 lb. of "Albone" 50 (hydrogen peroxide) or 400 lb. of "Solozone" (sodium peroxide).

Add water to bring the total

volume up to 2500 gal. and continue agitating for another 5 minutes to insure complete mixing. Test the solution for alkalinity and peroxide concentration.

In charging the pulper, the following procedure is typical.

Run in the desired quantity of hot water (160 to 180°F.) followed in order by metered quantities of alkali solution and peroxide solution; add a weighed charge of waste paper.

Chemicals used in a typical pulping operation are as follows:

| | Pounds per Ton of Stock |
|---|---|
| Epsom Salt | 0.8 |
| Sodium Silicate | 40-120 |
| "Albone" 50 (Hydrogen Peroxide) | 25* |
| Caustic Soda | 60* |

To improve the brightness of the recovered fiber, it is sometimes desirable to apply a peroxide bleach to the stock. In that case, a thickener may be used for this washing step to dewater the pulp to a higher consistency and so obtain a higher bleaching efficiency in the peroxide bleaching step.

In the peroxide bleaching step, the recovered stock, thickened to about 12% consistency and heated at 120 to 150°F., is mixed with the peroxide bleaching solution, held for 1 to 3 hours, and either washed free from alkali or neutralized with an acid ($SO_2$ or $SO_2$ + $H_2SO_4$). A typical bleaching formula is:

| | Pounds per Ton of Recovered Stock |
|---|---|
| Epsom Salt | 1 |
| Sodium Silicate | 100 |
| "Albone" 50 (Hydrogen Peroxide) | 8-16† |
| Caustic Soda | 10-20† |

The best conditions for applying the peroxide recovery process to waste newsprint are somewhat different from those described for waste magazine stock. Because of the higher ground-wood content in waste newsprint (approximately 85% compared with an approximate average of 35% in magazine stock), lower alkalinities and lower temperatures (135°F.) are preferred; also, the washing facilities must be adequate for handling the lower - freeness stock.

The alkalinity requirements and the temperature conditions

---

* 30 lb. of "Solozone" and 30 lb. of caustic soda are equivalent to 25 lb. of "Albone" 50 and 60 lb. of caustic soda.

†"Solozone" may be used to replace all or part of the "Albone" and caustic soda.

in this process are closely inter-related with the pulping and cooking operations. The typical formula may, therefore, require some modification for best adaptation to the local conditions in a specific mill.

|                  | Pounds per Ton of Newsprint |
|------------------|-----|
| Epsom Salt       | 1   |
| Sodium Silicate  | 100 |
| "Albone" 50      | 25† |
| Caustic Soda     | 15† |

### Stiff Saturated Paper Sheet
### U. S. Patent 2,558,634

#### Formula No. 1

| Total Pulp           | 10,000.0 |
|----------------------|----------|
| Dry Pulp             | 100.0    |
| Total GR–S Latex     | 37.0     |
| Dry-Rubber Content   | 14.0     |
| "Indulin" A          | 7.0      |
| Sodium Hydroxide     | 0.5      |
| Water                | 28.0     |
| Alum (Dry Basis)     | 8.0      |

Considerable variations may be made in the quantities of rubber and "Indulin," depending on the properties required in the finished product.

† 25 lb. of "Albone" and 15 lb. of caustic soda may be replaced by the equivalent amounts of "Solozone" and sulfuric acid; also, proper proportions of "Albone" and "Solozone" may be used without caustic soda or sulfuric acid.

#### No. 2

| Total GR–S Latex     | 263.00 |
|----------------------|--------|
| Dry-Rubber Content   | 100.00 |
| "Indulin" A          | 50.00  |
| Sodium Hydroxide     | 3.20   |
| Water                | 192.00 |
| Ammonium Sulfate     | 7.50   |
| 27% Ammonium Hydroxide | 0.75 |
| Water                | 50.00  |

#### No. 3

| Total Neoprene Latex | 200.00 |
|----------------------|--------|
| Dry-Rubber Content   | 100.00 |
| "Indulin" A          | 50.00  |
| Sodium Hydroxide     | 2.70   |
| Water                | 194.00 |
| Ammonium Sulfate     | 6.40   |
| 27% Ammonium Hydroxide | 0.64 |
| Water                | 67.00  |

#### No. 4

| Total GR–S Latex     | 263.0 |
|----------------------|-------|
| Dry-Rubber Content   | 100.0 |
| "Indulin" A          | 25.0  |
| Sodium Hydroxide     | 1.6   |
| Water                | 96.0  |
| Ammonium Sulfate     | 3.7   |
| 27% Ammonium Hydroxide | 0.4 |
| Water                | 25.0  |

These mixtures may be diluted with water to consistencies required to secure the desired amount of impregnant retained by an immersed sheet. Excess impregnant is removed by passing through squeeze rolls, after which the sheet is dried.

Similar products may be manufactured by incorporation of "Indulin"–latex mixtures into pulp before forming the sheet. A solution containing 20% "Indulin" A and 1 to 2% sodium hydroxide is prepared. This is mixed with enough latex to give two to four times as much rubber solids as "Indulin." After thorough blending with beaten pulp of about 1% consistency, alum is added to precipitate the "Indulin" and rubber in and around the fibers; then the sheet is formed in the usual way on a paper machine, dried, and calendared.

---

### Coating for Strawboard
### U. S. Patent 2,711,156

| | |
|---|---|
| China Clay | 19.0 |
| Barium Sulfate | 11.0 |
| Caustic Soda | 5.5 |
| Hydroxyethylcellulose | 4.3 |
| Water | 60.2 |

Apply by a roller. Remove the excess by a doctor blade and spray with 10% sulfuric acid at 35°C.

### Very White Paper Coating
### U. S. Patent 2,710,285

| | |
|---|---|
| Refined Carnauba Wax | 100 |
| Oleic Acid | 15 |
| Isopropanolamine | 6 |
| Alphabenzyl-$\beta$-Methyl Umbelliferone | 1 |
| Distilled Water | 900 |

---

### Prevention of Pitch Formation in Manufacturing Wood Pulp
### Japanese Patent 2102 (1954)

| | |
|---|---|
| Xylol | 60 |
| Ethylene Dichloride | 22 |
| Polyethyleneglycol Monooleate | 8 |
| Water | 10 |

The pulp-cooking liquor is treated with 0.2 to 0.39% of this solution.

---

### Bleaching Wood Pulp
### U. S. Patent 2,707,145

Add to ground wood pulp, at about 176°F. and 3% consistency, 0.1% of the trisodium salt of ethylenediamine tetraacetic acid followed by 1% of an aqueous solution of sodium hydrosulfite, by weight of the dry pulp. During the additions, agitate the pulp vigorously.

---

### Electrolytic Recording Paper
### U. S. Patent 2,692,228

| | |
|---|---|
| Catechol | 5 |

Potassium Nitrate        10
Oxalic Acid              ½
Polyethylene Glycol
  (1540)                 5
Water                    100

Impregnate the paper with this and dry to about 40% moisture content.

---

Red Fancy Paper Coating

Special Red Pulp
  (43% Solids)           188 lb.
Maroon Pulp
  (32% Solids)           312 lb.
Red Pulp
  (40% Solids)           212 lb.
China Red Pulp
  (18% Solids)           165 lb.
Cerise Lake              13 lb.
Casein Solution
  (18% Solids)           240 lb.
Wetting Agent
  (60% Solids)           3 qt.
Evener                   3 pt.
Foam Killer              1 pt.
Water                    10 gal.

---

Book-Paper High Finish

English Clay (66%
  Solids Slurry)         100 lb.
Satin-White Pulp
  (66% Solids)           52 lb.
Water                    10 gal.
Casein Solution
  (19% Solids)           20 gal.

5% Rhodamine Dye
  Solution               4 oz.
Fusel Oil                ½ pt.

---

Colored Paper Coating

Coating Clay             250 g.
Color Pigment Paste
  (50% Solids)           50 g.
Water                    210 cc.
Concentrated
  Ammonia                5 cc.
18% Casein               350 g.

Mix with a stirrer and then add in a very slow stream with the stirrer going at a good speed.

Formaldehyde-Water-
  Ammonia Mix*           40 cc.

Screen and coat.

The casein solution used is prepared as follows:

Casein, Dispersed in    100 g.
Water                    400 cc.

Heat in a double boiler to 110°F. and then add:

Borax, Dissolved in      10 g.
Water                    100 cc.

Stir for 20 minutes and then add:

28% Concentrated
  Ammonia                6 cc.

Stir for 20 minutes, remove from the heat, and stir until cold.

---

\* Formaldehyde              10
  Water            10
  Ammonia          10
Add in the order listed and mix well.

This should give an approximate volume of 560 cc.

---

### Stabilized Dry Rosin Size
#### U. S. Patent 2,471,714

Dry rosin size is made resistant to oxidation by addition of 0.25 to 1.00% phenothiazine based on the rosin used.

---

### Sizing for Photographic Paper
#### U. S. Patent 2,514,690

| | |
|---|---|
| Water | 1000 |
| Water-Soluble Melamine or Urea–Formaldehyde Resin | 8 |
| Starch | 40 |

Citric acid is added to yield a pH of 3 to 5. Paper of high α-cellulose content is sized with this.

---

### High Wet-Strength Paper Sizing
#### U. S. Patent 2,549,177

| | |
|---|---|
| Oxidized or Chlorinated Starch | 1200 |
| Water | 4800 |

Heat for 10 minutes at 80 to 85°F. and add:

| | |
|---|---|
| Glyoxal | 24-60 |

Heat for 5 minutes, dilute to 12 liters with water, and cool. Apply at 110 to 150°F.

---

### Waterproof Top Paper Sizing

| | |
|---|---|
| Casein Solution (Borax and Ammonia Cut, 18% Solids) | 450 lb. |
| Shellac Solution (Borax and Ammonia Cut, 20% Solids) | 110 lb. |
| Formaldehyde | 1 gal. |
| "Dowicide"–G | 2 lb. |

---

### Paper-Match Stiffening

#### Formula No. 1
#### U. S. Patent 2,495,575

| | |
|---|---|
| Paraffin Wax | 67 |
| Limed Rosin | 15 |
| "Opal Wax" | 7 |
| Carnauba Wax | 11 |

#### No. 2
#### U. S. Patent 2,647,048

| | |
|---|---|
| Paraffin Wax | 61 |
| Limed Rosin | 18 |
| Gilsonite | 18 |
| Synthetic Wax ("Acrawax" C) | 3 |

---

### Greaseproof Coatings for Paper

#### Formula No. 1

| | |
|---|---|
| Zein | 100 |
| Triethylene Glycol | 40 |
| "Carbowax" 1540 | 30 |
| Denatured Alcohol | 190 |

## No. 2

| | |
|---|---|
| Zein | 100 |
| "Staybelite" Resin | 40 |
| Palmitic or Stearic Acid | 20 |
| Oleic Acid | 10 |
| Denatured Alcohol | 210 |
| Pigments* | 50 |

## No. 3

| | |
|---|---|
| Zein | 100 |
| "Staybelite" Resin | 60 |
| Palmitic or Stearic Acid | 30 |
| Denatured Alcohol | 190 |

## No. 4

| | |
|---|---|
| Zein | 100 |
| "Staybelite" Resin | 50 |
| Oleic Acid | 40 |
| Nonsolvent Plasticizer** | 40 |
| Denatured Alcohol | 230 |
| Pigments* | 50 |

## No. 5

| | |
|---|---|
| Zein | 100 |
| "Staybelite" Resin | 40 |
| "Carbowax" 1540 | 25 |
| Oleic Acid | 25 |
| Denatured Alcohol | 190 |

## No. 6

| | |
|---|---|
| Zein | 80 |
| Ammonium Resinate Solids* | 20 |
| 28% Aqueous Ammonia | 15 |
| Water | 400 |

* Titanium dioxide, titanium dioxide, lithopone, calcium or zinc sulfide.
** Tricresyl phosphate, dioctyl phthalate, or "Saniticizer" B-16.

## No. 7

| | |
|---|---|
| Zein | 80 |
| Papermakers' Dry Size | 20 |
| Water | 400 |
| 10% Aqueous Caustic Soda | 5 |

## No. 8

| | |
|---|---|
| Zein | 80 |
| Ammonium Resinate Solids* | 20 |
| 28% Aqueous Ammonia | 10 |
| Water | 600 |
| 10% Aqueous Caustic Soda | 7 |

## No. 9

| | |
|---|---|
| Zein | 95 |
| "Carbowax" 4000 | 5 |
| Water | 600 |
| 10% Aqueous Caustic Soda | 7 |

## No. 10

| | |
|---|---|
| Medium-Viscosity Polyvinyl Alcohol | 120 lb. |
| Water | 880 lb. |
| Butanol | 2 qt. |
| Preservative | 2 lb. |

## No. 11
### U. S. Patent 2,544,509

| | |
|---|---|
| Low-Viscosity Glue | 10.00 |

* "Dresinol" 205 or 215.

## Paper Coatings

*General Recommendations.* In formulating a pigmented acrylic coating it is recommended that the clay slurry be made under full agitation to insure good dispersion. The clay should also be screened. If the system is one of a protein-acrylic type or a starch-acrylic type, the protein or starch should be added before the acrylic. The last and final step should be the addition of acrylic with a minimum of agitation. This procedure markedly reduces the formation of foam.

The addition of a wax emulsion to the coating helps eliminate any sticking that may develop. An excess of wax emulsion should not be used, since it detracts from the receptivity to printing and resistance to grease.

Casein, starch and alpha protein should be prepared in the manner recommended by the various manufacturers: In casein and alpha protein systems it has been observed that ammonia-cuts result in coatings which offer better resistance to water, and tend to form less than when sodium hydroxide is used.

*Clay Dispersion.* In order to prepare an acrylic coating, a satisfactory means of dispersing the pigment must be employed. The dispersion of the clay pigment in an acrylic system is dependent on both the dispersing agent and the mechanical methods used to disperse the pigment. The clay dispersion should be adjusted to a pH of approximately 9 with caustic soda (0.1 to 0.3% based on weight of clay). Predispersed clays require less caustic than other type clays. "Calgon" and "Quadrafos" are effective dispersing agents for clay systems. It is recommended that the caustic and dispersing agents be dissolved in the water before adding the clay.

*Size Press.* Size press formulations based on a "Rhoplex" emulsion can vary rather widely both as to pigment composition and binder content. The use of an acrylic binder has made it possible to operate the size press at higher coating weights, thus effectively utilizing the size press as a coater. A typical size press formulation is presented below:

| | |
|---|---|
| Clay | 100 |
| Protein or Casein (Solids Basis) | 10 |
| "Rhoplex" B-15 (Solids Basis) | 10 |

Sodium Hydroxide 0.15
"Calgon" or
"Quadrafos" 0.3
Total Solids 30 to 40%

The normal viscosity of systems used in the size press varies from 100 to 800 cp., depending on solids. Coating weight applications with this formulation will generally range from 3 to 5 lb. per ream (3,000 sq. ft.).

Attempts to use "Rhoplex" systems at 60 to 65% solids have been unsuccessful. The failure can be attributed to the lack of sufficient water in the coating to permit adequate flow during the drying process. Acrylic coatings, in general, tend to exhibit fast rates of water release. Notable success has been obtained with a starch-"Rhoplex" formulation in which "Penford" Gum #300 has been substituted for the protein. If unusually good resistance to water is required a small amount of formaldehyde, which has been previously mixed with ammonia, should be added. However, it should be cautioned that any appreciable amount of formaldehyde will cause thickening to occur.

*Calender Stack.* In numerous efforts to coat on the calender stack, patterning has presented a problem. A formulation which has been developed in the laboratory minimizes the patterning effect, and also shows desirable flow properties necessary for this type of a coating.

Clay 100
"Penford" Gum #300
(Solids Basis) 10
"Rhoplex" B-15 (Solids
Basis) 5
"Calgon" or
"Quadrafos" 0.3
Sodium or Ammonium
Hydroxide 0.15
Total Solids 28 to 30%

*Roll Coaters, Air-Knife Coaters and Brush Coaters.* The acrylic coatings are well adapted for application on the roll coater, air-knife coater and brush coater. These coaters normally do not require high solids coatings to achieve the desired coating weights. Usually the coated sheets obtained from this type of coater are supercalendered, or brush finished. When coatings are mechanically finished, the properties of the acrylics contribute appreciably to the high gloss and smooth surfaces.

The outstanding mechanical stability of the "Rhoplex" coatings is of great importance in view of the cascading, pumping, and shearing to which the coatings are subjected. A typical starting formulation for coaters of this type is as follows:

| | |
|---|---|
| Clay | 100 |
| "Rhoplex" B-15 (Solids Basis) | 8 |
| Protein or starch (Solids Basis) | 8 |
| "Calgon" or "Quadrafos" | 0.3 |
| Sodium or Ammonium Hydroxide | 0.15 |
| Total Solids | Approximately 40% |

*Knife-Blade Coaters.* One of the most desirable qualities of a "Rhoplex" binder in a knife-blade coating is the high degree of mechanical stability which it exhibits when subjected to the high shear of the knife blade. The ability to formulate high solids coatings with "Rhoplex" B-15 or "Rhoplex" B-60K is important in this type of coater. The incorporation of protein or starch contributes substantially to the flow characteristics of the coating in this case.

*Coatings Resistant to Grease.* "Rhoplex" B-15 and "Rhoplex" B-60K are good film-forming materials when dried at room temperatures. However, these acrylic emulsions may not be used alone because the resulting films are too tacky for normal use. To eliminate this tacky condition the use of clay or a water-soluble resin, "Amberlac" 165, incorporated into the system, has proven successful. Both methods have found commercial acceptance.

*Clay Coatings.* Slightly opaque, greaseproof coatings for paperboard, which are non-blocking and can withstand scoring, can be made by using a combination of "Rhoplex" B-15 and an ordinary coating clay. A suggested formulation is as follows:

| | |
|---|---|
| Clay | 70 |
| Water | 30 |
| "Calgon" or "Quadrafos" | 0.2 |
| "Rhoplex" B-15 (46% Solids) | 82 |

To prepare this formulation, the clay should first be dispersed in water. The "Rhoplex" B-15 is then added with constant stirring. The resulting coating contains approximately 60% total solids, and has a viscosity of approximately 70 cp. (Brookfield, 60 r.p.m., 25°C.) Viscosity may be controlled by the addition of dilute caustic soda; increasing pH of the system will result in a higher viscosity. Coating weights of 5 to 7 lb. per 3,000 sq. ft. are desirable, and a series of lighter coats proves more effective than one heavy coat. Greaseproof coatings of this type show good resistance to turpentine and vegetable oils, both flat and after scoring. They are resistant to blocking when placed face-to-face at 210°F. and 50% relative humidity under a pressure of 1 lb. per sq. in. for 16 hours. At 180°F. (100% relative humidity) under similar conditions of pressure and time the coatings also show no tendency toward blocking.

*"Amberlac" 165 Greaseproof Coatings.* "Amberlac" 165 is the ammonium salt of a synthetic resin complex, supplied as an aqueous solution. It may be used as an additive to a film-forming resin, although it does not form a film itself. The principal function of "Amberlac" 165 is to prevent blocking (sticking) of clear thermoplastic coatings. "Amberlac" 165 may be used in ratios of 100 parts of "Rhoplex" emulsion to 30 to 40 parts of "Amberlac" 165 (solids on solids). If "Rhoplex" B-15 is used, the coating will not be quite as resistant to blocking as when "Rhoplex" B-60K is used; but coatings based on "Rhoplex" B-15 will show better flexibility and resistance to scoring. The preparation of a "Rhoplex"-"Amberlac" 165 coating is quite simple, and merely involves adding the "Amberlac" 165 to the "Rhoplex" emulsion, followed by the addition of a wax emulsion. A recommended formulation is given below:

| | |
|---|---|
| "Rhoplex" (Solids) | 100 |
| "Amberlac" 165 (Solids) | 40 |
| Microcrystalline Wax Emulsion (Solids) | 14 |

Total Solids to this emulsion should be adjusted to about 25 to 30% for best operability.

The incorporation of the wax emulsion prevents picking on driers and calenders and "drag" in the coating, thereby permitting better slip characteristics in the automatic pasting machines. The resistance to scuff of the final coating is greatly improved by the wax.

*Color Formulations.* The compatibility of the "Rhoplex" emulsions with a wide variety of colored pigments makes them particularly useful in formulating colored coatings. A formulation for a colored coating, showing a very good covering power and intensity of color, is presented below:

| | |
|---|---|
| Clay | 78.5 |
| Titanium | 19.6 |
| Monastral Blue WDBP-192-E | 1.9 |
| Rhoplex B-15 (Solids) | 16 |
| Sodium Hydroxide | 0.2 |
| "Calgon" or "Quadrafos" | 0.1 |

At a 60% total solids level this formulation should be satisfactory for application by air-knife, size press and other types of coating equipment where a relatively low viscosity system is required. The viscosity can be increased by adding small amounts of protein, or by increase the solids content. Numerous other pigments can also be used with equal success in the above formulation. The wet rub resistance of the colored coating compares to that of a regular clay-titanium coating. The best resistance to wet rub is obtained when the coatings are mechanically finished.

Metallic coatings containing "Rhoplex" B-15 or "Rhoplex" B-60K have been successfully prepared. In particular, "Rhoplex" B-15, an emulsion of pH 6 to 6.8, readily lends itself to the formulation of metallic coating systems.

### Operational Suggestions

*Foam Control.* Three very good defoamers for Rhoplex coatings are "Foamex" tributyl phosphate and "Nopco" 1497-V. The defoamers should be added in small quantities (0.5 to 1% based on the total solids of the coating). Additional quantities of defoamer should be added only when the foam condition so demands.

The elimination of free falls, cascading, and air inlets into pumps will keep the foam to a minimum. The coating formulation should not be severely agitated in the mixer after the "Rhoplex" emulsion has been added. When making small batches of coating in large kettles, the mixer should be used in an intermittent motion once the acrylic resin has been added to the system. This technique tends to minimize the formation of foam.

*Control of Exposed Shallow Areas.* Shallow areas of coating should be avoided—if this is impossible then the exposed area can be hooded to prevent the evaporation of water at the film surface. The acrylic resins tend to release water rapidly; therefore, shallow and slow moving areas in the coating tend to skim over.

*Eliminate Top Coatings.* Studies reveal that water-soluble polymers should not be applied over a "Rhoplex"-bound coating, because the function of the acrylic is to provide ink receptivity and print characteristics desirable in a coating. If materials such as carboxymethyl cellulose, polyvinyl alcohol and starch are applied over an acrylic coating then the purpose of using a synthetic binder is defeated. However, zinc sulfate can be added at the calendar stack in place of water, particularly if the coating is composed of "Rhoplex" B-15 or "Rhoplex" B-60K blended with casein or protein. The zinc sulfate improves the resistance of the coating to water.

*Broke Recovery.* Broke recovery is not normally considered a troublesome problem. However, when the binder is solely acrylic, and the coating weight is extremely heavy the coating tends to be released from the fibers in agglomerates, ra⁺her than in discrete particle form. Small amounts of protein or starch present in the coating tend to eliminate this condition and broke recovery is no longer a problem. A small amount of bentonite or similar material, added to the broke beater, is also effective.

*Clean-Up.* A "Rhoplex" emulsion, once permitted to dry on a kettle or floor, is very difficult to remove. The emulsion should be removed before it has dried. The ordinary solvents will not dissolve a "Rhoplex" film. Alcohol will tend to soften the film. Should it be necessary to clean rollers or any part of the coating equipment, alcohol is recommended. Pigmented acrylic coatings containing insolubilized protein or starch offer no difficulty in cleanup.

## CHAPTER XVI

# PHOTOGRAPHY

Photographic Developer

Formula No. 1
*U. S. Patent 2,840,471*

| | |
|---|---|
| Sodium Sulfite, Anhydrous | 100.0 |
| Hydroquinone | 10.0 |
| Glycine | 10.0 |
| 1-Phenyl-3-Pyrazolidone | 0.4 |
| Sodium Borate | 3.0 |
| Boric Acid | 3.5 |
| Potassium Bromide | 1.0 |
| Water | To make 1000.0 |

No. 2
*U. S. Patent 2,757,091*

| | |
|---|---|
| "Elon" | 96-150 gr. |
| Sodium Sulfite | 160-196 g. |
| Sodium Bisulfite | 3-7 g. |
| Hydroquinone | 55-150 gr. |
| Glycine | 4-14 g. |
| Sodium Carbonate | 75-322 gr. |
| Acetone | 155-644 gr. |
| Triethanolamine | 155-644 gr. |
| Water | To make ½ gal. |

Photographic Developer
(Stable to Air Oxidation)
*U. S. Patent 2,902,367*

| | |
|---|---|
| Hydroquinone | 12 |
| Sodium Sulfite | 2 |
| Sodium Carbonate, Monohydrate | 36 |
| Boric Acid | 18 |
| Sodium Hydroxide | 12 |
| Potassium Bromide | 1 |

Sensitive Photographic
Developer
*East German Patent 12,504*

| | |
|---|---|
| p-Hydroxyphenylglycine | 0.40 |
| Trisodium Phosphate | 5.00 |
| Sodium Sulfite | 5.00 |
| Potassium Bromide | 0.50 |

| | | | |
|---|---|---|---|
| Sodium Hydroxide | 1.60 | Potassium Bromide | 10 g. |
| Dodecylpyridinium | | Borax | 1 g. |
| Bromide | 0.06 | Boric Acid | 5 g. |
| Water | 1000.00 | Boron Oxide | 3 g. |

For use dissolve in 1 liter of water.

Rapid Photographic Developer

### Formula No. 1
#### U. S. Patent 2,877,116

| | |
|---|---|
| Water | 1 l. |
| Sodium Sulfite | 37.8 g. |
| Pyrogallol | 25.2 g. |
| Amidol | 2.07 g. |
| Sodium Carbonate | |
| (Mono) | 49.6 g. |
| Potassium Bromide | 3.0 g. |

### No. 2
#### U. S. Patent 2,882,152

| | |
|---|---|
| Water | 1000.0 cc. |
| Sodium Sulfite | 12.6 g. |
| Pyrogallol | 25.2 g. |
| 2,4-Diamino- | |
| phenol HCl | 2.0 g. |
| Sodium Carbon- | |
| ate, $H_2O$ | 49.6 g. |
| Potassium Brom- | |
| ide | 3.0 g. |
| Mix A* | 0.5-5.0 ml. |

### Color Photography Bleach
#### U. S. Patent 2,843,482

| | |
|---|---|
| Potassium Ferro- | |
| cyanide | 20 g. |

* Mix A

| | |
|---|---|
| Hydrazine | 10 g. |
| Triethanolamine | 90 g. |

### Fast Photograph Washing Solution
#### U. S. Patent 2,860,978

| | |
|---|---|
| Sodium Sulfite | 20.0 |
| Sodium Bisulfite | 5.0 |
| Tetrasodium Ethylene- | |
| diaminetetraacetic | |
| Acid | 0.5 |
| Water | To make 1000.0 |

### Stable Single-Powder Developer
#### U. S. Patent 2,893,865

| | |
|---|---|
| "Metol" | 1.5 |
| Sodium Sulfite | 80.0 |
| Hydroquinone | 3.0 |
| Borax | 3.0 |
| Potassium Bromide | 0.5 |
| Lithium Hydroxide | 1.2 |
| Potassium Metabi- | |
| sulfite | 6.4 |

### Combined Photographic Developer & Fix
#### U. S. Patent 2,897,080

| | |
|---|---|
| Sodium Sulfite | 309 |
| Hydroquinone | 155 |
| Sodium Hydroxide | 129 |
| Potassium Bromide | 103 |

| Sodium Thiosulfate | 896 |
| Methyl-p-Amino- | |
| phenol Sulfate | 56 |

### Stayflat Photographic Film

| Gelatin | 3 oz. |
| Glycerin | 3 oz. |
| Carbolic Acid | 6 drops |
| Water | To make 32 oz. |

The gelatin and glycerin are mixed with 24 oz. of cool water and allowed to stand at room temperature for 30 minutes before heating to 120°F. until the gelatine dissolves. The acid and remaining water to make 32 oz. of solution are then added. The mixture should be applied to the glass or metal (aluminum) support at a temperature of 90°F. and the coated surface allowed to set for about 24 hours to congeal and condition the gelatin.

### Reconditioning Exposed & Developed Film
### U. S. Patent 2,846,334

*Mixture A*—5 g. of camphor is dissolved in 5 oz. of alcohol at 78°F. then added to 5 oz. of eucalyptus oil.

*Mixture B*—4 parts of glycerin is added to 6 parts of distilled water at 95°F.

The film is subjected to the vaporized solutions in an evacuated chamber at temperatures of 90 to 100°F. for a period of approximately three hours. Lower temperatures require proportionately longer periods. This does not require the film to be uncoiled prior to treatment (often resulting in additional cracking and breaking). Film reconditioned by this procedure is ready for use, and is unmarred by uneven deposits of the treating solution due to the vaporized state in which it is introduced.

### Barium Radiographic Contrast Agent

#### Formula No. 1

| Barium Sulfate U.S.P. | 80.00 |
| Talc U.S.P. | 18.49 |
| Soluble Saccharin | 0.01 |
| Sodium Alginate | 1.50 |

#### No. 2

| Barium Sulfate | 48.15 |
| Sodium Carboxymethyl- | |
| cellulose | 0.65 |
| Sodium Alginate | 0.68 |
| Pectin | 0.15 |
| Sodium Saccharin | 0.05 |
| Sucrose | 1.00 |

| Light Sensitive Fluid (Diazotype) U. S. Patent 2,861,008 | | 2,3-Dihydroxynaph-thalene-6-Sulfonic Acid | 2.0 |
|---|---|---|---|
| p-Diazodimethylani-line-Zinc Chloride Double Salt | 2.0 | Tartaric Acid | 2.0 |
| | | Saponin | 0.2 |
| | | Water | 100.0 |

## Chapter XVII

# POLISHES

Polyethylene Emulsions

Formula No. 1

| Emulsifiable "A-C" | |
|---|---|
| Polyethylene | 40 |
| Oleic Acid | 8 |
| Morpholine | 8 |
| Water | 184 |

Melt the "A-C" Polyethylene and add the oleic acid, avoiding overheating. With the melt temperature at 248-266°F., add the morpholine. (Care must be taken to prevent boiling out the morpholine.) There is a sufficient excess to allow for minor evaporation. Avoid open flames since morpholine has a flash point of 100°F.

While "A-C" Polyethylene is being melted, heat the water (bring to a boil and turn off heat just before using).

With rapid stirring (but below speed at which air is whipped in), slowly add the melt to the water (1-3 minutes). During the addition, the temperature of the melt should be 239-257°F. and the water 203-210°F. (not boiling). Add the melt at a steady rate to the top of the vortex formed by the stirring action (the melt stream should not hit the beaker side or the stirrer shaft). With proper addition the melt will spiral down the vortex and be emulsified enroute. If stirring speed is too low or rate of addition too high the melt will accumulate in the vortex and impair the quality of the finished product. Cover, and with moderate stirring allow to cool to 104-122°F. Make up to weight to replace loss of water due to evaporation.

No. 2

| "A-C" Polyethylene 629 | 40 |
| Oleic Acid | 8 |
| Morpholine | 8 |
| Water | 184 |

Melt the polyethylene in the oleic acid. Heat to 212-230°C*

and add morpholine slowly with stirring. Hold temperature at 208-210°F, add water (heated to 100°C.) slowly with rapid agitation. Cover and stir until 104-122°F. Add water to make up to 184.

* Avoid open flame.

## Anionic Emulsions
### (Wax to Water Method)

|  | No. 3 | No. 4 | No. 5 | No. 6 | No. 7 | No. 8 |
|---|---|---|---|---|---|---|
| "A-C" Polyethylene 629 | 40 | 40 | 40 | 40 | 40 | 40 |
| Oleic Acid | 4 | 6 | 5 | 5 | 6 | 7 |
| 3-Methoxy Propylamine | 3 | — | — | — | — | — |
| 2-Amino 2-methyl, 1-propanol | — | 3 | — | — | — | — |
| Morpholine | — | — | 7 | — | — | — |
| Monoethanolamine | — | — | — | 4 | — | — |
| Diethanolamine | — | — | — | — | 5 | — |
| Triethanolamine | — | — | — | — | — | 7 |
| Water | To make up to desired concentration | | | | | |

Other amines are also quite satisfactory such as:

    diethyl ethanolamine
    2-amino, 1-butanol
    2-amino, 2-ethyl, 1, 3-propanediol
    2-amino, 2-methyl, 1, 3-propanediol
    Tris (hydroxymethyl) aminomethane

Other fatty acids which may be used are:

    stearic                    soya
    palmitic                   linseed
    myristic                   tall oil

"A-C" Polyethylene 629 may also be emulsified with potassium or sodium soaps. Borax may be used in combination with amine soaps.

## Nonionic Emulsions

### (Wax to Water Method)

| | No. 15 | No. 16 | No. 17 | No. 18 | No. 19 | No. 20 | No. 21 | No. 22 | No. 23 |
|---|---|---|---|---|---|---|---|---|---|
| "A-C" Polyethylene 629 | 40 | 40 | 40 | 40 | 40 | 40 | 40 | 40 | 40 |
| "Michelene" D. S. | 8 | — | — | — | — | — | — | — | — |
| "Alrosol" | — | 8 | — | — | — | — | — | — | — |
| "Monamine" ACO-100 | — | — | 5 | — | — | — | — | — | — |
| "Pluramine" S-100 | — | — | — | 4 | — | — | — | — | — |
| "Permaline" A-100 | — | — | — | — | 8 | — | — | — | — |
| "Nopcogen" 14-11 or 14-L | — | — | — | — | — | 8 | — | — | — |
| "Hyonic" FA-40 | — | — | — | — | — | — | 8 | — | — |
| "Ninol" 737 | — | — | — | — | — | — | — | 8 | — |
| "Ninol" 2012E | — | — | — | — | — | — | — | — | 8 |
| Water | To make up to desired concentration | | | | | | | | |

## Cationic Emulsions
### (Wax to Water Method)

|                          | No. 9 | No. 10 | No. 11 | No. 12 | No. 13 | No. 14 |
|--------------------------|-------|--------|--------|--------|--------|--------|
| "A-C" Polyethylene 629   | 40    | 40     | 40     | 40     | 40     | 40     |
| "Armac" T                | 8     | —      | —      | —      | —      | —      |
| "Armac" HT               | —     | 8      | —      | —      | —      | —      |
| "Armac" C                | —     | —      | 8      | —      | —      | —      |
| Acetic Acid              | —     | —      | —      | 2      | 1      | 1      |
| "Ethomeen" 18-12         | —     | —      | —      | 8      | —      | —      |
| "Alro" Amine O           | —     | —      | —      | —      | 7      | —      |
| "Alro" Amine S           | —     | —      | —      | —      | —      | 6      |
| Water                    | To make up to desired concentration |

Several formulations are included in the following pages to illustrate some of the ways "A-C" Polyethylene 629 may be used. Many variations will be immediately apparent to the experienced formulator. However, it is strongly recommended that the basic laboratory emulsification procedure described in the preceding section should be tried before modifications are made.

---

### Heavy Duty Floor Polish

"A-C" Polyethylene 629 in combination with a resin and a leveling agent gives a dry-bright floor polish which has high gloss, water resistance, and exceptional durability. The formulations using "Ubatol" 2003 as the resin and "Durez" 15546 as the leveling agent are suggested as high quality industrial polishes.

The emulsions and solution are prepared separately to 15% total solids and blended.

### "A-C" Polyethylene 629 Emulsion

| "A-C" Polyethylene 629 | 40 parts |
|------------------------|----------|
| Oleic Acid             | 6        |
| Morpholine             | 7        |
| Water                  | 254      |

### Resin Emulsion I

| "Ubatol" 2003 (40% total solids) | 100.0 |
|----------------------------------|-------|
| Dibutylphthalate                 | 3.3   |
| "KP" 140                         | 2.6   |
| Water                            | 201.0 |

Mix together at room temperature the dibutylphthalate, "KP" 140, and the "Ubatol" 2003. Stir the mixture moderately for one-half hour. Then add the water and stir a few minutes.

*Resin Emulsion II*

| "Resyn" 78-3021 | |
| --- | --- |
| (35% solids) | 100 |
| Water | 133 |

*Leveling Agent Solution*

| "Durez" 15546 Resin | 30 |
| --- | --- |
| Ammonium | |
| Hydroxide (28%) | 5 |
| Water | 165 |

Heat water to 185°F., add one-half of the ammonium hydroxide, and with good agitation add the "Durez" 15546. Temperature should be maintained at 185-190°F. Add the remainder of the ammonium hydroxide and stir until solution of the "Durez" 15546 is obtained.

The following are the polish proportions.

Formula No. 1
(For Buffability)

| "A-C" Polyethylene 629 | |
| --- | --- |
| emulsion | 5 |
| Resin Emulsion 1 or 11 | 4 |
| "Durez" 15546 solution | 6 |

No. 2
(For Scuff Resistance)

| "A-C" Polyethylene 629 | |
| --- | --- |
| emulsion | 4 |
| Resin Emulsion 1 or 11 | 7 |
| "Durez" 15546 solution | 4 |

The finished product is obtained by blending the desired amount of leveling solution with the wax emulsion. Recommended proportions are 25 or 30 parts of leveling solution to 75 or 70 parts of emulsion. For greater water resistance ammonia cut resins may be used. Best leveling and gloss in this formulation are obtained with borax cut resins particularly "Shanco" L-1001.

These proportions may be varied considerably to bring out the qualities most desired by the formulator. Increasing the A-C Polyethylene emulsion improves polishability and film flexibility. Increasing the resin and the leveling agent improves the hardness and dirt pick-up. Excellent leveling is attained with these three components over a wide range of proportions.

Another suggestion for a heavy duty floor polish is as follows:

No. 3

| "A-C" Polyethylene 629 | 30 |
| --- | --- |
| "Beckacite" PX361 | 30 |
| Oleic Acid | 9 |
| Morpholine | 9 |
| Water | 382 |

Melt wax, resin and oleic acid with agitation until all the resin is dissolved. Cool to 257°F. Add morpholine and hold at 257°F. Add wax mixture at 257°F. to water heated to 203°F. with agitation. Agitate while cooling to room temperature.

Add 70-80 parts of the wax emulsion to 30-20 parts of the leveling solution.*

---

High Shellac Floor Polish

The following formula is representative of a good, high Shellac type dry-bright polish containing "A-C" Polyethylene 629.

### Borax Cut Shellac Solution

| | |
|---|---|
| Shellac, Bleached dewaxed | 108.75 |
| Borax, 5 Mol | 21.50 |
| Water | 709.75 |
| | 840.00 |

In a heated kettle, raise the temperature of the water to 160°F. and add the borax with stirring. Continue stirring and heating and add the shellac in granular form. Stir until solution takes place, cool the bath and make up to weight for loss of water due to evaporation.

---

* Leveling Solution.

| | |
|---|---|
| "Durez" 15546 | 119.0 |
| Ammonia | 26.4 |
| Water | 648.6 |

### "A-C" Polyethylene 629 Wax Emulsion

| | |
|---|---|
| "A-C" Polyethylene 629 | 90 |
| "Durez" 219 Resin | 30 |
| "Neofat" 42-12 | 12 |
| 2-Amino-2-methyl-1-propanol | 12 |
| Water | 706 |
| | 850 |

### Finished Product

| | |
|---|---|
| Borax Cut Shellac Solution | 65 |
| Wax Emulsion | 35 gal. |
| | 100 |

---

Carnauba Base Floor Polish

This polish will serve to demonstrate the value of "A-C" Polyethylene 629 in a high wax type dry-bright formulation. The polish dries to a hard, flexible film of excellent gloss and good leveling. A relatively small amount of "A-C" Polyethylene 629 is used, resulting in gloss and leveling superior to straight carnauba base polishes. Preparation of a typical formulation containing 15% total solids is as follows:

### Wax Emulsion

| | |
|---|---|
| Carnauba | 30 |
| "A-C" Polyethylene 629 | 10 |
| Oleic Acid | 7 |

| | |
|---|---|
| Morpholine | 6 |
| Water | 260 |

*Leveling Agent Solution*

| | |
|---|---|
| "Shanco" L-1001 | 30 |
| Borax, 5 Mol | 13 |
| Water | 244 |

Heat water to 194°F., add the borax with agitation until dissolved. Maintaining the temperature at 194°F., add the Shanco resin and stir until the solution is achieved.

## No Rub Polishes

| Formula | No. 1 | No. 2 | No. 3 | No. 4 | No. 5 |
|---|---|---|---|---|---|
| "A-C" Polyethylene 629 | 22.5 | 22.5 | 15 | 15 | 15 |
| "PE"–100 | 22.5 | 22.5 | 15 | — | — |
| "Petronauba" C | — | — | — | 15 | 15 |
| "C-700" or "C-1035" | 14 | 14 | 15 | 15 | 15 |
| "Durez" 219 | 41 | 41 | — | — | — |
| Carnauba No. 3 | | | | | |
| NC Ref. (Pure) | — | — | 55 | — | 55 |
| Carnauba No. 1 Yellow | — | — | — | 55 | — |
| Oleic Acid | 4-6 | 4-6 | 12 | 12 | 12 |
| Triethanolamine | 4 | — | — | — | — |
| Morpholine | — | 2-2.5 | 11 | 11 | 11 |
| KOH (85%) | 2 | 2 | — | — | — |
| Ammonia Water (26-28%) | — | 3 | — | — | — |
| Borax | 3 | — | 4 | 4 | 4 |
| Water | 750 | 750 | 800 | 800 | 800 |

Formulas No. 1 and 2 are produced by the "wax to water" system of emulsification as follows:

Melt the waxes, "A–C" Polyethylene 629 and resin at temperatures up to 350°F. Cool to 250°F. and add the oleic acid with agitation. When the temperature reaches 205 to 210°F., add the KOH and borax (no borax in formula 2) in a hot saturated solution followed by the TEA. Cook for 20 minutes at 205 to 210°F. (for formula No. 2, add the KOH solution and cook for 15 minutes; add the morpholine followed by the ammonium hydroxide with equal parts of hot water, and cook for an additional 5 minutes). Pour the wax-mix

(205 to 210°F.) into one half of the water at 205 to 210°F. with rapid agitation. Add the remaining water (cold) while cooling to room temperature.

These formulations should be free from scum and sediment. Formula No. 2 will have excellent water resistance; and both formulas will give very high gloss, good buffing properties, good slip resistance, and long wear with a minimum of discoloration. Excellent leveling can be obtained with 20% of a leveling resin solution. These formulations are stable at high solids concentration, and will have good stability with high concentrations of leveling resin. "Ubatol" polystyrene can be added to these emulsions with good results.

The procedure for manufacturing formulas 3, 4 and 5 is as follows:

Melt the waxes, "A-C" Polyethylene 629 and oleic acid together and bring the temperature to 205 to 210°F. Add the morpholine followed by the borax in a hot saturated solution with agitation. Begin adding half of the water (205 to 210°F.) to the wax-mix (205 to 210°F.) slowly until the emulsion inverts to the "oil in water" type. The remaining hot water and cold water can be added to a rapid rate. Cool to room temperature with slow stirring.

---

### Silicone Furniture Polish

#### Formula No. 1

| | |
|---|---|
| Silicone Fluid | 2.2 |
| Oxidized Microcrystalline Wax | 3.6 |
| Mineral Spirits | 94.2 |

#### No. 2

| | |
|---|---|
| Microcrystalline Wax | 5.0 |
| Silicone "DC-200" | 0.5 |
| "Ethomeen" 18/12 Acetate | 3.0 |
| Mineral Spirits | 10.0 |
| Water | 81.5 |

Heat the wax, silicone, mineral spirits and "Ethomeen" 18/12 acetate until the wax melts. Add the water to the wax very slowly at first with stirring. Then add the remainder of the water while continuing agitation. This formulation dries with a good gloss which upon buffing takes on a high luster.

---

### Wax Paste Polish

#### Formula No. 1

| | |
|---|---|
| Carnauba Wax | 30.0 |
| Beeswax | 30.0 |
| Naphtha | 50.0 |

| | |
|---|---|
| Triethanolamine | 4.3 |
| Stearic Acid | 8.0 |
| Water | 65.0 |

### No. 2

| | |
|---|---|
| Carnauba Wax | 30.0 |
| Beeswax | 30.0 |
| Naphtha | 50.0 |
| Monoethanolamine | 1.9 |
| Stearic Acid | 8.0 |
| Water | 65.0 |

### No. 3

| | |
|---|---|
| Carnauba Wax | 30.0 |
| Beeswax | 30.0 |
| Naphtha | 50.0 |
| Morpholine | 2.6 |
| Stearic Acid | 8.0 |
| Water | 65.0 |

-----

## Liquid Cream Wax Polish

### Formula No. 1

| | |
|---|---|
| Carnauba Wax | 12.0 |
| Beeswax | 6.0 |
| Naphtha | 70.0 |
| Triethanolamine | 4.8 |
| Stearic Acid | 8.0 |
| Water | 180.0 |

### No. 2

| | |
|---|---|
| Carnauba Wax | 12.0 |
| Beeswax | 6.0 |
| Naphtha | 70.0 |
| Monoethanolamine | 2.1 |
| Stearic Acid | 8.0 |
| Water | 180.0 |

### No. 3

| | |
|---|---|
| Carnauba Wax | 14.0 |
| Beeswax | 4.0 |
| Naphtha | 25.0 |
| Monoethanolamine | 2.0 |
| Stearic Acid | 8.0 |
| Water | 240.0 |

### No. 4

| | |
|---|---|
| Carnauba Wax | 12.0 |
| Beeswax | 6.0 |
| Naptha | 70.0 |
| Morpholine | 3.0 |
| Stearic Acid | 8.0 |
| Water | 180.0 |

-----

## Automobile Polish*

### Formula No. 1

| | |
|---|---|
| Carnauba Wax | 9.0 |
| Beeswax | 8.0 |
| Naphtha | 75.0 |
| Triethanolamine | 2.7 |
| Stearic Acid | 7.0 |
| Water | 75.0 |

### No. 2

| | |
|---|---|
| Carnauba Wax | 9.0 |
| Beeswax | 8.0 |
| Naphtha | 75.0 |
| Monoethanolamine | 1.2 |
| Stearic Acid | 7.0 |
| Water | 75.0 |

* About 25 lb. of water-absorbing abrasive, such as bentonite, can be added to produce a paste polish; 60 lb. of an oil-absorbing abrasive, such as tripoli, makes a liquid polish.

### No. 3

| | |
|---|---|
| Carnauba Wax | 9.0 |
| Beeswax | 8.0 |
| Naphtha | 75.0 |
| Morpholine | 1.7 |
| Stearic Acid | 7.0 |
| Water | 75.0 |

A steam- or hot-water-jacketed kettle is preferred for making wax polishes, as a satisfactory temperature must be maintained to prevent caking of the wax along the sides of the kettle and to avoid discoloration by overheating the wax. A paddle-type, hand-operated stirrer or a low-speed, large-bladed propeller is also suggested for successful operation. Since morpholine has a flash point of 100°F., it should not be added to the mixture in the presence of open flames. If the wax is melted by means of a gas burner, the gas should be turned off during the addition of the morpholine.

Melt the waxes and stearic acid, add the amine, and maintain the temperature at about 90°C. Add the naphtha slowly and stir until a clear solution is obtained and the temperature is 90 to 95°C. *Avoid the use of open flames.*

The method of adding the abrasive depends on the type used. An oil-absorbing abrasive, such as tripoli, should be well mixed with the hot naphtha solution of waxes just before the water is added. An abrasive that absorbs water, such as bentonite, is best stirred into the finished emulsion.

Heat the water to boiling, add it to the naphtha solution, and stir vigorously until a good emulsion is obtained. Continue stirring slowly until the emulsion has cooled to room temperature.

The proportions of waxes can be changed as desired, depending on the ease of polishing required and the hardness of the final film. A high-melting hydrocarbon wax can be used in place of all or part of the beeswax with good results. When the primary use of the automobile polish is for polishing rather than as a cleaning and polishing combination, it will be more satisfactory without an abrasive.

---

### Liquid Floor Polish

#### Formula No. 1

| | |
|---|---|
| "Estawax" 20 | 4-5 |
| Paraffin Wax (140°F. AMP) | 7-6 |
| Stoddard Solvent | 89 |

## No. 2

| | |
|---|---|
| "Estawax" 25 | 4-5 |
| Paraffin Wax | |
| (140°F. AMP) | 7-6 |
| Stoddard Solvent | 89 |

Heat the solvent to 180°F. and add the molten waxes. Cool with stirring and pour into containers at 100 to 140°F. Do not stir rapidly when the mixture approaches the pouring temperature. Rapid agitation will affect the crystal formation adversely and may cause separation of the solvent.

---

## Water-Emulsion Floor Waxes

### Formula No. 1

| | |
|---|---|
| a "Duroxon" J–324 | 39.0 |
| "Shanco" 300 Resin | 39.0 |
| Prime Yellow | |
| Carnauba Wax | 29.0 |
| Oleic Acid | 11.0 |
| Morpholine | 7.5 |
| Borax | 4.5 |
| Potassium Hydroxide | 0.4 |
| Water | To 12% solids |

Add the melted wax to water.

| | |
|---|---|
| b "Durez" 15546 | |
| | 12% Ammonia cut |

Mix 80 parts of a with 20 parts of b.

### No. 2

| | |
|---|---|
| a "Duroxon" J–324 | 39.0 |
| "Shanco" 300 Resin | 39.0 |
| "Duroxon" H–110 | 29.0 |
| Oleic Acid | 5.0 |
| Morpholine | 11.0 |
| Borax | 4.5 |
| Potassium Hydroxide | 0.4 |
| Water | To 12% solids |
| b "Durez" 15546 | |
| | 12% Ammonia cut |

Mix 80 parts of a with 20 parts of b.

### No. 3

| | |
|---|---|
| a "Duroxon" J–324 | 20.0 |
| Oleic Acid | 2.0 |
| 3-Methoxy- | |
| propylamine | 1.5 |
| Water | To 13% solids |
| b "Ludox" | |
| | Reduced to 13% solids |
| c Manila Loba C | |
| | Resin Dispersion 13% |

Mix 50 parts of a with 25 parts of each b and c.

### No. 4

| | |
|---|---|
| "Duroxon" J–324 | 20 |
| "Shanco" 300 Resin | 20 |
| Oleic Acid | 4 |
| Morpholine | 6 |
| Water | To 15% solids |

### No. 5

| | |
|---|---|
| "Duroxon" H–110 | 20 |
| "Durez" 219 Resin | 20 |
| 2-Amino 2-Methyl | |
| 1-Propanol | 6 |
| Water | To 13% solids |

## No. 6

| a "Duroxon" H–110 | 12.0 |
|---|---|
| Oleic Acid | 2.4 |
| Morpholine | 1.6 |
| Water | To 15% solids |
| b "Ubatol" 2001 | 50 |
| Water | 50 |

Add the melted wax to the water. Mix 60 parts of a with 40 parts of b.

## No. 7

| "Duroxon" H–110 | 50 |
|---|---|
| Morpholine | 6 |
| Water | To 12% solids |

---

## Liquid Solvent Wax

### Formula No. 1

| "Duroxon" R-21 | 5 |
|---|---|
| Paraffin Wax | |
| (126/130°F. AMP) | 5 |
| Mineral Spirits or | 90 |
| Mineral Spirits and | 80 |
| Turpentine | 10 |

Gel-point: Less than –4°F.

### No. 2

| "Duroxon" R-21 | 3.4 |
|---|---|
| "FT" Wax 300 | 3.3 |
| Paraffin Wax | |
| (126/130°F. AMP) | 3.3 |
| Mineral Spirits | 90.0 |

Gel-point: Less than –4°F.

### No. 3

| "Duroxon" R-21 | 10 |
|---|---|

| Mineral Spirits | 90 |
|---|---|

Gel-point: Less than –4°F.

## No. 4

| "Duroxon" R-21 | 5 |
|---|---|
| Paraffin Wax | |
| (126/130°F. AMP) | 5 |
| Mineral Spirits | 50 |
| Turpentine | 40 |

Gel-point: Less than 14°F.

## No. 5

| "Duroxon" R-21 | 7.5 |
|---|---|
| Paraffin Wax | |
| (126/130°F. AMP) | 7.5 |
| Mineral Spirits or | 85.0 |
| Mineral Spirits and | 75.0 |
| Turpentine | 10.0 |

Gel-point: 29°F.

## No. 6

| "Duroxon" R-21 | 7.5 |
|---|---|
| Paraffin Wax | |
| (126/130°F. AMP) | 7.5 |
| Mineral Spirits | 60.0 |
| Turpentine | 25.0 |

Gel-point: 32°F.

## No. 7

| "Duroxon" R-21 | 5 |
|---|---|
| "FT" Wax 300 | 5 |
| Paraffin Wax | |
| (126/130°F. AMP) | 5 |
| Mineral Spirits | 85 |

Gel-point: 35.6°F.

## No. 8

| "Duroxon" R-21 | 7.5 |
|---|---|

Paraffin Wax
(126/130°F. AMP) 7.5
Mineral Spirits 50.0
Turpentine 35.0
Gel-point: 39.5°F.

### No. 9

"Duroxon" R-21 10
Paraffin Wax
(126/130°F. AMP.) 10
Mineral Spirits 80
Gel-point: 39.5°F.

### No. 10

"Duroxon" R-21 5
"FT" Wax 300 5
Paraffin Wax
(126/130°F. AMP) 5
Mineral Spirits 75
Turpentine 10
Gel-point: 41°F.

### No. 11

"Duroxon" R-21 7.5
Paraffin Wax
(126/130°F.) 7.5
Silicone Oil (350 cstks.) 2.0
Mineral Spirits 83.0

### No. 12

"Duroxon" R-21 3.5
Silicone Oil (350 cstks.) 3.5
Stoddard Solvent 20.0
Mineral Spirits 73.0

The waxes are melted together at a temperature of 212 to 225°F. While agitating strongly, the mineral spirit, respectively the blend of mineral spirit and turpentine is slowly added in a steady stream. Then the heat is shut off and agitation is continued while cooling the mass to room temperature.

Where "FT" Wax 300 is part of the formula, the waxes are heated in the presence of approximately 20 to 30% of the total quantity of solvent until a clear solution results. For this purpose temperatures of approximately 200 to 212°F. are recommended. Only when such a clear solution is achieved, the balance of solvent is added. This procedure can be recommended as a matter of general practice in order to prevent any separation of wax components or premature crystallization.

### No. 13

"Duroxon" E–321 5
Paraffin Wax
(126/130°F. AMP) 5
Mineral Spirits 90

Products with a higher viscosity can be obtained by adding "Duroxon" J–324 to the composition.

### No. 14

"Duroxon" E–321 5
"Duroxon" J–324 1

Paraffin Wax
   (126/130°F. AMP)    4
Mineral Spirits         90
A further viscosity increase can be obtained without change of solids content if, in the place

of mineral spirits fresh spirits of gum turpentine is used. When blends of turpentine and mineral spirits are used, the viscosity depends on the proportion of turpentine.

|  | No. 15 | No. 16 | No. 17 | No. 18 |
|---|---|---|---|---|
| "Duroxon" E–321 | 8 | 7 | 10 | 9 |
| "Duroxon" J–324 | – | 1 | – | 1 |
| Paraffin Wax (126/130° F. AMP) | 7 | 7 | 10 | 10 |
| Mineral Spirits | 85 | 85 | 80 | 80 |

No. 19

| "Duroxon" E–321 | 9 |
|---|---|
| Microcrystalline Wax (150/160°F.) | 1 |
| Paraffin Wax (126/130°F. AMP) | 10 |
| Mineral Spirits | 80 |

Small quantities of nonionic emulsifiers may be added to liquid solvent wax formulations. Such emulsifiers prevent agglomeration of crystalline particles and improve gel - formation. Emulsifiers of the type "Igepal" CO–880, 'Emulphor" ON–870, "Hoechst" 2106, "Atlas" G–3960, and others suitable for this purpose. They are melted together with the waxes.

The liquid wax dispersions described in the preceding para-

graph may be prepared according to the following procedure: Heat the waxes until a clear melt results. For this, a temperature of 212 to 230°F. is recommended. Then start the agitation and add the solvent in a steady stream, making sure that the temperature never drops below 185°F. Where limitations in plant equipment do not permit the melting of the waxes at the temperatures indicated, the melt with the solvent should be held at 185 to 190°F. with agitation for a certain length of time in order to assure complete solution of the highest melting wax components and any polyethylene in the formula. This step will also prevent premature crystallization of a part of the wax components which would later result in separation and settling. After cooling the wax to room temperature with constant agitation, it may be passed through a homogenizer for further increasing its smoothness.

---

### Floor Polish Paste

| | |
|---|---|
| "Duroxon" R–11 | 6 |
| "Durmont" 500 Refined Montan Wax | 4 |
| Carnauba Wax No. 3 North Country | 4 |
| Paraffin Wax 143/150°F. Fully Refined | 11 |
| Mineral Spirits | 75 |

The waxes and paraffin are melted together and then the slightly prewarmed solvent is added with good stirring. Solution should be complete. Otherwise, reheat slightly. Then cool with agitation to a temperature of 110 to 115°F. and pour into cans.

---

### Automobile Cleaner-Polish

| | |
|---|---|
| "Duroxon" R–11 | 10.0 |
| "Durmont" 500 | 5.0 |
| Carnauba Wax No. 3 North Country | 5.0 |
| Silicone Oil 350 centistokes | 5.0 |
| "Snow Floss" | 10.8 |

### No. 292 Air Floated

| | |
|---|---|
| Cream Tripoli | 1.2 |
| Mineral Spirits | 63.0 |

Melt the waxes; in a separate container the silicone oil is mixed with the mineral spirits and heated to approximately 120°F. This solution is added slowly with agitation to the wax melt. Solution must be complete, otherwise, reheat slightly. Then, while agitating, add the "Snow Floss" and the Tripoli. Cool the solution to 120°F. with continued agitation. Then pour into cans and allow to cool undisturbed.

---

### Bright Drying Floor Wax Emulsion

#### Formula No. 1

| | | |
|---|---|---|
| a | "Duroxon" J–324 | 150 |
| | Oleic Acid | 10 |
| | Morpholine | 17 |
| | Monoethanolamine | 3 |
| | Water To make | 1000 |
| | (Appx. 16% solids) | |
| b | "Shanco" L–1001 | 160 |
| | Ammonia (28%) | 36 |
| | Water | 804 |

#### Final Composition

| | |
|---|---|
| a | 85 parts (by volume) |
| b | 15 parts (by volume) |

This product can be readily made by conventional procedures. When the "water-to-wax" method of manufacture is used, it can be modified by reducing the amount of amine (morpholine and monoethanolamine) recommended for *a*. It is also possible to employ other leveling resins than the recommended "Shanco" L–1001.

### No. 2

| | | |
|---|---|---|
| *a* | "Duroxon" J–324 | 100.0 |
| | Oleic Acid | 14.0 |
| | Morpholine | 14.3 |
| | Monoethanolamine | 3.1 |
| | Water      To make | 712.0 |
| | (16% solids) | |
| *b* | "Ubatol" U–2003 | |
| | @ 40% | 100.0 |
| | Plasticizer KP–140 | 2.6 |
| | Dibutyl Phthalate | 3.3 |
| | Water | 46.6 |
| *c* | "Durez" 15546 resin | 140.0 |
| | Ammonia (28%) | 21.8 |
| | Water | 838.2 |

*Final Composition* (Add in the order listed)

| | |
|---|---|
| *a* | 36.5 |
| Water | 26.0 |
| *b* | 23.5 |
| *c* | 14.0 |

It is recommended that *a* be prepared by the "wax-to-water" method. This emulsion should be almost completely transparent. Best leveling is usually obtained after the final composition has been allowed to stand undisturbed for at least 24 hours.

---

### Silicone Polishing Cloth
*German Patent 941,309*

| | | |
|---|---|---|
| *a* | Methylpolysiloxane | |
| | Oil | 12 |
| | Isopropanol | 6 |
| | Triethylamine | 7 |
| | Oleic Acid | 1 |
| *b* | Cresol Soap Solution | |
| | (1%) | 175 |

Mix *a* and pour into *b*. Impregnate soft cotton cloth with above for 15 minutes; squeeze and dry.

---

### Mineral Oil Emulsion Polish

| | |
|---|---|
| Mineral Oil (Light) | 40.0 |
| "Ethofat" 60/15 | 2.5 |
| "Ethofat" 60/20 | 2.5 |
| Water | 55.0 |

Dissolve the "Ethofat" 60/15 and "Ethofat" 60/20 in the mineral oil using heat if necessary. The water is then added to the oil with agitation.

---

### Aerosol Polish

| | |
|---|---|
| "A-C" Polyethylene 629 | 3.0 |
| Silicone ("Dow" | |
| DC–200, 50 CS) | 1.8 |
| Naphtha | 50.2 |
| "Genetron" 12 | 45.0 |

Melt the "A-C" Polyethylene 629, add the silicone and bring

to 230°F. Heat naphtha to 158-176°F. and add slowly keeping temperature at 185 to 194°F. (solution should be clear and homogeneous). Continue agitation and cool to room temperature, charge to aerosol containers, cool, and pressure fill with "Genetron".

To apply, spray light coat on clean surface and buff to high gloss.

---

### Aerosol Waxless Polish
### U. S. Patent 2,856,297

| | |
|---|---|
| Lauric Isopropylamide | 0.5 |
| Methylene Chloride | 25.0 |
| Trichloromonofluoro-methane | 25.0 |
| Difluorodichloro-methane | 50.0 |
| Dimethylpolysiloxane | 2.0 |

---

### Auto Cleaner Polish

| | |
|---|---|
| "A-C" Polyethylene 629 | 13.0 |
| Silicone ("Dow" DC–200, 500 CS) | 8.0 |
| Stearic Acid | 7.0 |
| Morpholine | 1.7 |
| Water | 150.0 |
| Naphtha | 100.0 |
| J. M. "Snow Floss" | 20.0 |

Melt the "A-C" Polyethylene and stearic acid together and add the silicone. Cool to 221 to 230°F. and add the morpholine.

Heat the naphtha to 158 to 176°F. and add slowly with stirring, holding the temperature at 185 to 194°F. (solution should be clear and homogeneous). Add the water (185 to 194°F.) with moderate agitation. Finally add the "Snow Floss" and cool with agitation to room temperature. To apply, rub in well to assure removal of surface film, allow to dry and wipe off.

---

### Ball Bearing Polish
### U.S.S.R. Patent 109,150

| | |
|---|---|
| Spindle Oil | 15-35 |
| Kerosene | 15-35 |
| Stearin | 2-5 |
| Graphite Powder | 1-3 |
| Emery Powder M-14 | 24-7 |

---

### Metal Polish

| | Formula No. 1 | No. 2 |
|---|---|---|
| Polyethylene-glycol 1500 | 35 | 50 |
| "Tergitol" Nonionic NPX | 3 | 3 |
| Citric Acid | 5 | — |
| Sodium Chloride | 5 | — |
| Bentonite | 8 | 9 |
| "MultiCel" 000 | 19 | 21 |
| Water | 25 | 29 |

Stir polyethyleneglycol 1500, water, and "Tergitol" nonionic NPX until a clear solution is obtained. If polish No. 1 is being formulated, add the citric acid and sodium chloride and stir until dissolved. Then, for both polishes, add the "MultiCell" 000 and the bentonite and stir until a smooth paste is obtained.

### No. 3

| "Carbowax" | |
| --- | --- |
| Polyethyleneglycol 1500 | 35 |
| "Tergitol" Anionic 7 | 3 |
| Citric Acid | 5 |
| Sodium Chloride | 5 |
| Bentonite | 8 |
| "MultiCel" 000 | 19 |
| Water | 25 |

Mix "Carbowax", water, and anionic 7 and stir until a clear solution is obtained. Add the citric acid and sodium chloride and stir until dissolved. Then add the "MultiCel" 000 and the bentonite and stir until a smooth paste is obtained.

---

### Chemical Polishing of Steel
*Japanese Patent 2817 (1956)*

| Hydrogen Peroxide (30%) | 4 |
| --- | --- |
| Hydrofluoric Acid (40%) | 4 |
| Hydrochloric Acid (d. 1.18) | 1 |

Immerse steel for 10 minutes, then wash with water.

### Chemical Polishing of Aluminum
*Japanese Patent 2962 (1956)*

| Phosphoric Acid | 100 cc. |
| --- | --- |
| Potassium Nitrate | 5-20 g. |
| Copper Sulfate | 0.05-1 g. |

---

### Alkaline Aluminum Cleaner

| Anhydrous Sodium Metasilicate | 30 |
| --- | --- |
| Alkyl Aryl Sodium Sulfonate (85%) | 10 |
| Trisodium Phosphate Dodecahydrate | 35 |
| Soda Ash | 20 |
| Tetrasodium Pyrophosphate Anhydrous | 5 |

---

### Metal Cleaner

| Sodium Metasilicate, Pentahydrate | 34.5 |
| --- | --- |
| Sodium Phosphate, Monobasic | 12.0 |
| Trisodium Phosphate, Dodecahydrate | 33.5 |
| "Tergitol" Nonionic NPX | 5.2 |
| Sodium Alkyl Aryl Sulfonate | 14.8 |

---

### Silver Cleaner

| "Tergitol" Nonionic NPX | 6.5 |
| --- | --- |
| "Carbowax" Polyethyleneglycol 400 | 4.0 |
| Ammonium Carbonate | 2.6 |
| "Ivory" Soap | 1.5 |

Chalk 6.5
"MultiCel" 000 26.4
Water 52.5

Dissolve the soap in part of the water, heating to obtain solution. Add the rest of the water and cool to room temperature. Add the ammonium carbonate, stir until dissolved, and then stir in nonionic NPX and "Carbowax" polyethyleneglycol 400. Add the abrasives and stir until thoroughly mixed and a smooth paste is obtained. The polish will become somewhat stiffer on standing several days. The amount of abrasives can be varied to obtain the desired viscosity.

---

### Silver Polish (Dip)
*Italian Patent 503,144*

Thiourea 8.0
Hydrochloric Acid 1.0
Wetting Agent 0.3

---

### Paste Polish

| | Formula No. 1 | No. 2 | No. 3 | No. 4 |
|---|---|---|---|---|
| "Epolene" N | 20 | 20 | 18 | 15 |
| Paraffin Wax | 8 | 4 | 4 | 8 |
| Beeswax | — | 4 | 4 | — |
| Carnauba | — | — | 2 | 5 |
| "DC–200" (silicone oil) | 2 | 2 | — | — |
| Turpentine | 30 | — | — | — |
| "Amsco" 46 spirits | — | 40 | 32 | — |
| VM&P naphtha | 20 | — | — | — |
| "Solvesso" 100 | — | 30 | 40 | 32 |
| "Stoddard" solvent | 20 | — | — | 40 |

In the preparation of these polishes, the waxes and solvents are heated to approximately 200°F. or until a clear solution is obtained. The mixture is then cooled with agitation until the first sign of cloudiness after which the mixture is poured into a container and allowed to solidify. It has been found that homogenization of the mixture just after the first sign of cloudiness tends to give a much smoother paste.

No. 5

| | |
|---|---|
| "A-C" Polyethylene 629 | 15 |
| Carnauba wax | 5 |
| Paraffin Wax | 5 |
| Turpentine | 25 |
| Naphtha | 50 |

Melt the "A-C" Polyethylene and the waxes together and cool to 212°F. Heat the turpentine to 122 to 140°F. and add to the melt with stirring. Heat the naphtha to 122 to 140°F. and add with stirring. Continue agitation until the polish cools to 131°F. and pour into container.

Apply in an even film and buff to a high gloss.

---

Antislaking Buffing Composition

Formula No. 1
*U. S. Patent 2,847,290*

| | |
|---|---|
| Vienna Lime | 65 |
| Stearic Acid | 15 |
| Acidless Tallow | 15 |
| N-Tallow Trimethyldiamine | 5 |

No. 2
*U. S. Patent 2,850,369*

| | |
|---|---|
| Vienna Lime | 77 |
| Stearic Acid | 14 |
| Acidless Tallow | 6 |
| N-Tallow-N,N',N'-tris (hydroxyethyl) trimethylene diamine | 3 |

Lime Buffing Composition
*U. S. Patent 2,899,289*

| | |
|---|---|
| Lime | 70-80 |
| Stearic Acid | 10-20 |
| Petrolatum | 1-6 |
| Tallow | 2-5 |
| Tertiary Amine | 0.1-5 |

---

Abrasive Vehicle (Oil)
*U. S. Patent 2,889,215*

| | |
|---|---|
| Diesel Oil | 82.99 |
| Lard Oil | 12.97 |
| Sodium Dodecyl-benzene Sulfonate | 2.04 |
| #1 cup Grease | 2.00 |

---

Razor Strop Compound
*U. S. Patent 2,766,128*

| | |
|---|---|
| Sodium Benzoate | 10 fl. oz. |
| Water | 10 fl. oz. |
| Gum Arabic | 2 oz. |

---

Insecticidal Floor Wax

Formula No. 1
(For Ants)

| | |
|---|---|
| a Carnauba Wax or Blend | 45.0 |
| Technical Chlordane | 2.5 |
| "Carbitol" | 7.0 |
| "Tween" 80 | 12.0 |
| Water | 410.0 |
| b Standard Shellac or Resin Dispersion* | 94.0 |

## No. 2
### (For Flies)

| | |
|---|---|
| a Carnauba Wax or | |
| Blend | 45.0 |
| 98% DDT | 2.5 |
| Butyl "Cellosolve" | 7.0 |
| "Tween" 80 | 12.0 |
| Water | 410.0 |
| b Shellac Dispersion* | 94.0 |

## No. 3
### (For Bugs)

| | |
|---|---|
| a Carnauba Wax or | |
| Blend | 40.0 |
| Beeswax | 5.0 |
| 95% Lindane | 2.5 |
| "Carbitol" | 7.0 |
| "Tween" 80 | 12.0 |
| Water | 410.0 |
| b Shellac Dispersion* | 94.0 |

## Opaque-White Nonrubbing Floor Wax

In preparing opaque-white nonrubbing carnauba wax dispersions, use light-colored ingredients, a minimum amount of dispersing agents, and a mutual solvent.

---

\* Shellac Dispersion

| | |
|---|---|
| Bleached Dewaxed Shellac | 50.0 |
| 28% Aqueous Ammonia | 7.2 |
| Water | 373.8 |

It is standard practice to use about 1 part of shellac dispersion to 5 parts of wax dispersion.

## Formula No. 1

| | |
|---|---|
| Carnauba Wax No. 1 | 47.5 |
| White Oleic Acid | 4.0 |
| Soap Flakes | 3.3 |
| "Carbitol" | 7.1 |
| Triethanolamine | 2.0 |
| Water | 411.1 |

## No. 2

| | |
|---|---|
| Carnauba Wax No. 1 | 47.3 |
| White Oleic Acid | 4.0 |
| Soap Flakes | 3.3 |
| Butyl "Cellosolve" | 7.0 |
| Triethanolamine | 2.0 |
| Water | 411.4 |

## No. 3

| | |
|---|---|
| Carnauba Wax No. 1 | 47.3 |
| White Oleic Acid | 6.0 |
| Butyl "Cellosolve" | 7.0 |
| Morpholine | 3.7 |
| Water | 411.0 |

## No. 4

| | |
|---|---|
| Carnauba Wax No. 1 | 47.3 |
| White Oleic Acid | 6.0 |
| Butyl "Cellosolve" | 7.0 |
| 2-Amino 2-Methyl 1-Propanol | 3.3 |
| Water | 411.4 |

## No. 5

| | |
|---|---|
| Carnauba Wax No. 1 | 47.5 |
| Butyl "Cellosolve" | 7.1 |
| "Tween" 80 | 9.5 |
| Water | 410.9 |

## No. 6

| | |
|---|---|
| Carnauba Wax No. 1 | 35.6 |
| "Mekon" Y–20 | 11.9 |
| Butyl "Cellosolve" | 7.1 |
| "Tween" 80 | 9.5 |
| Water | 411.0 |

## No. 7

| | |
|---|---|
| Carnauba Wax No. 1 | 35.6 |
| Bleached Beeswax | 11.9 |
| Butyl "Cellosolve" | 7.1 |
| "Tween" 80 | 9.5 |
| Water | 411.0 |

## No. 8

| | |
|---|---|
| Carnauba Wax No. 1 | 35.6 |
| "Mekon" Y–20 | 8.0 |
| Bleached Beeswax | 4.0 |
| "Carbitol" | 7.0 |
| "Tween" 80 | 9.5 |
| Water | 411.0 |

Wax dispersions are prepared by adding boiled water, containing the amine (if used) to the melted wax mixed with the other ingredients. Where soap flakes are used, they are dispersed in the melted wax before the water solution is added.

---

### Floor-Wax Emulsion

#### Formula No. 1

| | |
|---|---|
| "Chlorowax" 70 | 17.0 |
| "Crown" Wax 23 | 66.0 |
| Oleic Acid | 8.3 |
| Morpholine | 9.0 |
| Water | 730.0 |

Melt the waxes together and add the oleic acid and morpholine with stirring. Maintain the gel at 200 to 210°F. and add hot water at 205 to 210°F., slowly at first, with rapid stirring. After the gel inverts to an oil-in-water emulsion, you may add the water more rapidly. After half the water has been added, discontinue the heat and cool the batch as rapidly as possible while the remainder of the water is added at room temperature, with slow agitation.

### No. 2

| | |
|---|---|
| "Chlorowax" 70 | 7.7 |
| Carnauba Wax | 5.8 |
| "Crown" Wax 23 | 64.0 |
| Oleic Acid | 7.7 |
| Triethanolamine | 9.7 |
| Borax | 5.2 |
| Water | 730.0 |

Use the procedure given under formula 1, but add the borax, dissolved in 25 lb. of boiling water slowly to the mixture of waxes and emulsifier.

---

### Nonrubbing Floor Wax

#### Formula No. 1
North-Country

| | |
|---|---|
| Carnauba Wax No. 3 | 4.65 |
| "Cardis" 319 Wax | 4.65 |
| "Triton" X–100 | 1.50 |

| | |
|---|---|
| Morpholine | 0.85 |
| Water | 88.35 |

Melt and mix and the waxes in a steam-jacketed kettle (or in a water bath, etc.), add the "Triton," and stir in until dissolved; then mix in the morpholine. Slowly add boiling water, a small portion at a time homogeneously absorbed, with constant agitation, until the first-formed water-in-oil emulsion suddenly inverts to oil-in-water. Then add the balance of the water more rapidly, with agitation. Let cool.

If making up a small amount, e.g., 5 gal. or less, hand stirring is sufficient for the agitation, but mechanical agitation is still to be preferred.

The final one third or one quarter of the water can be added cold and mixed in.

After the preparation is made up to volume, it is not necessary to continue the agitation while cooling.

Resin or shellac solutions (10 to 20%) can be stirred in hot or cold, as is the common practice with water-emulsion floor waxes.

### No. 2

| | |
|---|---|
| "Cardis" 314 | 55.0 |
| "Durez" 219 | 25.0 |
| Refined Carnauba | |

| | |
|---|---|
| Wax No. 3 | 25.0 |
| Oleic Acid | 15.0 |
| Morpholine | 5.0 |
| Borax | 8.0 |
| Caustic Potash | 0.4 |
| Water | To make 1000.0 |
| Shellac-Substitute Solution | 250.0 |

Melt the "Cardis," "Durez," carnauba wax, and oleic acid; cool to 210°F. Then add the morpholine and stir 10 minutes at 210°F. Add the borax and caustic potash dissolved in 30 ml. boiling water and stir 10 minutes.

Pour the hot wax blend into the total volume of water at 210°F. under strong agitation, cool, adjust the $pH$ at 8.5 to 9.0 with ammonia, if necessary, and add the shellac substitute solution (12% "Durez" 15546) in water.

The finished emulsion shows perfect water resistance, very good gloss, leveling, and wetting properties. The heat stability is very good. The freezing stability is fair and can be improved to perfection by using 3 g. of 2-amino 2-methyl propanol in addition to the indicated 5 g. morpholine. The water resistance of the last formulation is good for practical purposes.

## No. 3

| | |
|---|---|
| "Cardis" One | 60.0 |
| "Durez" 219 | 40.0 |
| "Warco" 180 White | 20.0 |
| Oleic Acid | 8.0 |
| Morpholine | 9.0 |
| Borax | 8.0 |
| Caustic Potash | 0.4 |
| Water | To make 1000.0 |
| Shellac-Substitute Solution | 250.0 |

Melt the "Cardis" One, "Durez," "Warco," and oleic acid and cool to 210°F. Add the morpholine and stir 10 minutes at 210°F. Add the borax and caustic potash, dissolved in 30 ml. boiling water, and stir 10 minutes. Pour the hot wax blend into the total volume of water at 210°F. under strong agitation; cool, adjust the $p$H at 8.5 to 9.0 with ammonia, if necessary, and add the shellac-substitute solution (12% "Durez" 15546 in water).

## No. 4

| | |
|---|---|
| "Cardis" 314 | 57.0 |
| "Durez" 219 | 48.0 |
| Oleic Acid | 10.0 |
| Morpholine | 5.0 |
| Borax | 8.0 |
| Caustic Potash | 0.4 |
| Water | To make 1000.0 |
| Shellac-substitute solution ("Durez" 15546, 12% Solids) | 250.0 |

Melt the "Cardis," "Durez," and oleic acid and cool to 210°F. Add the morpholine and stir 10 minutes at 210°F. Add the borax and caustic potash, dissolved in 30 ml. boiling water, and stir 10 minutes at 210°F. Pour the hot wax blend in to the total volume of water at 210°F. under strong agitation, cool, adjust the $p$H at 8.5 to 9.0 with ammonia water, and add the "Durez" 15546 solution of 12% solid content, or as desired.

## No. 5

| | |
|---|---|
| "Cardis" One | 50.0 |
| "Durez" 219 | 40.0 |
| "Warco" 180.White | 30.0 |
| Morpholine | 3.0 |
| Soap | 12.0 |
| Caustic Potash | 0.4 |
| Water | To make 1000.0 |
| Shellac-Substitute Solution | 250.0 |

Melt the "Cardis" One, "Durez," and "Warco" and cool to 210°F. Add the morpholine and stir 10 minutes at 210°F. Add the soap flakes and caustic potash dissolved in 40 ml. boiling water and stir 10 minutes.

Pour the hot wax blend into the total volume of water at 210°F. under strong agitation, cool, adjust the pH at 8.5 to 9.0 with ammonia, if necessary, and add the shellac-substitute solution (12% "Durez" 15546 in water).

### No. 6

| | |
|---|---|
| "Cardis" One | 50.0 |
| "Durez" 219 | 40.0 |
| "Warco" 180 White | 30.0 |
| Oleic Acid | 8.0 |
| Morpholine | 9.0 |
| Borax | 8.0 |
| Caustic Potash | 0.4 |
| Water          To make | 1000.0 |
| Shellac-Substitute Solution | 250.0 |

Melt the "Cardis" One, "Durez," "Warco," and oleic acid and cool to 210°F.

Add the morpholine and stir 10 minutes at 210°F. Add the borax and caustic potash, dissolved in 30 ml. boiling water, and stir 10 minutes.

Pour the hot wax blend into the total volume of water at 210°F., under strong agitation, cool, adjust the pH at 8.5 to 9.0 with ammonia, if necessary, and add the shellac-substitute solution (12% "Durez" 15546 in water).

The finished emulsion shows very good water resistance, gloss, and leveling properties. The films are highly scuff and wear resistant.

## Water-Emulsion Paste Waxes

### Formula No. 1

| | | |
|---|---|---|
| a | "FT Wax" 200 | 12 |
| | "Duroxon" C–60–A | 12 |
| | Stearic Acid | 5 |
| | "Lorol" 28 | 3 |
| | Mineral Spirits | 15 |
| b | Water at 212°F. | 49 |
| | Triethanolamine | 3 |
| | Borax | 1 |

### No. 2

| | |
|---|---|
| "Duroxon" H–110 | 10 |
| "Duroxon" C–60–A | 10 |
| "Alrosol" B | 1 |
| Morpholine | 4 |
| Water | 75 |

Pour hot water into the melted wax.

---

## Liquid Solvent Waxes

### Formula No. 1

| | |
|---|---|
| "Duroxon" R–21 | 5 |
| "FT Wax" 300 | 5 |
| Paraffin Wax (M.P. 133–135°F.) | 5 |
| Mineral Spirits | 85 |

### No. 2

| | |
|---|---|
| "Duroxon" R–11 | 6.5 |
| Mineral Spirits | 93.5 |

---

## Solvent-Type Paste Waxes and Shoe Polishes

### Formula No. 1

| | |
|---|---|
| "Duroxon" R–11 | 6 |

| Crude Carnauba Wax | |
|---|---|
| No. 3NC | 4 |
| Beeswax | 1 |
| Crude Paraffin Wax | |
| (M.P. 143–150°F.) | 19 |
| Turpentine | 35 |
| Mineral Spirits | 35 |

Pouring Temperature: 120 to 130°F.

### No. 2

| "Duroxon" J–324 | 10 |
|---|---|
| "FT Wax" 300 | 10 |
| Paraffin Wax (M.P. 133–135°F.) | 10 |
| Mineral Spirits | 70 |

## Stable Wax–Solvent Floor Polish

### Formula No. 1

| Carnauba Wax | 6 |
|---|---|
| Petrolatum | 12 |
| Beeswax | 12 |
| Turpentine | 70 |

### No. 2

| Carnauba Wax | 6 |
|---|---|
| Petrolatum | 12 |
| Beeswax | 12 |
| Turpentine | 65 |
| "Aroclor" 1242 | 5 |

### No. 3

| Carnauba Wax | 6 |
|---|---|
| Petrolatum | 12 |
| Beeswax | 12 |
| Turpentine | 60 |
| "Aroclor" 1242 | 10 |

After preparation, the products are poured, while liquid, into glass bottles and placed in storage.

---

## Liquid Solvent Floor Wax

| "Duroxon" R–21 | 3.4 |
|---|---|
| White "FT Wax" 300 | 3.3 |
| Paraffin Wax (M.P. 133–135°F.) | 3.3 |
| Mineral Spirits | 90.0 |

Melt the waxes with 20% of the solvent; then stir in the heated solvent; finally cool, with agitation, to room temperature. This remains fluid at low temperatures.

---

## Buffing Compound
### U. S. Patent 2,681,274

| Silica Sand (200 Mesh) | 75 |
|---|---|
| Polyglycol Distearate | 20 |
| Beeswax | 2 |
| Sodium Bicarbonate | 3 |
| Tartaric Acid | 6 |

The articles buffed with this compound are immersed in water, acidified with hydrochloric or sulfuric acid, to complete cleaning.

---

## Metal Abrasive

| Alumina | 74 |
|---|---|
| Stearic Acid | 26 |

Heat and mix until uniform.

## CHAPTER XVIII

# PYROTECHNICS AND EXPLOSIVES

### Low Sensitivity Blasting Explosive
#### U. S. Patent 2,860,041

| | |
|---|---|
| Ammonium Nitrate | 51.8 |
| Sodium Nitrate | 16.1 |
| Aluminum Powder | 2.7 |
| Sodium CMC | 1.0 |
| Pecan Meal | 1.0 |
| Mineral Oil | 0.4 |
| Zinc Oxide | 0.3 |
| Nitrostarch (20% Moisture) | 33.4 |

### Electric Detonator Delay
#### U. S. Patent 2,749,226

| | |
|---|---|
| Red Lead | 91-95 |
| Silicon* | 9-5 |

\* Not less than 90% less than 5 microns and all of it less than 20 microns.

### Solid Rocket Fuel Igniter
#### U. S. Patent 2,849,300

| | |
|---|---|
| Ammonium Nitrate | 40-96 |

| | |
|---|---|
| Carbon | 2-40 |
| Potassium Dichromate | 2–40 |

### Nonluminous Igniter for Projectiles

#### Formula No. 1
#### U. S. Patent 2,716,599

| | |
|---|---|
| Bismuth Oxide | 45-85 |
| Manganese | 15-55 |
| Stearic Acid | 10 |
| Graphite | 10 |

#### No. 2
#### U. S. Patent 2,714,061

| | |
|---|---|
| Barium Oxide | 60-7 |
| Antimony Trisulfide | 21.3-35 |
| Asphalt | 1-4 |
| Graphite | 1-4 |

### Signal Flare
#### Swedish Patent 130,801

| | |
|---|---|
| Strontium Nitrate | 57 |
| Magnesium Powder | 34 |

Polyvinyl Chloride     3
Dibutyl Phthalate     6
This lights easily in the cold and is explosion proof.

---

### Smoke Mixture
*U. S. Patent 2,842,502*

Sulfamic Acid     50-80
Potassium Chlorate
or
Ammonium Chlorate    50-20

---

### Yellow Smoke
*German Patent 918,196*

Potassium Chlorate    18
Lactose     7
Pyroxylin, Powdered    11
Sodium Bicarbonate    14
Sudan Yellow     50

---

### Projectile Tracer Smoke
*U. S. Patent 2,823,105*

Strontium Oxide     70
Calcium Resinate     10
1-Methylaminoanthra-
quinone     15
Magnesium Powder    5
Use 5 to 80 g. per projectile. Produces a red smoke. For black smokes replace 1-methylamino-anthraquinone by anthracene and magnesium powder by catechol.

---

### Bullet Tracer
*U. S. Patent 2,899,291*

Magnesium     31.8
Strontium Nitrate    30.9
Strontium Tartrate    27.3
Hexachlorobenzene    4.6
Charcoal     0.9
Stearic Acid     4.5

## Chapter XIX

# RUBBER, RESINS, PLASTICS, AND WAXES

High Sealing Strength Wax
U. S. Patent 2,846,375

| Paraffin Wax | 25-45 |
|---|---|
| Microcrystalline Wax | 10-45 |
| Petrolatum | 30-50 |

Nonblocking Paraffin Wax
U. S. Patent 2,761,851

| Paraffin Wax (M.P. 130-2°F.) | 99-95 |
|---|---|
| Polyethylene (M.W. 2000-1500) | 1-5 |

Melt together at 220°F.

Free Flowing Powdered Wax
U. S. Patent 2,777,776

Coat a powdered wax with powdered tricalcium orthophosphate (0.25 to 2%) by tumbling.

Defoamer for Melted Wax
U. S. Patent 2,796,355

| Potassium Laurate | 5 |
|---|---|
| Glyceryl Monolaurate | 95 |

Use 0.1 lb. of above to 100 lb. wax.

Antioxidant for Paraffin Wax
U. S. Patent 2,741,563

| Butylhydroxyanisole | 3-6 |
|---|---|
| Propyl Gallate | 0.5-2 |
| Citric Acid | 0.5-2 |

Stabilizer (Oxidation) of Paraffin Wax

Formula No. 1
U. S. Patent 2,860,064

| Dodecyl Gallate | 0.00005-0.1 |
|---|---|
| Octadecyl Citrate | 0.00005-0.1 |

No. 2

| tert-Butylhydroxyanisole | 0.00005-0.1 |
|---|---|
| Octadecyl Citrate | 0.00005-0.1 |

286

Refractory Coating for
Wax Molds
*U. S. Patent 2,852,399*

| | |
|---|---|
| a Butyl Titancete | 4 |
| Xylene | 16 |

Mix *a* to dissolve. Then mix

| | |
|---|---|
| b Zircon or Sillimanite (200-300 Mesh) | 80 |

Mix to dissolve. Then mix with
with *b* and dry.

---

Sootless Candles
*U. S. Patent 2,807, 524*

Addition of 0.2% dicyclopentadienyl iron to candle composition produces a candle that burns without soot.

---

### Crayons

| | Formula No. 1 | No. 2 | No. 3 | No. 4 |
|---|---|---|---|---|
| "Duroxon" B-120 | 50 | 20 | 20 | – |
| "Duroxon" C-60 | – | – | – | 20 |
| "FT" Wax 200 | 10 | – | 20 | 20 |
| "FT" Wax 300 | – | 40 | – | – |
| Paraffin Fully Refined 125/129°F. | – | – | 40 | 40 |
| Chrome Oxide Green | 40 | – | – | – |
| Brown Iron Oxide | – | 40 | – | – |
| Red Iron Oxide | – | – | 20 | – |
| Bordeaux Red | – | – | – | 20 |

---

### Low Density Nonshrinking Foamed Urethane

*Prepolymer Formulation*

| | Formula No. 1 | No. 2 | No. 3 | No. 4 |
|---|---|---|---|---|
| DB Castor Oil | 305 | – | – | – |
| "Estynox" 300 | 53 | – | – | – |
| "Flexricin" 9 | – | 276 | – | – |
| "Flexricin" 15 | – | – | 290 | – |
| "Flexricin" 20 | – | – | – | 311 |
| Tolylene Diisocyanate 80/20 | 242 | 324 | 310 | 289 |
| Reaction Temperature | 110°C | 130°C | 130°C | 130°C |

Charge the castor polyol into a three-neck, one-liter flask fitted with a stirrer, thermometer and condenser with drying tube attached. Slowly add the tolylene diisocyanate to the charge with moderate agitation over a thirty minute period. Then heat the re- actants to the prescribed temperature for one hour. Cool to 65°C. and package in 1 qt. tinned cans.

| Foam Formulation | Formula No. 1 | No. 2 | No. 3 | No. 4 |
|---|---|---|---|---|
| Prepolymer | 100.00 | 100.00 | 100.00 | 100.0 |
| Dimethyl Siloxane "Dow-Corning" DC-200 | .50 | .50 | .50 | .5 |
| Buffered Diethylethanolamine* | 4.40 | 4.40 | 4.40 | 4.4 |
| Distilled Water | .35 | 1.15 | 1.04 | .8 |

Laboratory Procedure for Foam Preparation: The dimethyl siloxane is hand mixed into the prepolymer. To this mixture is added the buffered diethylethanolamine together with the water. The blend is manually mixed to a point where incipient foaming becomes observable (about 45 seconds) after which it is poured into a waxed paper container to foam. The foam is allowed to cure at room temperature.

---

Polyvinyl Chloride Handbag and Novelty Sheeting

| "Pliovic" G90V | 100 |
|---|---|
| Primary Plasticizer | 11 |

*Diethylethanolamine ... 42
HCl (36.5%) ... 24
Distilled Water ... 34
Adjust pH to 9.8
Water Content 49.1%

| Low-Temperature Plasticizer | 10 |
|---|---|
| "Paraplex" G60 | 4 |
| Extender Plasticizer ("HB-40") | 15 |
| Secondary Plasticizer | 10 |
| "Atomite" | 20 |
| "Mark" XI | 2 |
| "Mark" XX | 1 |

---

Polyvinyl Chloride Calendered Flooring

| "Pliovic" G90V | 100 |
|---|---|
| Dioctyl Phthalate | 25 |
| "Paraplex" G40 | 25 |
| Calcimum Carbonate | 80 |
| "Silene" | 20 |
| "Tribase"* | 3 |
| Lead Stearate* | 3 |
| Color | 2 |

---

* For lead-free systems, 3 parts of "Synpron" CS138 and 1 part of stearic acid should be substituted for this stabilizing system.

## Polyvinyl Chloride Garden Hose

### Formula No. 1
#### (Clear)

| | |
|---|---|
| "Pliovic" G90V | 100.0 |
| Dioctyl Phthalate | 28.0 |
| Diisooctyl Adipate | 20.0 |
| "Paraplex" G60 | 2.0 |
| "Mark" XV | 3.0 |
| "Synpron" 450 | 1.0 |
| Stearic Acid | 0.5 |
| Color | Variable |

### No. 2
#### (Opaque)

| | |
|---|---|
| "Pliovic" G90V | 100 |
| Dioctyl Phthalate | 28 |
| Diisooctyl Adipate | 20 |
| "Paraplex" G60 | 2 |
| "Atomite" | 20 |
| "Mark" XI | 2 |
| "Mark" XX | 1 |
| Color | Variable |

### Refrigerator Gasket

| | |
|---|---|
| "Pliovic" G90V | 100 |
| "Paraplex" G50 | 40 |
| Dioctyl Phthalate | 20 |
| "Santicizer" 141 | 20 |
| "Atomite" | 80 |
| "DS 207" | 4 |
| Lead Carbonate | 5 |

### Polyvinyl Chloride Blown Film

| | |
|---|---|
| "Pliovic" G90V | 100.0 |
| Dioctyl Phthalate | 16.0 |
| "Santicizer" 141 | 10.0 |
| "Paraplex" G60 | 2.0 |
| "Mark" XV | 3.0 |
| "Synpron" 450 | 1.0 |
| Stearic Acid | 0.5 |

### Translucent or Clear Films
#### (10 mils and under)

### Formula No. 1

| | |
|---|---|
| Polyvinyl Chloride Resin | 100 |
| "Santicizer" 107 | 50 |
| Stabilizer 1 | |

### No. 2

| | |
|---|---|
| Polyvinyl Chloride Resin | 100 |
| "Santicizer" 107 | 25 |
| "Santicizer" 141 | 25 |
| Stabilizer 1 | |

This offers improved processing characteristics and light stability over formula 1 and is flame resistant.

### No. 3

| | |
|---|---|
| Polyvinyl Chloride Resin | 100 |
| "Santicizer" 107 | 20 |
| "Santicizer" 141 | 20 |
| "Santicizer" 160 | 10 |
| Stabilizer 1 | |

This is more economical than formula 2 and is more solvent and grease resistant.

## No. 4

| | |
|---|---|
| Polyvinyl Chloride | |
| Resin | 100 |
| "Santicizer" 107 | 20 |
| "Santicizer" 141 | 20 |
| Dioctyl Adipate | 10 |
| Stabilizer | 1 |

This offers excellent low temperature flexibility, flame resistance, and light stability, as well as good processing characteristics.

## No. 5

| | |
|---|---|
| Polyvinyl Chloride | |
| Resin | 100 |
| "Santicizer" 140 | 15 |
| "Santicizer" 160 | 15 |
| "Paraplex" G–60 | 7 |
| Dioctyl Adipate | 13 |
| Cadmium-Type Stabilizer | |

This is an economical, quality formulation, using "Paraplex" G–60 as a stabilizing plasticizer.

## No. 6

| | |
|---|---|
| Polyvinyl Chloride | |
| Resin | 100 |
| "Santicizer" 140 | 15 |
| "Santicizer" 107 | 10 |
| "Santicizer" 160 | 15 |
| Dioctyl Adipate | 10 |
| Stabilizer | 1 |

This is an economical formulation of good general properties.

## No. 7

| | |
|---|---|
| Polyvinyl Chloride | |
| Resin | 100 |
| "Santicizer" 141 | 35 |
| "Paraplex" G–60 | 15 |
| Cadmium-Type Stabilizer | |

This is a quality formulation with excellent flame resistance, light stability, and ease of processing.

## No. 8

| | |
|---|---|
| Polyvinyl Chloride | |
| Resin | 100 |
| "Santicizer" 140 | 25 |
| "Paraplex" G–60 | 10 |
| Dioctyl Adipate | 15 |
| Cadmium-Type Stabilizer | |

The low-temperature flexibilizing plasticizers selected will determine the quality and economies of this formulation.

The following low-temperature plasticizers were tested in formula 8 in place of dioctyl adipate and the low temperature flexibility obtained was as follows:

| | |
|---|---|
| "Hylene" D | –23.5°C. |
| Dioctyl Sebacate | –24.9°C. |
| "Flexol" 4GO | –19.8°C. |
| "Flexol" TOF | –21.1°C. |
| "Baker's" P–4 | –20.7°C. |
| "Baker's" P–8 | –15.3°C. |
| "Plastolein" 9250 | –23.4°C. |
| "Plasticizer" SC | –22.8°C. |
| "Harflex" 500 | –21.3°C. |

## Opaque Vinyl Sheeting

### Formula No. 1

| Polyvinyl Chloride | |
|---|---|
| Resin | 100.0 |
| "Santicizer" 107 | 15.0 |
| "Santicizer" 141 | 17.5 |
| "Paraplex" G–50 | 7.5 |
| Dioctyl Sebacate | 10.0 |
| Filler | 15.0 |
| Stabilizer 1 or 2 | |

This yields top-quality, flame-resistant, nonmigrating films of good oil resistance and excellent low-temperature flexibility.

### No. 2

| Polyvinyl Chloride | |
|---|---|
| Resin | 100 |
| "Santicizer" 160 | 20 |
| "Santicizer" 140 | 15 |
| "Plastolein" 9250 | 15 |
| Filler | 15 |
| Stabilizer 1 or 2 | |

This gives an economical flame-resistant product.

### No. 3

| Polyvinyl Chloride | |
|---|---|
| Resin | 100 |
| "Santicizer" 160 | 35 |
| "Plastolein" 9250 | 15 |
| Filler | 15 |
| Stabilizer 1 or 2 | |

This has good processing characteristics and is economical.

## Plastic Insulation

### Formula No. 1
(For Calendering Clear)

| "Geon" 404 | 100 |
|---|---|
| Tin Stabilizer | 3-5 |
| Dibasic Lead Stearate | 0-0.25 |

### No. 2
(For Calendering Opaque)

| "Geon" 404 | 100 |
|---|---|
| Dibasic Lead Phthalate | 5-10 |
| Dibasic Lead Stearate | 0.5-0.75 |

### No. 3
(For Extrusion and Molding)

| "Geon" 404 | 100 |
|---|---|
| Dibasic Lead Phthalate | 5-10 |
| Calcium Stearate | 2 |

### No. 4
(For Wire Insulation)

| "Geon" 404 | 100 |
|---|---|
| Dibasic Lead Phthalate | 10 |
| Dibasic Lead Stearate | 2 |
| Calcium Stearate | 2 |

---

## Polyvinyl Chloride ("Pliovic")
## Latex Backing and Coating

### Formula No. 1

| "Pliovic" Latex 300 | |
|---|---|
| (50% Solids) | 200.0 |

Dioctyl Phthalate

Emulsion*                62.4

3% "Methocel"

Solution (25 cp.)        10.0

No. 2

"Pliovic" Latex 300

(50% Solids)            120

"Chemigum" Latex

245CHS (40%

Solids)                 100

3-5% "Methocel"

Solution (25

(cp.)        Sufficient for

             suitable coating

             viscosity

Compounds based on "Pliovic" latex 300 can be applied to backing materials by any of the usual coating methods. Brushing, spraying, or saturating techniques may also be employed. After coating, water must be removed from the film by drying at 212°F. or less.

To achieve satisfactory physical properties, it is necessary to fuse the dried film. Fusion can be obtained at 275°F. However, shorter fusion cycles of 1 to 5 minutes are possible by heating at 300 to 450°F.

---

* Dioctyl Phthalate     100.0
  "Santomerse" D          2.0
  Water                  54.0

Flameproof Asphalt

U. S. Patent 2,667,425

Chlorinated Paraffin

(< 50% Chlorine)    5-40

Antimony Oxide         5-40

Hydrated Lime         1½-15

Inorganic

Filler        To make 100

Melt 15% of this with asphalt and mix until uniform.

---

Asphalt Tile

Formula No. 1

(Dark Red)

"Cumar" Resin T–15     37.6

Ground Limestone       12.7

Asbestos (7R)          33.5

Iron Oxide              3.8

"Paraflux"             10.7

Aluminum Distearate     1.7

No. 2

(Light Red)

"Cumar" Resin T–3      42.0

Ground Limestone       19.1

Asbestos (7R)          27.8

Iron Oxide              3.1

"Hercolyn"              6.5

Aluminum Distearate     1.5

No. 3

(White for Mottling)

"Cumar" Resin T–3      42.5

Ground White

Limestone             35.5

Asbestos (7R)          10.5

| Zinc Oxide | 4.0 |
|---|---|
| "Hercolyn" | 6.0 |
| Aluminum Distearate | 1.5 |

The limestone should all pass through 40 mesh and be 80% retained on 100 mesh.

---

## Reducing Viscosity of Molten Asphalt

"Wax S–1167" has the property of reducing the viscosity of hot - melt asphalts. Electrical units, such as transformers, frequently use asphalts as the potting compound. A reduction in the viscosity of such compounds is of definite interest. The viscosity of an asphalt which is normally applied at 375 to 400°F. is reduced by 5% "Wax S–1167" from 1700 centistokes to 800 centistokes.

---

## Noninflammable Transparent Pyroxylin Film
### Dutch Patent 65,793

| Pyroxylin | 200 |
|---|---|
| Zinc Bromide | 6 |

Dissolve these in 1 liter of a solution of:

| Butanol | 4 |
|---|---|
| Methyl Ethyl Ketone | 2 |
| Toluene | 1 |
| Xylene | 1 |

## Plasticizer for Polyethylene

Polyethylene is difficult to plasticize without exudation of the plasticizer and a hazy finished product. The addition of 2% diglycol stearate S, incorporated directly with the molding powders, has been reported to plasticize polyethylene effectively. The surface of the plastic is left smooth and a very clear bright product is obtained.

---

## Rubber-Surface Preservative
### Hungarian Patent 134,251

| Glycerin | 80 |
|---|---|
| Alcohol | 4 |
| Turkey Red Oil | 8 |
| Water | 8 |

---

## Molding Compound for Jigs
### U. S. Patent 2,656,281

| Asbestos Powder | 50.00 |
|---|---|
| Sodium Silicate | 25.00 |
| Soda Ash | 12.50 |
| Glycerin | 6.25 |
| Iron Oxide | 6.25 |

---

## Recovery of Inserts from Phenolic Moldings

Metallic inserts may be recovered undamaged from phenolic moldings by the following method:

Cut or break the molding into pieces and discard all excess material. Place the pieces in a wire-mesh basket and treat at 200 to 210°F. in the following solution:

| Potassium Hydroxide | 8 oz. |
| Sodium Hydroxide | 8 oz. |
| Sodium Cyanide* | 1 oz. |
| "Duponol" WA | 1/10 oz. |
| Water | 1 gal. |

The cured phenolic molding compound will rapidly disintegrate in the solution and the metallic inserts are readily removed. If the inserts are discolored by the treatment, they may be put through a bright dip and are then ready for reuse.

## Glazing Dull Plastic Casts

Dull-surfaced phenolic plastic moldings or casts can be given a durable, ceramiclike glaze finish with this simple brush-coating technique.

First, mix a small quantity of liquid phenol-formaldehyde casting resin with a few drops of catalyst. A good catalyst for this work can be made by mixing 1 part hydrochloric acid with 5 to 10 parts glycerin, by weight. Thin to coating consistency by

* Highly poisonous.

adding small quantities of alcohol to the resinous mixture. Then, using a good paint brush of proper size, coat the surfaces of the casting.

The brush coating can be rapidly solidified and fused on dull plastic surfaces with the heat from an infrared lamp. In addition to yielding a high-gloss finish, the coating will cover pinholes and other small defects on the cast plastic article.

## Electrician's Tape

### Formula No. 1
(Friction)

| Rubber | 42 |
| Carcass Reclaim | 12 |
| Barytes | 20 |
| Zinc Oxide | 20 |
| "Cumar" P–25 | 6 |

### No. 2
(Coat)

| Pale Crepe | 40 |
| Carcass Reclaim | 25 |
| Barytes | 18 |
| Zinc Oxide | 12 |
| "Cumar" P–25 | 4 |
| Wool Grease | 1 |

The degree of tack is controlled by the amount of mastication of rubber. The wool grease is included in the coat stock for easy stripping and the

adhesive properties of the tape
can be varied by increasing or
decreasing the amount of this
lubricant.

---

### V-Belt Base Compound

Formula No. 1
(Natural Rubber)

| | |
|---|---|
| Smoked Sheet | 100.00 |
| "Reogen" | 2.00 |
| Stearic Acid | 1.00 |
| Zinc Oxide | 5.00 |
| "AgeRite" Resin D | 1.00 |
| "AgeRite Hipar" | 1.00 |
| Palm Oil | 4.00 |
| "SRF" Black | 75.00 |
| Sulfur | 3.00 |
| "Altax" | 1.50 |
| "Unads" | 0.15 |
| Specific Gravity | 1.24 |

Press Cure
10 minutes at 307°F.

No. 2
(GR–S and Natural Rubber
Blend)

| | |
|---|---|
| Smoked Sheet | 75.00 |
| GR–S 100 | 25.00 |
| "Reogen" | 5.00 |
| Stearic Acid | 1.00 |
| Zinc Oxide | 3.00 |
| "AgeRite" Resin D | 1.00 |
| "AgeRite Hipar" | 1.00 |
| "Cumar" MH | 2.00 |
| "P–33" | 200.00 |
| Sulfur | 2.00 |

| | |
|---|---|
| "Altax" | 2.25 |
| "Ledate" | 0.25 |
| Specific Gravity | 1.36 |

Press Cure
10 minutes at 307°F.

---

### V-Belt Fabric

Formula No. 1
(Skim)

| | |
|---|---|
| Smoked Sheet | 60.00 |
| Whole-Tire Reclaim | 80.00 |
| "Reogen" | 2.00 |
| Stearic Acid | 2.00 |
| Zinc Oxide | 3.00 |
| "AgeRite" Resin D | 1.50 |
| "AgeRite Hipar" | 1.50 |
| "Cumar" MH | 2.00 |
| "Thermax" | 35.00 |
| Sulfur | 2.75 |
| "Altax" | 1.50 |
| "Ledate" | 0.10 |
| Specific Gravity | 1.14 |

Press Cure
10 minutes at 307°F.

No. 2
(Friction)

| | |
|---|---|
| Smoked Sheet | 60.00 |
| Whole-Tire Reclaim | 80.00 |
| "Reogen" | 5.00 |
| Stearic Acid | 2.00 |
| Zinc Oxide | 3.00 |
| "AgeRite" Resin D | 1.50 |
| "AgeRite Hipar" | 1.50 |
| "Cumar" MH | 2.00 |
| "Thermax" | 35.00 |

| | |
|---|---|
| Sulfur | 2.75 |
| "Altax" | 1.50 |
| "Ledate" | 0.10 |
| Specific Gravity | 1.135 |

Press Cure
    10 minutes at 307°F.

---

### V-Belt Cord Skim
### (Cushion)

| | |
|---|---|
| GR–S 10 | 42.00 |
| Smoked Sheet | 28.00 |
| Brown Crepe | 15.00 |
| Whole-Tire Reclaim | 30.00 |
| "Reogen" | 3.00 |
| Stearic Acid | 1.00 |
| Zinc Oxide | 3.00 |
| "AgeRite Hipar" | 1.00 |
| "P–33" | 20.00 |
| "MAF" Black | 20.00 |
| Sulfur | 2.25 |
| "Altax" | 2.25 |
| "Ledate" | 0.25 |
| Specific Gravity | 1.14 |

Press Cure
    10 minutes at 307°F.

---

### General-Purpose Automotive
### Compound

#### Formula No. 1
#### (GR–S)

| | |
|---|---|
| GR–S 25 | 100.0000 |
| "Bondogen" | 2.0000 |
| Stearic Acid | 1.0000 |

| | |
|---|---|
| Zinc Oxide | 5.0000 |
| "AgeRite" Gel | 1.0000 |
| "AgeRite" White | 0.2500 |
| "Cumar" MH | 20.0000 |
| Light Process Oil | 15.0000 |
| Sunproofing Wax | 4.0000 |
| "MAF" Black | 55.0000 |
| "Dixie" Clay | 20.0000 |
| Sulfur | 2.2500 |
| "Altax" | 1.5000 |
| "Cumate" | 0.1875 |
| Specific Gravity | 1.2100 |

Press Cure
    12 minutes at 307°F.

#### No. 2
#### (Natural Rubber Reclaim)

| | |
|---|---|
| Flat-Bark Crepe | 50.00 |
| Peel Reclaim | 90.00 |
| Refined-Tailings Reclaim | 23.00 |
| "Reogen" | 2.00 |
| Stearic Acid | 1.00 |
| Zinc Oxide | 3.00 |
| "AgeRite Stalite" | 1.00 |
| Hard Hydrocarbon | 7.50 |
| "Thermax" | 40.00 |
| Sulfur | 3.00 |
| "Altax" | 1.00 |
| "Ledate" | 0.15 |
| Specific Gravity | 1.17 |

Press Cure
    10 minutes at 307°F.

## Rubber-Covered Roll Compounds

### Formula No. 1
#### (For Suction Press Roll )

| | |
|---|---|
| Smoked Sheet | 100.00 |
| "Reogen" | 6.00 |
| Zinc Oxide | 100.00 |
| "AgeRite" Resin D | 1.00 |
| 'AgeRite Hipar" | 1.25 |
| "MPC" Black | 2.00 |
| "Dixie" Clay | 25.00 |
| Calcium Oxide | 3.50 |
| Sulfur | Variable |

### No. 2
#### (For Size and Wax Roll)

| | |
|---|---|
| Smoked Sheet | 100.0 |
| "Reogen" | 7.5 |
| Zinc Oxide | 82.5 |
| "AgeRite" Resin D | 1.0 |
| "AgeRite Hipar" | 1.0 |
| Lithopone | 82.5 |
| Calcium Oxide | 4.0 |
| Sulfur | Variable |

The amount of sulfur added varies with the hardness desired in the finished product.

---

## Floor Tiling

### Formula No. 1

| | |
|---|---|
| GR–S 25 | 100.0 |
| "Reogen" | 5.0 |
| Stearic Acid | 1.0 |
| Zinc Oxide | 5.0 |
| "AgeRite Stalite" | 1.0 |
| Ceresin Wax | 1.0 |
| Burgundy Pitch | 5.0 |
| "Cumar" MH | 20.0 |
| Heavy Calcined Magnesia | 15.0 |
| "Dixie" Clay | 80.0 |
| Calcium Silicate | 20.0 |
| "Asbestine" | 80.0 |
| Titanium Dioxide | 20.0 |
| Sulfur | 7.0 |
| "Altax" | 1.5 |
| "Unads" | 0.1 |
| Specific Gravity | 1.75 |
| Shore Hardness | 90.00 |

Press Cure
  8 minutes at 320°F.

### No. 2

| | |
|---|---|
| GR–S 25 | 100.0 |
| High-Styrene Copolymer 70/30 | 10.0 |
| Medium-Hard Cumarone–Indene Resin | 20.0 |
| Cumarone–Indene Resin (M.P. 10°C.) | 5.0 |
| Zinc Oxide | 5.0 |
| Stearic Acid | 0.5 |
| Paraffin | 1.0 |
| "Purecal" M | 300.0 |
| Hard Clay | 100.0 |
| Titanium Dioxide | 10.0 |
| Sulfur | 4.0 |
| Benzothiazyl Disulfide | 2.0 |
| Copper Diethyl Dithiocarbamate | 0.1 |

This low-cost base formula produces a high-grade, nonbrittle tile of excellent color. Both GR–S and titanium dioxide may be saved with no sacrifice in quality.

---

## Building-Wire Insulation

### Formula No. 1
### (GR–S Code*)

| | |
|---|---|
| GR–S 65 | 100.0 |
| "Bondogen" | 2.0 |
| Stearic Acid | 2.0 |
| Zinc Oxide | 5.0 |
| "AgeRite" White | 1.0 |
| "AgeRite" Resin D | 1.0 |
| Hard Hydrocarbon | 100.0 |
| Paraffin | 3.0 |
| Calcium Carbonate | 75.0 |
| "Dixie" Clay | 75.0 |
| "Thermax" | 100.0 |
| Sulfur | 3.0 |
| Litharge | 2.0 |
| "Altax" | 1.5 |
| Methyl "Zimate" | 1.5 |
| Specific Gravity | 1.44 |

### No. 2
### (GR–S RH–RW*; Heat and Moisture Resistant)

| | |
|---|---|
| GR–S 65 | 75.00 |
| GR–S 60 | 25.00 |
| "Reogen" | 5.00 |
| Stearic Acid | 1.00 |
| Zinc Oxide | 50.00 |
| "AgeRite" White | 1.00 |
| "AgeRite" Resin D | 1.00 |
| Hard Hydrocarbon | 30.00 |
| Ozokorite Wax | 2.00 |
| Sunproofing Wax | 2.00 |
| Calcium Carbonate ("Atomite") | 50.00 |
| Calcined Clay | 75.00 |
| Sulfur | 1.25 |
| Litharge | 5.00 |
| "Altax" | 2.50 |
| Methyl "Zimate" | 1.25 |
| Specific Gravity | 1.56 |

### No. 3
### (GR–S RH–RW*; Heat and Moisture Resistant)

| | |
|---|---|
| GR–S 65 | 75.0 |
| GR–S 60 | 25.0 |
| "Reogen" | 5.0 |
| Stearic Acid | 1.0 |
| Zinc Oxide | 50.0 |
| 'AgeRite" White | 1.0 |
| 'AgeRite" Resin D | 1.0 |
| Hard Hydrocarbon | 30.0 |
| Ozokerite Wax | 2.0 |
| Sunproofing Wax | 2.0 |
| Calcium Carbonate ("Atomite") | 50.0 |
| Calcined Clay | 75.0 |
| Sulfur | 1.0 |
| "Altax" | 1.0 |
| Methyl "Zimate" | 2.0 |
| "Cumate" | 0.5 |
| Specific Gravity | 1.55 |

---

* For continuous vulcanization.

## Cord Compound
### Formula No. 1
### (POSJ*)

| | |
|---|---|
| Smoked Sheet | 100.00 |
| "Reogen" | 2.00 |
| Stearic Acid | 2.00 |
| Zinc Oxide | 10.00 |
| "AgeRite" Resin D | 1.00 |
| "AgeRite" White | 0.50 |
| Sunproofing Wax | 2.00 |
| Water-Washed Clay | 75.00 |
| Calcium Carbonate | 50.00 |
| Red Oxide | 5.00 |
| "MPC" Black | 0.25 |
| Sulfur | 1.25 |
| "Altax" | 0.65 |
| "Bismate" | 0.65 |
| Specific Gravity | 1.45 |

### No. 2
### (Insulation; Neutral*)

| | |
|---|---|
| GR–S 65 | 75.00 |
| GR–S 60 | 25.00 |
| "Reogen" | 5.00 |
| Stearic Acid | 1.00 |
| Zinc Oxide | 10.00 |
| "AgeRite" Resin D | 1.00 |
| "AgeRite" White | 1.00 |
| "Cumar" MH | 20.00 |
| Sunproofing Wax | 4.00 |
| "Dixie" Clay | 100.00 |
| Calcium Carbonate | 75.00 |
| Sulfur | 2.00 |
| "Altax" | 2.00 |
| "Ledate" | 1.00 |
| Ethyl "Selenac" | 0.25 |
| Specific Gravity | 1.55 |

### No. 3
### (Jacket*)

| | |
|---|---|
| GR–S 100 | 100.0 |
| "Bondogen" | 2.0 |
| Stearic Acid | 1.5 |
| Zinc Oxide | 3.0 |
| "Rio" Resin | 3.0 |
| Hard Hydrocarbon | 30.0 |
| Sunproofing Wax | 4.0 |
| "MAF" Black | 40.0 |
| "EPC" Black | 40.0 |
| Sulfur | 2.0 |
| "Altax" | 1.0 |
| "Bismate" | 1.0 |
| Specific Gravity | 1.18 |

### Neoprene Compounds
### Formula No. 1
### (High-Tension Jacket*)

| | |
|---|---|
| Neoprene GN | 100.0 |
| "Altax" | 1.5 |
| Stearic Acid | 1.0 |
| "AgeRite" White | 0.5 |
| "Rio" Resin | 10.0 |
| Light Process Oil | 2.0 |
| Paraffin Wax | 5.0 |
| "Thermax" | 30.0 |
| "Dixie" Clay | 30.0 |
| Extra Light Calcined Magnesia | 4.0 |
| Zinc Oxide | 20.0 |
| "Permalux" | 0.5 |
| Specific Gravity | 1.54 |

* For continuous vulcanization.

## No. 2
### (Heavy-Duty Jacket†)

| | |
|---|---|
| Neoprene GN | 100.0 |
| "Altax" | 1.5 |
| Stearic Acid | 3.0 |
| "AgeRite Stalite" | 2.0 |
| Paraffin Wax | 8.0 |
| Light Process Oil | 8.0 |
| "EPC" Black | 40.0 |
| "P–33" | 25.0 |
| Extra Light Calcined Magnesia | 4.0 |
| Zinc Oxide | 7.5 |
| Specific Gravity | 1.39 |

## No. 3
### (Line-Wire Insulation*)

| | |
|---|---|
| Neoprene GN | 100.0 |
| "Altax" | 1.0 |
| Stearic Acid | 2.0 |
| "AgeRite Stalite" | 1.0 |
| "AgeRite" White | 1.0 |
| Petrolatum | 2.0 |
| Paraffin Wax | 4.0 |
| Light Process Oil | 17.0 |
| "MPC" Black | 25.0 |
| "Dixie" Clay | 90.0 |
| Extra Light Calcined Magnesia | 4.0 |
| Zinc Oxide | 5.0 |
| "Permalux" | 0.5 |
| Specific Gravity | 1.55 |

† For pan and drum cures.
* For continuous vulcanization.

## Calendered Upholstery Sheeting

| | |
|---|---|
| "Pliovic" DB80V | 100.0 |
| DOP | 20.0 |
| DOS | 12.0 |
| "Paraplex" G–50 | 10.0 |
| "Multifex" MM | 15.0 |
| "Plumb-O-Sil" C | 3.0 |
| Stearic Acid | 0.5 |

## General Purpose Calendered Film

| | |
|---|---|
| "Pliovic" DB80V | 100.0 |
| DOP | 30.0 |
| "Santicizer" 141 | 7.0 |
| DOS | 6.0 |
| "Mark" XI | 2.0 |
| "Mark" XX | 1.0 |
| "Monoplex" S–71 | 1.0 |
| Stearic Acid | 0.5 |

## Extruded Cove Base Molding

| | |
|---|---|
| "Pliovic" DB80V | 100 |
| Dioctyl Phthalate | 22 |
| "Paraplex" G50 | 5 |
| "Dutrex" 20 | 20 |
| "Baker's" P–4 | 5 |
| "Vopcolene" 75 | 4 |
| Lead Carbonate | 5 |
| Whiting | 50 |
| "Atomite" | 150 |
| "Acrawax" C, Atomized | 2 |
| Color | As required |

## Butyl Rubber Base Compound

| | |
|---|---|
| "Enjay" Butyl 218 | 100 |
| MPC Black | 50-90 |

Zinc Oxide 5
Stearic Acid 2
Sulfur 1¼
Tetramethyl Thiuram
  Disulfide 1½
Mercaptobenzothiazole 1
Cure 20 minutes at 320°F.

---

Non-Cracking Polyethylene
U. S. Patent 2,765,293
0.5 to 5% Ethylene distearamide ("Acrawax C") is incorporated in polyethylene before molding or drawing.

---

Moth Killing Plastic for Bags

Formula No. 1
U. S. Patent 2,861,965

Polyethylene 100.00
Camphor 5.05
Alcohol 0.05

No. 2
Polyvinyl Chloride 100.0
Dioctyl Phthalate 35.0
Cadmium Stearate 1.0
Calcium Stearate 0.4
Mineral Oil 2.0
Naphthalene 6.0

---

Plastic Lens Antifogging
Composition
U. S. Patent 2,889,298

Polyvinyl Alcohol 5.0
Glycerin 2-2.5
Lithium Chloride 0.2-0.3

Sodium Oleylmethyl
  Tauride 0.1-0.2
Alcohol
  (50%)  To make 100

---

Temperature Resistant Plastic
U. S. Patent 2,883,352

Phenolformaldehyde
  Resin 20-25
Hexamethylane-
  tetramine 2-4
Stearic Acid 2-5
Manganese Dioxide 20-40
Kaolin 5-15
Magnesium Oxide 20-40
Asbestos 5-20
This will stand temperatures of 250 to 300°C.

---

Cellulose Acetate Solvent
U. S. Patent 2,739,069

Methylene Chloride 83-93
Methanol 2-5
Butyl Alcohol 5-12

---

Etching Solvent for Cellulose
Acetate Printing Plate
U. S. Patent 2,891,849

Dimethyl Sulfoxide 1
Isopropanol 3-4

---

Castor Oil Alkyd Resin
Glycerin
  (100%
  basis) 18.6

Phthalic
Anhydride 35.9
AA Castor
Oil 50.0
———
104.5
–4.5 (water)
———
100.0

Charge all materials to the alkyd kettle. Heat to 435°F. and hold at 435°F. for an acid number of 40 (solids basis). Raise the temperature to 500°F. and hold at 500°F. for desired acid number and viscosity. Cool and reduce to 65% solids in xylol.

Improving Odor of
Petroleum Resins
U. S. Patent 2,857,361

Vanillin 0.05
Butylhydroxyanisole 0.20

These are added to petroleum resins in the above percentages to give a practically odorless coating when resin is applied as a coating.

"Dacron" ("Terylene") Solvent
German Patent 959,328

Trifluoroacetic Acid 10
Ethylene Dichloride 32

Benzol (20 to 50%) may be added to lower cost of above.

Bleedproof Decalcomania
U. S. Patent 2,758,035

Lacquer Backing
Nitrocellulose 12.8
Titanium Dioxide 34.2
Soybean Oil 6.8
Dibutylphthalate 6.8
Castor Oil 6.8
Butyl "Cellosolve" 17.1
Butyl Alcohol 6.8
Butyl Acetate 8.7

Blocking Film
Animal Glue 7.2
Glycerin 2.7
Ethyl Lactate 8.2
Water 65.6
Polyvinyl Acetate
Emulsion (55%) 16.3

Chewing Gum Base

"Vistanex" LM 100
Wax (135 EMP) 40
Petrolatum (165 EMP) 50
Powdered Sugar 333
Flavor 1

Awning-Coating Organosol

"Pliovic" AO 100.0
"Hercoflex" 150 40.0
"Paraplex" G–60 5.0
"Advance" L 1.5
Dibutyl Tin Maleate 1.5
"Victor" 85 2.0
Butyl "Cellosolve" 4.0

| | |
|---|---|
| "Apcothinner" | 45.0 |
| Titanium Dioxide | 1.5 |
| "Surfex" | 20.0 |

This formulation contains approximately 78% solids.

---

## Thin Unsupported Film

### Formula No. 1
#### (60% Solids)

| | |
|---|---|
| "Pliovic" AO | 100.0 |
| Di-2-Ethylhexyl Phthalate | 27.5 |
| Polyethylene Glycol 200 Monolaurate | 7.5 |
| Xylol | 65.6 |
| "Troluoil" | 43.8 |
| "Victor" No. 85 | 2.0 |
| Antimony Oxide | 6.5 |
| Titanium Dioxide | 8.0 |
| "Non–Fer–Al" or "Metro Nite" | 12.5 |

### No. 2
#### (87% Solids)

| | |
|---|---|
| "Geon" Resin 121 | 56 |
| Dioctyl Phthalate | 28 |
| Oil-Soluble Stabilizer | 3 |
| Diluent | 13 |

### No. 3
#### (75% Solids)

| | |
|---|---|
| "Pliovic" AO | 100.0 |
| Dioctyl Phthalate | 35.0 |
| "Apcothinner" | 47.0 |
| "Victor" 85 | 2.0 |

| | |
|---|---|
| "Carbowax" 1000 Distearate | 2.5 |
| "Advance" L | 3.0 |

### No. 4

| | |
|---|---|
| "Pliovic" AO | 100 |
| "Hercoflex" 150 | 26 |
| "Paraplex" G–60 | 7 |
| "Apcothinner" | 47 |
| "Aldo" 25 | 2 |
| "Hypodyne" No. 5 | 1 |
| "Advance" L | 3 |

### No. 5
#### (60% Solids)

| | |
|---|---|
| "Pliovic" AO | 100.0 |
| Dioctyl Phthalate | 27.0 |
| "Glyco" S–1118 | 6.0 |
| "Victor" 85 | 2.0 |
| Xylol | 57.8 |
| "Apcothinner" | 31.2 |

### No. 6

| | |
|---|---|
| "Marvinol" VR–10 | 100.0 |
| Dioctyl Phthalate | 30.0 |
| Dibutyl Tin Maleate | 1.0 |
| "Advance" No. 3 | 0.5 |
| Calcium Stearate | 0.3 |
| Carnauba Wax | 1.0 |

When starting, it has been found advisable to run with a large amount of bleed to prevent initial hang-up in the die. Once the stock is flowing freely throughout the extruder, the bleed can be cut back to a minimum. A small amount of bleed helps prevent stagnation of material in the extruder head. The die temperature for starting should be 20 to 30°F. lower than used for standard operating conditions. Once the tube is formed, operating the extruder near maximum speed seems to prevent stock stagnation.

In beginning work, it is best to use highly stabilized, overlubricated vinyl-compounds until the necessary operating techniques are worked out. The plasticizer content for very thin film is considerably lower than that used for heavier 4-gage film, which is usually about 45 parts in 100 parts resin. Films containing this quantity of plasticizer tend to be too soft when they are as thin as 2 mils and a lower plasticizer ratio is preferable for many applications.

|  | Start of Run | Main Part of Run |
|---|---|---|
| Die Temperature, °F | | |
| Core Pin | 335-350 | 395-425 |
| Die Face | 335-350 | 395-425 |
| Head Temperature, °F. | 350-360 | 380-400 |
| Barrel Temperature, °F. | | |
| First Section | 350-360 | 350-360 |
| Second Section | 320 | 320 |
| Screw Speed, rpm. | 40 | 40 |

### Calendered Vinyl Sheeting

| | |
|---|---|
| Vinyl Chloride Polymer ("Geon" 101) | 100.00 |
| Dioctyl Phthalate | 25.0 |
| Octyl Diphenyl Phosphate | 15.00 |
| Dioctyl Sebacate | 10.00 |
| Calcium Carbonate | 10.00 |
| Lead Silicate | 5.00 |
| Stearic Acid | 0.25 |
| Color Pigment | 2-5 |

All ingredients may be mixed in a Banbury mixer or on a rubber mill. For best results, grind the pigment and lead silicate in a portion of the plasticizer on a paint mill and then preblend all ingredients in a ribbon blender or similar device before Banbury mixing or milling.

This sheeting can be calendered to 0.02 in. thickness for upholstery and similar uses. It has flame resistance and good low-temperature flexibility.

## Vinyl Plastisol Compound

| | |
|---|---|
| "Geon" 121 | 100 |
| Dioctyl Phthalate | 40 |
| Octyl Diphenyl Phosphate | 10 |
| Dioctyl Sebacate | 15 |
| Calcium Carbonate | 15 |
| Barium-Cadmium Stabilizer ("Barca" 10) | 5 |
| Color Pigment | 2 to 5 |

Grind the calcium carbonate and pigment in a portion of the plasticizer on a paint mill or other suitable grinder. Put the plasticizers in a mixer and stir in the dispersed pigment and stabilizer. Add the resin slowly, stirring continuously. Deaerate and the plastisol is ready for coating. Fuse by heating the coating to 350°F.

This is suitable for heavy-weight fabric coating.

---

## Transparent Vinyl Films

### Formula No. 1
### (4 to 8 mil)

| | |
|---|---|
| Polyvinyl Chloride | 100 |
| Dioctyl Phthalate | 24 |
| "MPS–500" | 18 |
| Dioctyl Adipate | 10 |
| Stabilizer | 4½ |

This formulation is suggested for good low-temperature characteristics and all-around good utility in the many end uses where transparent vinyl films are required.

### No. 2
### (4 to 8 mil)

| | |
|---|---|
| Polyvinyl Chloride | 100 |
| Dioctyl Phthalate | 18 |
| "MPS–500" | 26 |
| "Flexol" 4 GO | 10 |
| Stabilizer | 4½ |

This is similar to formula 1, but with a saving in cost and small sacrifice in quality.

### No. 3
### U. S. Patent 2,599,573

| | |
|---|---|
| "Vinylite" (VYNU) | 150 |
| Tributyl Acetylcitrate | 18 |
| Dihexyl Phthalate | 30 |
| Polyethylmethacrylate | 2 |

---

## Antiseptic Plastic Sheeting
### British Patent 684,404

| | |
|---|---|
| Polyvinyl Chloride | 100 |
| Dibutyl Phthalate | 40 |
| "Cresantol" | 0.5–7.5 |
| "Aerosol" OT | 0.5–5 |

---

## Expanded Polyvinyl Plastic

### Formula No. 1
### U. S. Patent 2,626,968

| | |
|---|---|
| Polyvinyl Chloride | 90 |
| Dicyclohexyl Ammonium Nitrite | 10 |

Blend together on a heated

mill and foam by heating to 350°F.

### No. 2
*British Patent 692,142*

| | |
|---|---|
| Polyvinyl Chloride | 50 |
| Tritolyl Phosphate | 50 |

Treat on rubber mill with:

| | |
|---|---|
| Benzene Sulfonic Acid Hydrazide | 12 |
| Benzoic Acid | 4 |

Heat and cool under pressure in forms.

---

### Dipping Organosol

| | |
|---|---|
| "Pliovic" 40 | 100.0 |
| Dioctyl Phthalate | 35.0 |
| "Victor" 85 | 2.0 |
| "Carbowax" 1000 Distearate | 2.5 |
| "Apcothinner" | 24.0 |
| "Standard" 350 | 24.0 |
| "Surfex" | 20.0 |
| Aluminum Stearate | 1.5 |

It is recommended to dissolve the aluminum stearate by heating in part of the plasticizer. When cool, this solution is stirred into the formulation after grinding.

When equipment imposes limits on the baking and fusing cycles, the relative proportions of the volatile ingredients can be adjusted.

### Flooring Organosol

| | |
|---|---|
| "Pliovic" 40 | 100 |
| "Hercoflex" 150 | 15 |
| Diisobutyl Ketone | 3 |
| "Standard" 350 | 44 |
| "Aldo" 25 | 2 |
| "Hypodyne" No. 5 | 1 |
| "Advance" L | 3 |

---

### Plastic Upholstery Sheeting

| | |
|---|---|
| "Geon" 101 Resin | 100.0 |
| Dioctyl Sebacate | 15.0 |
| Octyl Diphenyl Phosphate | 15.0 |
| Polymeric Plasticizer | 7.0 |
| Dioctyl Phthalate | 17.0 |
| Calcium Carbonate | 15.0 |
| Basic Lead Carbonate | 3.0 |
| Barium-Cadmium Laurate | 1.0 |
| Stearic Acid | 0.5 |
| Phthalocyanine Blue | 2.0 |

---

### Molding and Extruding Plastic

#### Formula No. 1

| | |
|---|---|
| "Geon" Resin 101 | 44.5 |
| Dioctyl Phthalate | 15.5 |
| Secondary Plasticizer | 9.0 |
| Filler | 22.0 |
| Pigment-Type Stabilizer | 6.8 |
| Titanium Dioxide | 2.2 |

#### No. 2

| | |
|---|---|
| "Geon" Resin 101 | 63.0 |

| | |
|---|---|
| Dioctyl Phthalate | 18.5 |
| Secondary Plasticizer | 12.0 |
| Polymeric Plasticizer | 5.0 |
| Oil-Soluble Stabilizer | 1.5 |

### No. 3

| | |
|---|---|
| "Geon" Resin 101 | 64.5 |
| Dioctyl Phthalate | 32.5 |
| Oil-Soluble Stabilizer | 1.3 |
| Pigment-Type Stabilizer | 1.3 |
| Lubricant | 0.4 |

### No. 4

| | |
|---|---|
| "Geon" Resin 101 | 65.5 |
| Dioctyl Phthalate | 19.5 |
| Dioctyl Adipate | 11.0 |
| Pigment-Type Stabilizer | 3.4 |
| Oil-Soluble Stabilizer | 0.6 |

### Fire-Retardant Transparent Film
#### (4 to 8 mil)

| | |
|---|---|
| Polyvinyl Chloride | 100 |
| Dioctyl Phthalate | 29 |
| Tricresyl Phosphate | 6 |

| | |
|---|---|
| "MPS–500" | 17 |
| Stabilizer | 7½ |

### Stabilizers for Transparent Films

#### Formula No. 1

| | |
|---|---|
| "DS–207" | 0.5 |
| Salol | 1.0 |
| "Plumbosil" B | 3.0 |

#### No. 2

| | |
|---|---|
| "DS–207" | 0.5 |
| Dibutyl Tin Maleate | 2.0 |
| "Plumbosil" B | 2.0 |

#### No. 3

| | |
|---|---|
| "Paraplex" G–60 | 5.0 |
| "Plumbosil" B | 2.0 |
| "DS–207" | 0.5 |

Formula No. 3 is particularly recommended for the preceding fire-retardant composition. "Paraplex" G–60 is a combination plasticizer-stabilizer. For this reason, when it is used, the dioctyl phthalate can be reduced by 5 parts. This allows stabilization at a very slight cost.

CHAPTER XX

# SOAPS AND DETERGENTS

## Industrial Cleaning Sanitizers

|  | Low pH | Med. pH | High pH |
|---|---|---|---|
| "CBD"-70 | 28.5 | 28.5 | 28.5 |
|  | or | or |  |
| "CDB"-85 | 22.5 | 22.5 | — |
| Sodium Sulfate | 11.5–17.5 | 11.5–17.5 | — |
| Sodium Tripolyphosphate | 30.0 | 30.0 | 42.5 |
| Trisodium Phosphate | — | — | 29.0 |
| Disodium Phosphate | — | 30.0 | — |
| Monosodium Phosphate | 30.0 | — | -- |
| Available Chlorine | 20.0 | 20.0 | 20.0 |

---

### Machine Dishwashing Compound

| "CDB"–70 | 1.5 |
|---|---|
| Tetrasodium Pyrophosphate | 48.0 |
| Sodium Metasilicate Anhydrous or GD Silicate | 11.0 |
| Soda Ash | 39.5 |
| Available Chlorine | 1.0 |

### Quaternary Sanitizer

| "Tergitol" Nonionic NPX | 10 |
|---|---|
| Soda Ash | 30 |
| Quaternary Germicide | 10 |
| Trisodium Phosphate and/or Tetrasodium Pyrosphosphate | 50 |

Nonionic NPX and quarternary compound are sprayed into a mixer containing the salts and thoroughly mixed. Normal use concentration is a 2% water solution.

308

Household Germicidal Cleaners

| | Commercial Laundry Bleach | Household Laundry Bleach | | Sanitizer-Bactericide | Scouring Powder |
|---|---|---|---|---|---|
| | | No. 1 | No. 2 | | |
| "CDB"-60 | — | 8.3–11.6 or | — | 23.3 | — or |
| "CDB"-70 | — | 7.1–10.0 | 5.0 | — | 0.50 |
| "CDB"-85 | 16.7 | — | — | — | 0.40 |
| Sodium Sulfate | 53.2–68.2 | To make 100 | 76.9 | 76.7 | — |
| Sodium Tripolyphosphate | 15–30 | 10–40 | 10.0 | — | 5.0 |
| Alkyl Aryl Sulfonate | — | 6.7 | 4.0 | — | — |
| Alkyl Sulfate | — | — | — | — | 2.0 |
| G Silicate | — | — | 4.0 | — | — |
| Abrasive | — | — | — | — | To make 100 |
| Optical Bleach | 0.1 | 0.1 | 0.1 | — | — |
| Available Chlorine | 15.0 | 5–7 | 3.5 | 14.0 | 0.35 |

Detergent Sanitizer

| | |
|---|---|
| Alkyldimethylbenzyl | |
| Ammonium Chloride | 5 |
| Tetrasodium EDTA | 3 |
| Synthetic Detergent | 92 |

Use 1 oz. to 1 to 2 gal. water.

Low Temperature
Wash & Bleach
*British Patent 802,035*

| | |
|---|---|
| Soap | 10.0 |
| Tetrasodium | |
| Pyrophosphate | 2.0 |
| Sodium Perborate | 0.7 |
| Malonitrile | 0.1 |

This can be used at 50 to 80°C.

Dishwashing Machine Detergent

| | |
|---|---|
| Tri-2-Ethylhexyl | |
| Phosphate | 0.75 |
| "Tergitol" Nonionic | |
| NPX or XD | 0.75 |
| Tetrasodium | |
| Pyrophosphate | 15.00 |
| Sodium | |
| Tripolyphosphate | 20.00 |
| Soda Ash | 20.00 |
| Sodium Metasilicate | 30.00 |
| Sodium Silicate | 11.00 |
| "Silene" EF | 2.50 |

Liquid Dishwashing Compound

| | |
|---|---|
| Alkyl Aryl Sulfonate | |
| Slurry | 20-25 |
| Alkyl Aryl Polyether | |
| Alcohol | 8-12 |
| Lauric Diethanolamide | 3-8 |
| Propyleneglycol | 2-3 |
| Perfume and | |
| Coloring | To suit |

Liquid Detergent
( Heavy Duty )

| | Formula No. 1 | No. 2 |
|---|---|---|
| Alkyl Aryl Sulfonate | — | 40.0 |
| Nonyl Or Octyl Phenol Ethylene Oxide (9 to 12 moles) | 15-20 | — |
| Lauric Diethanolamide (65% amide) | — | 4.0 |
| Propyleneglycol | 20-30 | — |
| Sodium Xylene Sulfonate | — | 5.0 |
| Tetra Potassium Pyrophosphate | 18-20 | 20.0 |
| Sodium Metasilicate | — | 7.5 |
| Optical Bleach | — | 0.1 |
| Water, Perfume and Color | | To make 100 |

## No. 3

| | |
|---|---|
| Alkyl Aryl Sulfonate | 16.00 |
| Aliphatic Ethylene Oxide Ether, 85% Active (Non-ionic) | 1.00 |
| Lauric Diethanolamide (65% amide) | 4.00 |
| Propyleneglycol | 1.00 |
| Sodium Toluene Sulfonate | 7.50 |
| Tetra Potassium Pyrophosphate | 18.00 |
| Sodium Metasilicate | 5.00 |
| Carboxy Methyl Cellulose (medium vis.) | 0.50 |
| Optical Brightener | 0.03 |
| Water | 47.00 |
| Perfume and Dye | To suit |

The CMC is first wetted out by dispersing it in the propylene glycol. This mix is slowly added to the water at about 120°F. under efficient agitation. The phosphate and toluene sulfonate are then added, followed by the alkyl aryl sulfonate—the amide and the non-ionic. The silicate is then added followed by the brightener, dye and perfume.

## High Foaming Liquid Detergent

| | |
|---|---|
| Ammonium Salt of Sulfated "Surfonic N-40" | 50 |
| Diethanolamide of Lauric Acid | 10 |
| Alcohol | 25 |
| Perfume | 1 |
| Water | 14 |

## Low-Foaming Laundry Detergent

| Formula | No. 1 | No. 2 |
|---|---|---|
| "Surfonic N-95" | 10.0 | 15 |
| Sodium Dodecylbenzene Sulfate | 5.0 | 10 |
| Sodium Triphosphate | 35.0 | 40 |
| Tetrasodium Pyrophosphate | 30.0 | — |
| Trisodium Phosphate | 17.5 | — |
| Carboxymethylcellulose | 2.5 | 2 |
| Sodium Carbonate | — | 33 |

## Salt Water Laundry Detergent

| | |
|---|---|
| "Renex" 31 | 13 |
| Borax (−10 mol.) | 61 |
| Sodium Sulfate, Anhydrous | 25 |
| CMC, Low viscosity | 1 |

## Automatic Washer Detergent

| | |
|---|---|
| "Tergitol" Nonionic NPX | 12 |
| Sodium Tripolyphosphate | 20 |
| Sodium Sulphate | 20 |
| Modified Soda | 25 |

| | |
|---|---|
| Sodium Silicate | 16 |
| "Cellosize" Hydroxyethyl-Cellulose WP-3 | 2 |
| "Silene" EF | 2 |
| Water (Plus dye if desired) | 3 |

## Liquid Dishpan Detergent

| | Formula No. 1 | No. 2 |
|---|---|---|
| "Tergitol" Nonionic NPX | 5 | 3 |
| Triethanolamine Salt of Dodecylbenzene Sulfonic Acid | 45 | 30 |
| Lauric Acid Amide of Diethanolamine | 10 | 7 |
| Ethanol | 15 | 10 |
| Water and Perfume | 25 | 50 |

## Dry Cleaning Concentrates

For efficient action a dry cleaning bath must contain sufficient water to remove "sweet" spots without affecting the sizing or body of the fabric. The use of synthetic detergents, or a mixture of them, in dry cleaning concentrates is a convenient way of carrying water into the cleaning bath. More water can be brought into the cleaning bath this way than by using soap and a solvent coupling agent.

The following is one suggested formulation:

| | |
|---|---|
| "Tergitol" Anionic 7 | 44.0 |
| "Tergitol" Nonionic NPX | 22.0 |
| Oleic Acid | 3.5 |
| Monoethanolamine | 1.8 |
| Sodium Nitrite | 0.8 |
| Naphtha | 27.9 |

Mix anionic 7 and nonionic NPX. Add the sodium nitrite, amine, and oleic acid and stir until all of the nitrite is dissolved. Stir in the naphtha. If the mixture is not clear, add a small amount of oleic acid to produce clarity. This formulation contains about 30% water but will not couple any added water into the naphtha cleaning bath.

## Hosiery Scour

| | |
|---|---|
| "Ethofat" 242/25 | 0.7 |
| TSPP | 1.5 |
| Sodium Metasilicate | 0.5 |

## Rug Shampoo

### Formula No. 1

| | |
|---|---|
| Oleic Acid | 28.0 |
| Coconut Oil Fatty Acids | 21.0 |
| Isopropanol (99%) | 30.0 |
| Triethanolamine | 14.0 |
| Monoethanolamine | 6.8 |
| "Tergitol" Nonionic NPX | 5.0 |
| Water | 15.0 |

Mix the oleic acid, coconut oil fatty acids, and isopropanol. Add the amines and NPX and stir until thoroughly mixed. Add the water, which will produce a clear liquid. This formula is based on a combining weight of 210 for coconut oil fatty acids and 282 for oleic acid. The proportions should be changed according to the combining weight of the particular fatty acids used. The triethanolamine may be replaced by using an additional 6.3 lb. of monoethanolamine to increase the detergent properties of the shampoo.

Dry shampoos based on NPX have excellent detergent properties and are easy to use. A very active dry powder cleaning agent can be made by dispersing a perchlorethylene-naphtha solution of the following concentrate on an inert carrier such as wood flour:

### No. 2

| "Tergitol" Nonionic | |
|---|---|
| NPX | 62.5 |
| Monoethanolamine | 2.2 |
| Oleic Acid | 10.3 |
| Perchlorethylene | 25.0 |

### Aerosol Rug and Upholstery Cleaner

| | | |
|---|---|---|
| a | Coconut Fatty Acids | 14.2 |
| | Isopropyl Alcohol | 4.8 |
| | Triethanolamine | 9.5 |
| | Perfume | To suit |
| b | "Veegum" | 4.8 |
| | Water (Deionized) | 61.7 |
| c | Dichlorodifluoro-methane | 5.0 |

Melt the fatty acids, add alcohol, triethanolamine and perfume, mix. Add "Veegum" slowly to water, mixing until smooth. Add a slowly to b with slow agitation until smooth. Add c. Pressure fill the propellent.

---

### Electric Lamp-Reflector Cleaner

| | |
|---|---|
| Glacial Acetic Acid | 1 |
| Methyl Alcohol | 2 |
| Oxalic Acid | 1 |
| Lactic Acid (80%) | 1 |
| Butyl Acetate | ½ |
| Amyl Acetate | ½ |
| Glycerin | 8 |

The ingredients are mixed until the oxalic acid crystals dissolve. The glycerin is then added followed by butyl alcohol until the solution becomes clear. The solution is applied at room temperature to reflectors and tubes with a soft brush then rinsed with a clean sponge.

### Safety Glass Cleaner
*U. S. Patent 2,878,188*

| | |
|---|---|
| Sodium Dihydrogen Phosphate | 86.0 |
| Di-Sodium Hydrogen Phosphate | 3.5 |
| Benzyl Alcohol | 2.5 |
| Sodium EDTA | 1.0 |
| Sodium Dodecyl-benzene Sulfonate | 6.0 |
| "Triton" X-100 | 1.0 |

### Eyeglass Cleaner
*U. S. Patent 2,870,029*

| | |
|---|---|
| Paraffin Wax | 81 |
| Ferric Oxide (Rouge) | 3½ |
| Corn Starch | 15½ |

Melt the wax and mix in the other items; cool slowly to uniform mix.

### Glass Cleaner

| | |
|---|---|
| Isopropanol (99%) | 35.00 |
| "Carbitol" Solvent | 7.50 |
| "Tergitol" Anionic 7 | 0.25 |
| Water | 57.25 |

Mix the isopropanol, "Carbitol" solvent, and water, and then add anionic 7.

### Aerosol Glass Cleaner

| | |
|---|---|
| Isopropanol (99%) | 25.0 |
| "Carbitol" Solvent | 7.5 |
| "Tergitol" Anionic 7 | 0.3 |
| Water | 64.0 |
| "Tween" 80 | 0.9 |
| "Arlacel" 80 | 0.6 |
| Dichlorodifluoro-methane | 1.7 |

### Laboratory Glass Cleaner

| | |
|---|---|
| Water | 7 gal. |
| "Texapon" P | 1 teaspoon |

### Grease-Cutting Compound

A 1% solution of a 50-50 mixture of nonionic NPX and tetrasodium pyrophosphate is effective for cutting grease found in kitchen flues, around meatpacking plants, and in commercial garages.

### Removing Silicone Grease Lubricants

Triethylamine or other organic amines readily dissolve silicone greases even at room temperature. Hence, they can be used to remove silicone greases that are employed as lubricants for stopcocks and ground glass surfaces.

### Rust Removal from Cement Patios

Dissolve one part sodium citrate in six parts of water, add six parts glycerin. Mix with powdered whiting to form a heavy paste. Spread a thick coating of the mixture on the stained area and allow to dry. Repeat treatment if necessary. Badly stained areas may require a week or more for complete removal. This mixture is also effective in removing stains on siding caused by rusty nails.

### Drain Cleaner
*U. S. Patent 2,816,012*

| | |
|---|---|
| Caustic Soda | 54.50 |
| Sodium Nitrate | 30.45 |
| Salt | 10.80 |
| Aluminum Alloy (containing 1-1.5% copper) | 4.25 |

### Hard Surface Cleaner

| | |
|---|---|
| "Tergitol" Nonionic NPX | 15 |
| Tetrasodium Pyrophosphate | 10 |
| Sodium Metasilicate | 5 |
| Soda Ash | 40 |
| Sodium Sulfate | 30 |

A more dusty powder can be made by replacing part of the sodium sulfate with "puffed" borax.

### Paint and Woodwork Cleaner

| | |
|---|---|
| "Tergitol" Nonionic NPX | 5 |
| Tetrasodium Pyrophosphate | 1 |
| Water | 94 |

Add nonionic NPX and tetrasodium pyrophosphate to the water and stir until a clear solution is obtained. The tetrasodium pyrophosphate content can be increased if desired. The formulation can be diluted with 1 or 2 volumes of water and still be an effective cleaner for painted walls and woodwork.

### Hardwood Floor Cleaner

| | |
|---|---|
| "Tergitol" Anionic 7 | 3.0 |
| "Tergitol" Nonionic NPX | 1.5 |
| Cleaners' Naphtha | 95.5 |

Mix anionic 7 and nonionic NPX until a clear solution is obtained. Add all or part of the naphtha as desired. Naphtha cleaners should never be used on asphalt tiles.

### Diode Assembly Cleaner
*British Patent 791,887*

| | |
|---|---|
| Sulfuric Acid | 20 cc. |
| Ammonium Sulfate | 125 g. |
| Water | 80 cc. |

Use at 95°C. for 10 minutes.

### Metal Parts Cleaner

| | |
|---|---|
| Pine Oil | 62.0 |
| Oleic Acid | 10.8 |

| Triethanolamine | 7.2 |
|---|---|
| "Dowanol" EB | 20.0 |

Dilute with an equal volume of a 50:50 mixture of naphtha and white spirit. Other formulations may be diluted with water.

---

### Aerosol Oven Cleaner

| Scale Wax | 2.0 |
|---|---|
| Ethylcellulose | 2.0 |
| Methylene Chloride | 39.0 |
| Methanol | 9.1 |
| "Cellosolve" Acetate | 9.6 |
| Trichlorethylene | 13.3 |
| "Freon" 12 | 25.0 |

---

### Auto Radiator Cleaner
### U. S. Patent 2,802,788

| Disodium EDTA | 18 |
|---|---|
| Sodium Tripolyphosphate | 59 |
| Sodium Bisulfite | 20 |
| Polyglycol 1600 Monooleate | 3 |

Use 2 to 4% of this mixture in water.

---

### Decontaminating Air or Walls from Poisonous Metals
### U. S. Patent 2,774,736

| EDTA | 10 |
|---|---|
| Sodium Metabisulfite | 90 |

Dissolve in water and spray on walls or pass contaminated air through a vessel containing this solution.

---

### Engine Degreasers

### Formula No. 1

| Petroleum Naphtha (290-350°F.) | 75.00 |
|---|---|
| Perchloroethylene | 24.98 |
| Inhibitor | 0.02 |

### No. 2

| Medium Solvency Aromatic Naphtha (200-300°F.) | 96 |
|---|---|
| Orthodichlorobenzene | 4 |

The problem of degreasing either heavy surface films of oil and grease or varnishlike deposits calls for specially compounded cleaners. Unlike the other types of degreasers, these cleaners are diluted prior to use with petroleum naphtha or kerosene and their application is followed by a high pressure water spray. This flush - off technique is prompted by the need to avoid direct skin contact with the compound and in consideration of the most economical means of rinsing away large amounts of grease and oil. Ideally, the degreaser is applied by brush or spray, allowed to remain in contact with the soiled surface for

15 to 30 minutes, following which the dirt and cleaner are flushed off the surface with water.

### No. 3
*U. S. Patent 2,107,287*

| | |
|---|---|
| Tall Oil Soap | 75.44 |
| Kerosene | 17.5-33 |
| Sodium Cresylate | 12.5-23 |

### No. 4

| | |
|---|---|
| Crude Pine Fatty Acid Soap | 39 |
| Alkali Cresylate | 12 |
| Kerosene | 19 |
| Amyl Alcohol | 12 |
| Water | 18 |

---

### Engine Cylinder (Carbon) Cleaner

### Formula No. 1
*U. S. Patent 2,904,458*

| | |
|---|---|
| Xylol | 32-40 |
| Glycol Monomethyl Ether | 30-32 |
| Trichlorobenzene | 20-24 |
| Butylamine | 10-12 |

### No. 2
*French Patent 1,113,500*

Add 0.3 g. per liter of triisopropylamine to gasoline. Its continual use prevents deposits of lead and carbon.

### No. 3
*Australian Patent 207,895*

| | |
|---|---|
| Gasoline | 88-97 |
| Amyl Nitrate | 1.5-6.5 |
| Lubricating Oil (S.A.E. 10-20) | 1-10 |

### No. 4

| | |
|---|---|
| Methylene Chloride | 6.4 gal. |
| "Dowanol" P-Mix | 3.2 gal. |
| Water | 0.4 gal. |
| Potassium Oleate | 2.4 lb. |
| Paraffin (M.P. 47-50.6°C.) | 0.8 lb. |

---

### Amber Cleaner and Brightener
*U.S.S.R. Patent 110,980*

| | |
|---|---|
| Toluene | 60 |
| Ethyl Acetate | 20 |
| Methanol | 20 |

Steep amber in above; drain and polish mechanically.

---

### Solvent for Boric Acid Flux Residue
*U.S.S.R. Patent 117,331*

An aqueous solution of potassium bisulfate is used.

---

### Removing Insoluble Calcium Salt Deposits from Fossils

| | |
|---|---|
| EDTA | 3 |
| Sod. Tripolyphosphate | 3 |
| "Igepol" C | 1 |
| Water | 93 |

## Cleaning Old Paintings

First, turpentine is swabbed onto the picture to protect the paint. Then, with a different swab, "Tergitol" Anionic 4 is applied in a rolling motion. Finally, every last trace of "Tergitol" Anionic 4 is removed by wiping with swabs soaked in naphtha. Swabs are made by rolling cotton on the ends of applicators. The entire operation is performed with great care. Only small areas are cleaned at a time. "Tergitol" Anionic 4 may be used full strength or may be diluted to as little as a 1 or 2% solution. The concentration that works best can be determined by experimenting along the narrow edge of the picture that will be concealed by the frame.

---

## Newspaper De–inker

| | |
|---|---|
| B.W. Sodium Silicate | 5.0 |
| Caustic Soda | |
| (76%–Sodium Oxide) | 1.5 |
| Hydrogen Peroxide | |
| (130 volumes) | 2.0 |
| "Ethofat" 242/25 or | |
| 60/20 | 1.0 |

---

## Liquid White Wall Tire Cleaner

| | |
|---|---|
| "Emcol P10–49" | 1 |
| "Emcol 5130" | 7 |

| | |
|---|---|
| Trisodium Phosphate | 4 |
| Water | 88 |

---

## Syndet Soap Bar

| | |
|---|---|
| Alkyl Benzene Sulfonate | |
| (65 to 85% Active) | 41.0 |
| Sodium Lauryl Sulfate | 15.0 |
| Tallow–Coco Soap | |
| (80%:20% ratio) | 30.0 |
| Calcium Stearate | 6.0 |
| Titanium Dioxide | 1.5 |
| "Polyox" WSR–35 or | |
| 205 | 0.2 |
| Water, Perfume | To suit |

---

## Metallic Soap

### (Dry Process)
*U. S. Patent 2,890,232*

| | | |
|---|---|---|
| *a* | Stearic Acid | 100.0 |
| | Calcium Oxide | 15.0 |
| *b* | Water | 4.7 |

Heat *a* to 93°C. and add *b*.
Mix well until reaction ends.

---

## Chimney Soot Cleaner
*French Patent 1,001,058*

| | |
|---|---|
| Sodium Nitrate | 23 |
| Sulfur | 25 |
| Salt | 50 |
| Zinc Dust | 2 |

Ignite in Chimney.

---

## Lipstick Stain Remover

| | |
|---|---|
| "Tergitol" Nonionic | |
| NPX | 1.0 lb. |

"Carbitol" Solvent
(low gravity) 2.0 gal.
Cleaners' Naphtha 8.0 gal.
Isopropanol (99%) 1.5 gal.

Avoid open flames when mixing or using this formula. Dissolve nonionic NPX in the low-gravity "Carbitol" solvent, add the naphtha and isopropanol, and stir until completely mixed. From ¼ to ½ of the naphtha may be replaced with carbon tetrachloride to give a faster drying solution and to reduce the fire hazard.

Stains are removed by placing the soiled fabric on a folded, clean cloth (stain against cloth). Daub the stain with another cloth, a portion of which is saturated with the solution. Continue daubing (not rubbing) with the solution until the stain is removed, moving the fabric at intervals so that it does not remain on the stained portion of the folded cloth. The cleaned area can be rinsed with straight naphtha or carbon tetrachloride, but it is not necessary.

Skin Nicotine Stainer Remover
*British Patent 745,177*

| | |
|---|---|
| Gum Arabic | 1 |
| Kaolin | 7 |
| Detergent | 3 |

| | |
|---|---|
| Calcium Hydroxide | 3 |
| Calcium Thioglycollate | 3 |
| Pumice Powder | 33 |
| Glycerin | 15 |
| Water | 35 |

Waterless Hand Cleaner

Formula No. 1

| | |
|---|---|
| Kerosene (Deodorized) | 35.8 |
| "Neo-Fat" 94-04 | 5.0 |
| "Ethomid" HT/15 | 3.0 |
| "Ethofat" O/15 | 5.0 |
| Caustic Soda (50%) | 1.3 |
| Water | 49.9 |

Add the "Neo-Fat" and "Etho"-chemicals to the kerosene and heat to 50°C. The required amount of water and caustic soda is heated to 50°C. and added to the kerosene solution with constant stirring. The mixture will set to a stiff gel.

No. 2

| | |
|---|---|
| Methylcellulose 4000 cp. | 15.0 |
| Alcohol | 5.0 |
| Lanolin | 0.5 |
| Glycerin | 6.0 |
| "Methylparaben" | 0.2 |
| Perfume | 0.4 |
| Water | To make 100.0 |

No. 3

| | |
|---|---|
| Water | 50.0 |
| "Carbopol" | .5 |

| NaOH (10%) | .1 |
| "Ethomeen" C-25 | .5 |
| Mineral Oil | 33.3 |
| Isopropyl Palmitate | 16.7 |

No. 4

| a Stearic Acid | 3.0 |
| Oleic Acid | 4.0 |
| "Renex" 690 or | |
| "Renex" 30 | 10.0 |

| Deodorized Kerosine | |
| or Mineral Spirits | 51.0 |
| b Sodium Hydroxide | 0.5 |
| Water | 31.5 |

Weigh part *a*. Heat to melt stearic acid, then cool to room temperature. Weigh and mix part *b*. Add *b* to *a*, mixing well until a gel forms.

## Industrial Hand Cleaner

| | Formula No. 1 | No. 2 | No. 3 | No. 4 |
| --- | --- | --- | --- | --- |
| Kerosene | 34.5 | 35.1 | 37.0 | — |
| White Oil | — | — | — | 34.5 |
| Oleic Acid | 10.7 | 10.8 | 11.4 | 10.7 |
| Triethanolamine | 5.3 | 2.7 | 2.8 | 2.6 |
| Monoethanolamine | — | 1.1 | 1.2 | 1.1 |
| "Tergitol" Anonionic 4 | 2.6 | 2.6 | — | — |
| "Tergitol" Nonionic NPX | — | — | 4.5 | 4.3 |
| Propylene Glycol | 4.3 | 4.3 | 4.5 | 4.3 |
| Powdered Sodium EDTA | — | — | 1.9 | — |
| Water | 42.6 | 43.4 | 36.7 | 42.5 |

Mix the kerosene or oil and oleic acid. Dissolve the amines in the water and stir in "Tergitol" surface active agent and propylene glycol. In formula 3, add "Versene" and stir until it is dissolved.

Add Formula No. 1 to No. 2 or No. 2 to No. 1, depending on conveniece, and stir until a smooth cream is obtained. No heating is required.

Lanolin may be added to the kerosene to reduce the defatting action of the kerosene on the skin and the surfactant content can be increased to make the cleaner easier to remove with water.

Creams 2, 3 and 4 are better detergents than cream 1 because of the presence of monoethanolamine as well as triethanolamine soap. When first made, cream 1 is slightly less viscous than the other creams but it becomes about as viscous as the others after standing several hours. This should be an advantage in packaging cream No. 1.

## Preventing Discoloration in Laundry Operations

"Tetrine"* chelates calcium and magnesium more effectively than the complex phosphates and, in addition, chelates iron, copper, nickel, and other heavy metals when present. Thus, metallic soaps cannot deposit on the goods when sufficient "Tetrine" is used.

In the soap tank, depending on conditions, 3 to 20 lb. of sodium "Tetrine" liquid conc. are used per 100 lb. of dry soap.

In the first rinse, at 3-gr. hardness per gallon of water, 2 oz. sodium "Tetrine" liquid conc. are used per 100 lb. of clothes washed. The quantity should be reduced if suds from regenerated soap overflow the machine.

In the sour, a 10% solution of sodium "Tetrine" liquid conc., acidulated with laundry sour to a $pH$ of 4.5 to 3.5, will remove heavy iron stains without degradation of coton fibres. Application of heat hastens the reaction. The goods should be thoroughly rinsed after treatment.

## Chelation of Calcium and Magnesium

Sodium "Tetrine", soap, and alkali, in combination, prevent the precipitation of calcium and magnesium soaps to a much higher degree than would seem possible from consideration of the equivalent quantities involved. In many instances, the precipitation of calcium and magnesium soaps is prevented by the presence of as little as one tenth of the quantity of sodium "Tetrine" that would appear to be necessary from stoichiometric calculation. The calcium and magnesium chelates formed by "Tetrine" are much more stable than the compounds formed when these same metallic elements combine chemically with the complex phosphates. For this reason, the "Tetrines" are much more effective at extreme dilutions than are the complex phosphates.

Sodium "Tetrine" is stable under all conditions of temperature, concentration, alkali, and

* Sodium ethylenediamine tetraacetate.

acid that may be encountered in laundering practice. Thus, when sodium "Tetrine" is added to a stock soap tank, it remains fully effective, even after several days at or just under the boil, whereas complex phosphates start to retrograde immediately after being put into solution and continue to revert to the simple phosphates from which they are made until all of their sequestering properties are destroyed.

At elevated temperatures, complex phosphates are completely broken down to simple phosphates in a matter of hours. For these reasons, it has been found in laundering practice that sodium "Tetrine", dollar for dollar, will produce much better results, as far as prevention of precipitation of calcium and magnesium are concerned, than do the complex phosphates.

Chelation of Iron, Nickel, Copper, and Other Heavy Metals

Sodium "Tetrine" chelates all heavy metals normaly encountered in laundering. It also chelates iron, nickel, copper, and other heavy metals, when present, preferentialy to calcium and magnesium. This is especially important for preventing the formation of iron, copper, and nickel soaps. Copper and nickel compounds are often present in minute quantities on the surface of "Monel" metal wash wheels due to the action of sodium hypochlorite and the sour on the metal.

Iron, copper, and nickel soaps, when permitted to form, gradually build up a grayish deposit of metallic soap on the goods. Sodium "Tetrine" prevents this build-up. Thus, goods washed continuously with the aid of sodium "Tetrine" in the soap remain sparkling white and bright as compared with clothes washed in the conventional manner.

Removal of Iron Stains without Degradation of Cotton

Sodium "Tetrine", used with regular laundry sour, will remove heavy iron stains without fabric degradation. This does not mean that a small amount of sodium "Tetrine" added to the sour can be expected to remove very heavy iron stains. It does mean that these stains can be removed by working with a fairly concentrated bath of sodium "Tetrine" and sours and that the removal of the iron stain is accomplished without degeneration of the fabric, provided, of course, that the sour is thoroughly rinsed from the goods after treatment.

---

### Softening the Water in the Soap Tank

Many laundries, operating with water containing a moderate amount of hardwater ingredients, for example, 3 gr. per gal., and having no water-softening system, have found it advantageous to add complex phosphates to the soap tank for combating hardness. Some laundries, operating with the aid of water-softening systems, recognizing that these systems do not remove the last traces of hardness and little, if any, of the iron and that the soil in clothes often contains hard-water ingredients, have added smaller quantities of the complex phosphates to stock soap tanks to take care of the smaller quantities of hardness present in the wash wheels. Such laundries find that sodium "Tetrine," used in place of complex phosphates, produces substantially better results.

Conditions vary so greatly from plant to plant that a specific recommendation is almost impossible to make under an unknown set of conditions. However, the working range for 3-gr. water is 3 to 20 lb. of sodium "Tetrine" per 100 lb. of soap.

---

### The First Rinse

Addition of sodium "Tetrine" to the first rinse has beneficial results. The regenerated soap produces substantial suds, giving extra sudsing to the goods at no sacrifice of time. Calcium and magnesium and, more important, iron, nickel, and copper compounds present in the bleach bath are chelated, thus preventing the deposition of metallic soaps on the goods.

When 3-gr. hardness per gallon water is encountered, sufficient sodium "Tetrine" should be added to the first rinse to assure the addition of 2 oz. of sodium "Tetrine" per 100 lb. of clothes washed. In some cases, this quantity must be reduced to prevent the development of excessive suds. As soap should always be balanced by alkali in laundering, we recommend the addition of 2 oz. of stock solution of silicate per 2 oz. of sodium "Tetrine" liquid to provide the proper alkali balance.

## The Sour

When used in the soap tank and the first rinse, sodium "Tetrise" will chelate and hold in solution the soluble iron compounds present in the water, thus preventing the precipitation of iron soap on the goods. "Tetrine" dissolves iron stains slowly on the alkaline side. Laundry operations are so rapid that the goods are not in contact with the alkaline solution long enough to permit the very small amount of "Tetrine" in the solution to completely solubilize heavy iron stains. Continuous use of "Tetrine" in the soap and first-rinse baths will gradually remove the iron soap deposited on the goods and will gradually brighten the fabric that has been dulled with iron.

When heavy iron stains must be removed, "Tetrine" in an acid medium is used as it works much more rapidly on the acid side as a solubilizing agent for heavy metals. Initially 1 part of sodium "Tetrine" can be used per 10 parts of water and then acidulated with any sour available to a $pH$ of 4.5 to 3.5. This solution will rapidly solubilize heavy iron stains and will have no deleterious affects on the fibre, if the acid is thoroughly rinsed from the goods after use. Application of heat speeds up removal of the stain and does not have a bad effect on the goods.

---

## Hand-Dishwashing Compound

### Formula No. 1

| | |
|---|---|
| Soda Ash | 40 |
| Sodium Bicarbonate | 20 |
| Tripolyphosphate | 15 |
| "Oronite" D–40 | 25 |

### No. 2
### (Light Duty)

| | |
|---|---|
| "Kreelon" 8D or 8G | 10-20 |
| Modified Soda | 70-85 |
| Sodium Metasilicate Pentahydrate | 4-8 |
| "Carbose" D (65% Sodium Carboxy Methyl Cellulose) | 1-2 |

### No. 3
### (Heavy Duty)

| | |
|---|---|
| "Kreelon" 8D or 8G | 20-40 |
| Sodium Polyphosphate | 15-25 |
| Modified Soda | 33-49 |
| Sodium Metasilicate Pentahydrate | 5-10 |
| "Carbose" D (65% Sodium Carboxy Methyl Cellulose) | 1-2 |

No. 4

| "Kreelon" CD | 25 |
|---|---|
| Modified Soda | 65 |
| Sodium Tripolyphosphate | 10 |

No. 5
(Light Duty)

| "Kreelon" CD | 20-30 |
|---|---|
| Modified Soda | 70-80 |

No. 6
(Heavy Duty)

| "Kreelon" CD | 45-55 |
|---|---|
| Sodium Polyphosphate | 15-25 |
| Modified Soda | 20-30 |

No. 7

| Sodium Dodecyl-benzene Sulfonate | 30 |
|---|---|
| "Ninol" AA62 | 15 |
| Water | To make 55 |

Germicidal Dishwasher

| "Nonic" 218 | 10 |
|---|---|
| "Hyamine" | 10 |
| Trisodium Phosphate | 3 |
| Water | 77 |

This works best at a pH of about 9.5.

Machine-Dishwashing
Compound
Formula No. 1
(For Hard Water)

| Anhydrous Sodium Tripolyphosphate | 55 |
|---|---|
| Anhydrous Sodium Metasilicate | 30 |
| Dense Soda Ash | 15 |

No. 2
(For Soft Water)

| Anhydrous Sodium Tripolyphosphate | 40 |
|---|---|
| Anhydrous Sodium Metasilicate | 50 |
| Dense Soda Ash | 10 |

No. 3
(Light Duty)

| "Kreelon" 4D | 15-25 |
|---|---|
| Modified Soda | 73-84 |
| "Carbose" | 1-2 |

No. 4
(Heavy Duty)

| "Kreelon" 4D | 45-55 |
|---|---|
| Sodium Polyphosphate | 15-25 |
| Modified Soda | 17-39 |
| "Carbose" | 1-3 |

No. 5

| Anhydrous Sodium Metasilicate | 40 |
|---|---|
| Anhydrous Tetrasodium Pyrophosphate | 40 |
| Anhydrous Sodium Carbonate | 18 |
| "Sterox" C.D. | 2 |

Detergent for Washing
Machines
(Controlled Suds)

The new automatic washers

perform better if there is not too much suds present. Excessive suds cushion the clothes as they are dropped back on top of the water and thus interfere with the mechanical action of the machine and also make rinsing more difficult.

| "Ethofat" 242/25 | 15-20 |
| Tetrasodium Pyro-phosphate | 35-45 |
| Anhydrous Sodium Sulfate or Sodium Bicarbonate | 40-45 |
| Low-Viscosity Car-boxy Methyl Cel-lulose | 2 |

This can be formulated in a dry mixer by adding all of the ingredients except the "Ethofat," starting the mixer, and then adding the "Ethofat" slowly. Some prefer to spray in the "Ethofat" while the mixer is in operation. Usually 0.25% of the product, on the basis of the water weight, is sufficient for home washing. Some formulators prefer to add about 20% of water to this formulation.

Liquid Dishwashing Compound

| "Nacconol" Z Flakes | 35.00 |
| 2-Methyl 2,4-Pentan-ediol | 7.50 |
| Water | 57.50 |
| Perfume | 0.25-0.50 |

## Marble Cleaner

| Sodium Carbonate | 2 |
| Pumic Stone | 1 |
| Finely Powdered Chalk | 1 |

Sift the mixture through a fine sieve. Mix well with water. Rub it well over the marble to remove any stain. Wash well with salt and water.

## Building Granite Cleaner

| Phosphoric Acid | 1 |
| Water | 3 |

Brush this solution on the granite. When the stain is decolorized, wash off the acid quickly.

## General Household Cleaners

### Formula No. 1

| "Kreelon" CD | 35-45 |
| Sodium Tripolyphos-phate | 15-35 |
| Modified Soda | 30-40 |

### No. 2

| Light Soda Ash | 30.0 |
| Sodium Metasili-cate, $5H_2O$ | 60.0 |
| "Oronite" | 7.5 |
| Pine Oil | 2.5 |

## No. 3

| | |
|---|---|
| "Kreelon" 8D or 8G | 15-25 |
| Sodium Tripolyphosphate | 15-35 |
| Modified Soda | 61-78 |
| Sodium Metasilicate Pentahydrate | 5-10 |
| "Carbose" D (65% Sodium Carboxy Methyl Cellulose) | 2-4 |

## Building-Maintenance Cleaner

### Formula No. 1

| | |
|---|---|
| "Kreelon" CD | 20-30 |
| Modified Soda | 60-65 |
| Polyphosphates | 10-15 |

### No. 2

| | |
|---|---|
| "Kreelon" 8D or 8G | 10-15 |
| Modified Soda | 68-79 |
| Sodium Polyphosphates | 10-15 |
| "Carbose" D (65% Sodium Carboxy Methyl Cellulose) | 1-2 |

## Sanitizing Detergent
### U. S. Patent 2,519,747

| | |
|---|---|
| Sodium Carbonate | 45.0 |
| Tetrasodium Pyrophosphate | 45.0 |
| Nonaethylene Glycol Monosoyate | 6.3 |
| Oleyl Dimethyl Ethyl Ammonium Bromide | 3.0 |

## Synthetic-Detergent Bar

| | |
|---|---|
| Tallow Soap | 47 |
| Dodecyl Benzene Sulfonate | 20 |
| "Nekal" A | 20 |
| "Aerosol" AY | 10 |
| Sodium Carboxy Methyl Cellulose | 2 |
| Perfume | 1 |

## Soapless-Detergent Toilet Bar

| | |
|---|---|
| "Lathanol" LAL | 33 |
| "Carbowax" 4000 | 66 |
| Perfume Oil | 1 |

Melt the "Carbowax" in a steam-jacketed kettle equipped with an agitator. Add the powdered "Lathanol" gradually to the melted "Carbowax," with slow agitation. Then add the perfume oil and cast in molds of any desired shape and size.

## Floor-Wax Remover

| | |
|---|---|
| Vegetable-Oil Soap | 20 |
| Ammonia | 5 |
| Water | 75 |

Use 2 to 8 oz. per gallon of water.

## Floor-Oil Absorbent
### U. S. Patent 2,601,862

| | |
|---|---|
| Dry Precipitated Calcium Carbonate | 90 |

| Lime | 2 |
|---|---|
| Magnesia | 8 |

## Toilet-Bowl Cleaner

### Formula No. 1
| Hydrochloric Acid | |
|---|---|
| (18°Bé.) | 25.0 |
| Nitrobenzol | 0.7 |
| "Triton" X–30 | 0.2 |
| Green Dye | To suit |
| Water | 74.0 |

### No. 2
| Sodium Bisulfite | 93.9 |
|---|---|
| Sal Soda | 2.0 |
| "Kreelon" | 4.0 |
| Isobornyl Acetate | 0.1 |

## Removing Discolorations from Sinks and Other Ceramics

| 10% Hydrochloric | |
|---|---|
| Acid | 3 oz. |
| "Bon Ami" | ½ can |

Mix the two thoroughly. Apply the paste to the discoloration, allow to stand 15 minutes. Rinse the paste off thoroughly with water. The item will be free of the objectionable stain. Repeat if necessary.

## Noncorrosive Metal Cleaner
### U. S. Patent 2,599,729
| Stabilized Trichloro- | |
|---|---|
| ethylene | 100 cc. |
| Triethanolamine | 2 g. |

| Monoethanolamine | 6 g. |
|---|---|
| Sodium Oleate | 1 g. |
| Water | 11 cc. |
| Benzyl Thiocyanate | 3 g. |

## Metal-Cleaner Emulsion
| Pine Oil | 62.0 |
|---|---|
| Oleic Acid | 10.8 |
| Triethanolamine | 7.2 |
| "Cellosolve" | 20.0 |

Before use dilute with an equal volume of:
| Naphtha | 50.0 |
|---|---|
| White Spirit | 50.0 |

## Flat-Bed Printing-Press Cleaner
| "Metso" 99 | 20 lb. |
|---|---|
| Boiling Water | 50 gal. |

Spray on press.

## Metal Base-Form (Printers') Cleaner
| "Metso" 99 | 1 lb. |
|---|---|
| Water | 1 gal. |

Soak for 2 hours in near boiling solution of this, then rinse, dry, and oil lightly.

## Spark-Plug Cleaner
### U. S. Patent 2,627,148
| "Bentonite | 0.25-2 |
|---|---|
| Water | 2.80-3 |
| Powdered Feldspar | 40-65 |

Diethyleneglycol
Monobutyl Ether
To make 100

---

Removing Rust from Watch
Parts

Formula No. 1

a Potassium Cyanide*    15
  Water                 30
b Potassium Cyanide*    15
  Soft Soap             15
Whiting                 30
Water

To make a stiff paste

Mix to dissolve. Moisten the rusted parts with a and then rub with b.

No. 2

Olive Oil               50
Flowers of Sulfur        5
Tripoli              10-30

No. 3

Tartaric Acid           30
Zinc Chloride           30
Mercuric Chloride*       2
Water                   50

No. 4

Lactic Acid              1
Aspic Oil                2

No. 5

Attach the rusted parts to a piece of zinc and immerse in diluted sulfuric acid until the rust disappears.

---

*Very poisonous.

Oiled Metal Cleaner

Sodium Metasilicate    46.0
Caustic Soda           27.6
Trisodium Phosphate    18.4
"Dresinate" X           8.0

Use 8 oz. of this per gallon water.

---

Baking-Pan Cleaner
U. S. Patent 2,525,079

Sodium Metasilicate    45
Soda Ash               32
Borax                  23

This is useful for removing silicone-resin coatings.

---

Silverware Cleaner
U. S. Patent 2,513,187

Calcium Hydroxide      53
Sodium Bicarbonate     45
Sodium Petroleum
  Sulfonate             2

Mix these dry and place them in an aluminum-foil envelope 32 sq. in. in area. For use, immerse in 2 to 4 qt. of water and immerse the silverware.

---

Silver-Tarnish Dip
Cleaner

Thiourea              30 g.
37.5% Hydrochloric
  Acid                15 cc.
Water    To make 10 fl. oz.

To perfume this mix, use:

| | |
|---|---|
| Benzaldehyde | |
| (F.F.C.) | ½ cc. |
| "Triton" X–100 | ½ cc. |

### Aluminum Cleaner

| | |
|---|---|
| Sodium Silicate | 10 |
| "Ninol" 1281 | 3 |
| Sodium Xylene Sul- | |
| fonate | 4-5 |
| Water | To make 100 |

### Internal-Combustion Engine Cleaner
#### U. S. Patent 2,671,036

| | |
|---|---|
| Lubricating Oil | |
| (SAE 30) | 50-75 |
| Linseed-Oil Soap | ½-1 |
| o-Cresol | ½-1 |
| Water | 50 |

After draining the oil, put this mixture in the crankcase. Allow the engine to run slowly for 1 hour and then drain.

### Engine-Carbon Remover

Cresylic acid emulsions have been specified for use as engine-carbon removers. The usual type of emulsion is made with large percentages of soap. These emulsions are quite stable at room temperature, but tend to break down at 55°C. which is the temperature at which the carbon-remover emulsion works with the greatest efficiency. The usual soap-type emulsions also tend to develop a great deal of foam, which is uncontrollable, particularly when used in automatic spraying machines.

The following formula overcomes these objections:

| | |
|---|---|
| Cresylic Acid | 88 |
| Diglycol Oleate S | 10 |
| Stearic Acid | 2 |

This is actually a soluble oil which can be emulsified readily at 55°C to give a highly efficient carbon remover. The emulsion formed by this soluble oil is not too stable at room temperature, but at 55°C. it does form an unusually stable emulsion which meets the requirements. Since there is no free soap in this formula, the formation of sodium cresylate is eliminated and thus, additional free cresylic acid is available for the active cleaning job. In addition, the foaming properties of this formulation are such as to cause little difficulty.

### Brake-Lining Cleaner
#### U. S. Patent 2,500,055

| | |
|---|---|
| Hardwood Ash | 50 |
| Fuller's Earth | 45 |
| Rosin | 5 |

Red Iron Oxide
(Optional) ½

---

Electrical Insulator Cleaner
*German Patent 802,941*

Quartz Powder        6 kg.
Hydrochloric Acid
(19°Bé.)            330 cc.
Corrosion Inhibitor  160 g.
Oxalic Acid          1900 g.
"Nekal" (Wetting
Agent)              500 g.
Methyl Orange        24 g.
Add enough water and mix
well to a paste before use.

---

Clear Emulsion Cleaner
*U. S. Patent 2,576,419*

Trichlorethylene     100 cc.
Triethanolamine
Oleate              2 g.
Monoethanolamine
Oleate              6 g.
Sodium Oleate        1 g.
Water                11 cc.

---

Silicone-Resin Coating
Remover

Formula No. 1
*U. S. Patent 2,525,079*

Sodium Metasilicate  35-55
Soda Ash             22-42
Borax                18-28
Dissolve in water.
This is used to clean silicone-

coated baking pans without damage to the metal.

No. 2
*U. S. Patent 2,662,837*

Sodium Hydroxide     0.1
Alcohol              5.2
Tetrachloroethane    94.7
Immerse the pan in this solution heated to 75°C. for 15 minutes. Then rinse with water.

---

Finger-Print Remover
(For Furniture, Etc.)
*U. S. Patent 2,517,636*

Spindle Oil          55
Sodium Petroleum Sul-
fonate              30
Sulfated Fish Oil    5
Water                5
Diacetone Alcohol    5

---

Polystyrene Cleaner

Denatured Alcohol    5.0
Ethyl Acetate        2.5
Butyl Alcohol        1.5
"Cellosolve"         1.0
This solution may be used to prepare the surface of polystyrene moldings for the application of any of the special lacquers for use on polystyrene.

Polystyrene moldings may be readily cleaned by wiping with a lintless cloth moistened with the solution.

## Brass and Copper Cleaner

| | |
|---|---|
| "Dresinate" X | 4-6 |
| Sodium Carbonate | 20-30 |
| Sodium Metasilicate | 25-30 |
| Trisodium Phosphate | 30-35 |
| Polyphosphate Se-questering Agent | 5-7 |
| Synthetic Wetting Agent | 1.0 |

## Driveway Cleaner

| | |
|---|---|
| "Dresinate" X | 10-12 |
| Sodium Carbonate | 5-20 |
| Sodium Orthosilicate | 40-50 |
| Trisodium Phosphate | 25-35 |
| Polyphosphate Se-questering Agent | 5 |

## Stable-Foam Detergent
U. S. Patent 2,691,636

| | |
|---|---|
| Sodium Alkylaryl Sulfonate | 10 |
| Dodecylamine Laurate | 1-2 |
| Sodium Tallow Alcohol Sulfate | 9 |
| Sodium Tripolyphosphate | 55 |
| Sodium Sulfate | 25 |

## Nondiscoloration Metal Detergent
U. S. Patent 2,698,302

| | |
|---|---|
| Sodium Dodecyl Benzene Sulfonate | 35.0 |
| Sodium Tripolyphosphate | 40.0 |
| Sodium Silicate | 3.0 |
| Carboxy Methyl Cellulose | 0.8 |
| Phenyl Thiourea | 0.1 |

## Aluminum Cleaner

| | |
|---|---|
| 85% o-Phosphoric Acid | 3 |
| Citric Acid C.P. | 4 |
| "Renex" 30 (Polyoxyethylene Alkyl Ether) | 2 |
| Methyl Ethyl Ketone | 3 |
| Water | 88 |

Apply with a sponge, soft brush, or rag. Leave on the surface no longer than 15 minutes. Rinse off with water. Do not use this compound on magnesium.

It may be found necessary to add a corrosion inhibitor, e.g., 1% sodium chromate.

## Removing Wax from Rubber Tile

| | |
|---|---|
| "Renex" 35 | 2 |
| Borax Powder | 8 |
| Water | 990 |

## Ceramic-Surface Cleaner
U. S. Patent 2,714,094

| | |
|---|---|
| Ammonium Bifluoride | 6-25 |
| Ammonium Chloride | 12-70 |
| Alkylaryl Sulfonate | 1 |

## Wall, Tile, and Floor Cleaner

| | |
|---|---|
| Alkylolamide (e.g., "Ninol") | 7 |
| Trisodium Phosphate | 2 |
| Sodium Tripolyphosphate | 2 |
| Sodium Lauryl Sulfate | 2 |
| Water | To make 100 |

## Paint-Brush Cleaner

| | |
|---|---|
| "Aerosol" C-61 | 5 |
| Butyl "Cellosolve" | 5 |
| Xylene | 90 |

Brushes and rollers may be cleaned immediately after use by being suspended in this mixture. In a few hours, practically all of the paint pigment settles in a layer beneath the bristles. The solvent mixture may be used repeatedly. When the pigment has settled, the brush may be removed and hung in circulating air to dry. For best results, the brush then should be washed in soapy water and rinsed. If the brush is to be used again immediately, it may be rinsed in solvent instead of drying and water wash.

## Dried Paint-Brush Cleaner

| | |
|---|---|
| "Aerosol" C-61 | 10 |
| Xylene | 45 |
| Butanol | 5 |
| Mineral Spirits | 40 |

The brush caked with hard dry paint is soaked in the mixture for a while and then worked up and down to aid the penetration into the caked paint. The soaking and working may be repeated several times if the paint is very hard and dry. When the old paint has soaked out of the brush, it is rinsed in clean, warm water, finally washed in soapy water and rinsed again, then hung up to dry. After drying, the bristles have the appearance and feel of those in a new brush.

## Paint and Tar Remover*

| | |
|---|---|
| Xylene | 140.0 |
| Trichlorethylene | 47.0 |
| Ethylene Dichloride | 61.0 |
| 99% Isopropanol | 33.0 |
| Oleic Acid | 40.0 |
| Sulfonated Castor Oil | 24.0 |
| Triethanolamine | 21.5 |

Mix the solvents, oleic acid, and sulfonated oil, add the amine, and stir to obtain a clear solution. Provide adequate ventilation and take special care to avoid inhaling the vapor and re-

---

* Avoid open flames when mixing or using this formula.

peated contact with the skin whenever chlorinate solvents are used.

This paint and tar remover combines four active solvents, yet is easily dispersed in water with which it forms a stable emulsion that is excellent for wool scouring.

---

### Lipstick-Stain Remover

Removal of lipstick stains from fabrics requires a grease solvent (naphtha or carbon tetrachloride) and a solvent (butyl "Carbitol") for the bromo-acid dyes. Other ingredients are added to assist in the wetting and detergent action of the solvents.

| | |
|---|---|
| "Tergitol" Wetting Agent 7 | 1.6 lb. |
| "Nacconol" Detergent NRSF | 0.4 lb. |
| Butyl "Carbitol" | 2.0 gal. |
| 99% Isopropanol | 1.5 gal. |
| Cleaner's Naphtha | 8.0 gal. |

Mix the "Tergitol" and "Nacconol" until the second is dissolved and a clear gel is obtained. Add the butyl "Carbitol," naphtha, and isopropanol and stir until a clear solution is obtained.

From one fourth to one half of the naphtha can be replaced with carbon tetrachloride to achieve faster drying and to reduce the fire hazard.

---

### Ink-Stain Removers

"Carbowax" compound 4000 is especially attractive as a vehicle for stain-removing agents, since the remover can be made in stick form. The stick can be applied easily and economically and makes a convenient package. These sticks have been found satisfactory for removing ink stains from all colors of rayon, cotton, wool, and silk without affecting the color of the fabric. The sticks will be found useful for removing iodine, rust, coffee and other stains as well.

---

### Ink-Stain Removing Stick

#### Formula No. 1

| | |
|---|---|
| "Carbowax" Compound 4000 | 60.0 |
| Triethanolamine | 21.5 |
| Oxalic Acid | 18.5 |

Melt the "Carbowax", add the triethanolamine and oxalic acid, and stir until the acid crystals have disappeared.

#### No. 2

| | |
|---|---|
| "Carbowax" Compound 4000 | 70 |
| Sodium Bisulfate | 25 |

"Carbitol" 6

Melt the "Carbowax", add the "Carbitol," and stir in the sodium bisulfite which has been powdered very finely.

In each case, stir until fairly viscous creams, just thin enough to pour, are obtained. Then pour into molds or metal containers, stirring well before each pouring to assure uniform distribution of the ingredients.

If a slightly softer stick according to formula 2 is desired, replace 7 to 10 lb. of the "Carbowax" with "Carbitol." Other stain-removing ingredients can be used with the "Carbowax" as the convenient and economical carrier.

### Spotting with Sticks

Lay the spotted fabric on a folded, clean cloth, with the soiled side of the fabric on the cloth. Using another small, clean cloth, apply water to the spot with a daubing action until no more color from the ink spot can be removed to the undercloth.

Hold the stick of formula 1 under hot water for an instant and daub the ink spot with the wet stick. This will release more color from the spot. Repeat this process until the spot is almost gone and no more color can be seen going from the spot on the folded cloth.

Hold the stick of formula 2 under hot water for an instant and daub the spot with the wet stick. Repeat several times, if necessary, until the spot disappears. Rinse well with water by daubing with a small cloth wet with water.

In removing ink stains from some white or light-colored rayons or silks, the blue color is removed but a deep pink spot may remain. This pink color can be lightened by repeated applications of the stick of formula 1, followed by the stick of formula 2, with a final thorough water rinse.

### No. 3
*U. S. Patent 2,393,865*

| | |
|---|---|
| Polyethylene Glycol 6000 | 60.0 |
| Triethanolamine | 21.5 |
| Oxalic Acid | 18.5 |

Melt the solid compound, add the triethanolamine and oxalic acid, and stir until the acid crystals have disappeared. When the mixture becomes a fairly viscous cream, pour into molds or metal containers.

## No. 4

| Polyethylene Glycol | |
|---|---|
| 6000 | 75 |
| Sodium Bisulfite | 25 |

Melt the solid compound, add "Carbitol" and stir in the sodium bisulfite (finely powdered). Stir until it becomes fairly viscous and pour into molds.

To remove ink from cloth, wet the stick of formula No. 1 with warm water and daub it on the ink spot. Follow with the stick of formula No. 2, wet with warm water. Rinse the cloth well with water.

---

## Water-Soluble Stain Remover

| Glycerin | 4 |
|---|---|
| Glacial Acetic Acid | ¾ |
| Methyl Alcohol | 1 |
| Lactic Acid | 1 |
| Amyl Acetate | 1 |

After making up the mixture, add 3 to 5 drops of either dilute hydrofluoric acid or a proprietary rust remover. Next add the smallest amount of butyl alcohol which will give a clear solution. The purpose of the butyl alcohol is to mix the wet and dry solvents and a complete and permanent mixture is not obtained as long as the solution remains turbid. After this, add ½ part of oleic acid, which will require several hours to dissolve in the mixture completely.

This formula is safe on rayon, silk, cotton, and acetate fabrics, but before use on acetate, it should be tested on a sample of acetate satin. It is a good stripping agent for tannin stains, such as those caused by coffee, tea, beer, coca-cola, ginger ale, and other soft drinks. If the tannin stain is well oxidized, however, it will be difficult to remove with this or any other known means. The solution is also said to be useful in removing many fruit, ink and dye stains, and to be reasonably safe on most dyed fabrics.

---

## Blood Stain Remover

| "Triton" X–100 (25% | |
|---|---|
| Solution) | 100 |
| Hexylene Glycol | 15 |
| Ammonium Hydroxide | 25 |

---

## Dry Cleaners' Water Repellent

| | | |
|---|---|---|
| a Aluminum Stearate | 2/3 | |
| Petrolatum | 2½ | |
| "Flexo Wax" C | 10 | |
| b Xylol | 5 gal. | |
| Naphtha | ½ gal. | |
| a (Above) | 2½ gal. | |

*c* Dry-Cleaning Sol-
  vent          10 gal.
*b* (Above)     1 gal.

Portion *a* is made by heating the components together and is then added to the solvents in *b*. In use, 1 part *b* is added to 10 parts of the dry cleaning solvent which is generally Stoddard solvent.

The water repellent is applied after the regular dry-cleaning cycle.

---

## Removing Tobacco Stains from Fingers

Hydrochloric Acid      4
Glycerin             100
Triple Rose Water    900

To use the lotion, saturate a piece of cotton with it and rub gently over the stained area. Several applications may be necessary before the desired results are obtained.

---

## Dry Rug Cleaner
### U. S. Patent 2,364,608

Diglycol Stearate    10 lb.
Water                22 gal.
"Stoddard" Solvent   10 gal.
Carbon Tetrachlor-
  ide                12 gal.
Petroleum Naphtha    4½ gal.

Deodorized Hydrocar-
  bon Solvent
  (B.P. 380–
  460°F.)           3½ gal.
Sodium Stearate
  Soap              13 lb.
Wood Flour          192 lb.

---

## Rug-Cleaning Shampoo

### Formula No. 1

*a* Oleic Acid            35
  "Trigamine"             15
  Water                  125
*b* Butyl "Cellosolve"     5
  Ethylene Dichloride     13
*c* Diethylene Glycol      15
  Isopropyl Alcohol       20

Mix *a* and *b* with a high-speed mixer. Add *c* slowly. This gives a clear solution which is readily emulsifiable in water.

### No. 2

Oleic Acid               28
Ethylene Dichloride      13
99% Isopropanol          14
Butyl "Cellosolve"        5
Triethanolamine          16
Water                   125

Mix the oleic acid, ethylene dichloride, isopropanol, and butyl "Cellosolve" and add the amine. Stir until thoroughly mixed and add the water. If the mixture is cloudy, add sufficient isopropanol to clear it.

An emulsion made of equal volumes of the soap and water is recommended for cleaning rugs and carpets.

Adequate ventilation should be provided and special care should be taken to avoid inhaling the vapors and repeated contact with the skin whenever chlorinated solvents are used.

### No. 3

| | |
|---|---|
| Oleic Acid | 28.2 |
| Coconut-Oil Fatty Acids | 21.0 |
| 99% Isopropanol | 30.0 |
| Triethanolamine | 14.2 |
| Monoethanolamine | 6.7 |
| "Terigitol" Wetting Agent 7 | 5.0 |
| Water | 15.0 |

Mix the oleic acid, fatty acids, and isopropanol. Mix the amines and "Tergitol" and stir until thoroughly mixed. Add the water which will produce a clear liquid.

This formula is based on a combining weight for coconut-oil fatty acids of 210. The proportion should be changed according to the combining weight of the particular fatty acid. All the triethanolamine may be replaced with an additional 6.3 parts of monoethanolamine to increase the detergent properties of the shampoo.

### No. 4

| | |
|---|---|
| "Tergitol" Wetting Agent 7 | 6.4 |
| "Nacconol" Detergent NRSF | 3.2 |
| "Carbitol" or 99% Isopropanol | 8.0 |
| Tetrasodium Pyrophosphate | 2.0 |
| Water | 90.0 |

Mix the "Tergitol" and "Nacconol" and stir at intervals until a clear gel is obtained. Add the "Carbitol" or isopropanol, then the water and tetrasodium pyrophosphate. Stir until a clear solution is obtained.

---

### Laundry Detergent

#### Formula No. 1

| | |
|---|---|
| "Santomerse" No. 1 | 50 |
| Soda Ash | 30 |
| Sodium Tripolyphosphate | 15 |
| Trisodium Phosphate | 5 |

#### No. 2

| | |
|---|---|
| Soap | 40 |
| "Santomerse" No. 1 | 10 |
| Tetrasodium Pyrophosphate | 20 |
| Modified Soda | 30 |

### No. 3

| | |
|---|---|
| "Ethofat" 242/25 | 18.0 |
| Low-Viscosity Carboxy Methyl Cellulose | 1.8 |
| Water | 13.2 |
| Tetrasodium Pyrophosphate | 15.0 |
| Trisodium Phosphate | 5.0 |
| Sodium Metasilicate | 10.0 |
| Soda Ash | 37.0 |

This can be made in a dry mixer by adding alkalis first and starting the mixer. The cellulose is dissolved in the water and sprayed on the mixing alkalis and then the "Ethofat" is sprayed in

This formula can also be used in the home laundry. Generally 0.2 to 0.3% of the finished product is used on the basis of the wash water.

### No. 4

| | |
|---|---|
| "Kreelon" CD | 45-55 |
| Soda Ash | 30-40 |
| Sodium Polyphosphate | 15-25 |

### No. 5

| | |
|---|---|
| Soap | 35-45 |
| "Kreelon" CD | 10-20 |
| Tetrasodium Pyrophosphate | 15-25 |
| Modified Soda | 30-40 |

### No. 6

| | |
|---|---|
| "Kreelon" 8D or 8G | 20-30 |
| Soda Ash | 40-62 |
| Sodium Polyphosphate | 15-25 |
| "Carbose" D (65% Sodium Carboxy Methyl Cellulose) | 2-4 |

### No. 7

| | |
|---|---|
| Soap | 35-45 |
| "Kreelon" 8D or 8G | 5-10 |
| Tetrasodium Pyrophosphate | 15-25 |
| Soda Ash | 17-44 |
| "Carbose" D (65% Sodium Carboxy Methyl Cellulose) | 1-3 |

### Laundry Detergent for Overalls and Greasy Clothing

Mix 1 gal. pine-oil disinfectant (coefficient 5) with 20 lb. soda ash. Screen and mix thoroughly with 200 lb. soda ash.

### Framed Sea-Water Soap

| | |
|---|---|
| "Nacconol" Z | 300 |
| Tallow Fatty Acids | 435 |
| Water | 78 |
| 33% Caustic Soda | 187 |
| Perfume Oil | 5 |

Run the melted fatty acids into a steam-jacketed soap crutcher. Add "Nacconol" with downward motion of the crutcher screw. Then add the water with continued agitation. Heat the mixture at 150 to 160°F. and add the caustic soda gradually, with agitation provided by upward motion of the crutcher screw. When saponification is complete and the soap mass is smooth and uniform, add the perfume oil and drop the mass into the frame. Allow to cool thoroughly, strip, and cut in cakes of the desired size.

---

### Milled Sea-Water Soap

| | |
|---|---|
| "Nacconol" Z | 30.00 |
| Toilet-Soap Base (12-14% Moisture) or Tallow Chip Soap (12-14% Moisture) | 69.00 |
| Perfume Oil | 1.00 |
| Oxidation Inhibitor | 0.05 |

Blend the ingredients in an amalgamator and feed slowly to a steel-roll or stone-roll soap mill. Then feed the milled soap to a standard soap plodder. Cut and press the extruded mass in the usual manner.

A rate of production somewhat below that of a standard toilet soap is to be expected.

Milling and plodding of this type of soap may be facilitated by replacing 10 parts of the soap stock with an equal amount of cornstarch.

### Transparent Soap

#### Formula No. 1

| | |
|---|---|
| Coconut Oil | 20.0 |
| Tallow | 13.0 |
| Castor Oil | 7.0 |
| Caustic Soda (38°Bé.) | 22.0 |
| Sugar | 14.0 |
| Crystal Soda | 0.5 |
| Water | 13.5 |
| Alcohol | 10.0 |

#### No. 2

| | |
|---|---|
| Coconut Oil | 19 |
| Tallow | 17 |
| Castor Oil | 8 |
| Caustic Soda (38°Bé.) | 23 |
| Glycerin | 4 |
| Sugar | 8 |
| Water | 6 |
| Alcohol | 15 |

---

### Filled Coconut-Oil Soap

For filling coconut soaps, potassium silicate is mixed with 10% potassium hydroxide of 38°Bé. In order to obtain a smooth surface of the soap, it is advisable to use one third of a 10% "Tylose" solution and two thirds of potassium silicate.

| | |
|---|---|
| Coconut Oil | 100 |
| Caustic Soda (38°Bé.) | 53 |
| Potassium Silicate (38°Bé.) | 32 |
| "Tylose" Solution HB 25 | 15 |

Unfilled Coconut Soap

| | |
|---|---|
| Coconut Oil | 70 |
| Tallow | 20 |
| Castor Oil | 10 |
| Caustic Soda (38-40°Bé.) | 50 |
| 10% "Tylose" Solution | 10 |

Filled Cold-Process Coconut Soap

| | |
|---|---|
| Coconut Oil | 80 |
| Tallow | 20 |
| Caustic Soda (38°Bé.) | 52 |
| Filling Solution (29°Bé.)* | 18 |
| "Tylose" Solution | 10 |

Shaving Soap

| | | |
|---|---|---|
| a | Beef Tallow | 25 |
| | Coconut Oil | 6 |
| | Stearin | 1 |
| | Lanolin | ½ |
| b | Caustic Soda (38°Bé.) | 7 |
| | Potassium Hydroxide (50°Bé.) | 8 |
| | Water | 2 |

* Filling Solution

| | |
|---|---|
| Water | 50 |
| Sugar | 15 |
| Potash | 7 |
| Sodium Chloride | 6 |
| Caustic Soda (38° Bé.) | 4 |

This solution is heated to the boiling point. It is preferable to prepare it the day before it is used that impurities may separate or settle down.

Heat *a* to 70°C. Then add *b* according to the half-boiled saponification method. Allow the resulting emulsion to stand for about 1 hour and then examine for complete saponification. If there is not enough caustic soda, add the necessary quantity. Neutralize the excess caustic soda with stearin. Add 1 to 2 kg. hot "Tylose" solution to the saponified mass.

All-Purpose Cleaner

Formula No. 1

| | |
|---|---|
| "Ninol" 1281 | 3.2 |
| 40% Potash Soap | 17.5 |
| Sodium Tripolyphosphate | 2.0 |
| Water | 77.3 |

Dissolve the phosphate in the water followed by the soap. Then stir in the "Ninol."

No. 2

| | |
|---|---|
| "Ninol" 1281 | 7.0 |
| Trisodium Phosphate | 4.0 |
| "Ultrawet" K. | 1.0 |
| Sequestering Agent* | 0.6 |

Powdered Detergent

| | |
|---|---|
| "Ultrawet" K | 37 |
| "Ninol" AA-62 | 5 |

* Tetrasodium ethylene diamine tetraacetate.

Sodium Tripolyphos-
phate                    40
Sodium Sulfate           10
Sequestering Agent*       8

---

Household Scouring Powder

Formula No. 1
Silica Flour (140
Mesh)                 80-90
Modified Soda            4-8
Tetrasodium Pyrophos-
phate                    1-3
Bentonite            0.05-1.0
"Kreelon" 4G             4-8

No. 2
Silica Flour or Vol-
canic Ash             84-92
Sodium Sesquicar-
bonate                  5-10
"Kreelon" 8G             3-6

---

Floor Cleaner
"Tergitol" Wetting
Agent 7                   4
"Nacconol" Detergent
NRSF                      1
Cleaner's Naphtha        95

Avoid open flames when handling or using naphtha cleaners.
Mix the "Tergitol" and "Nacconol" and stir until the "Nacconol" is dissolved and a clear
gel is obtained. Mix the gel with
sufficient naptha to produce a
solution. This can be stored as
a concentrate or the remainder
of the naphtha can be added immediately.

---

Floor or Wall Cleaner
"Ninol" 1281              7
Sodium Tripolyphos-
fonate                   2-3
Trisodium Phosphate      2
Sodium Xylene Sul-
fonate                   2-3
Water          To make 100

---

Janitor's Scrub Soap
"Nacconal" Z             22
Trisodium Phosphate       5
High-Viscosity Carboxy
Methyl Cellulose         21
Pine Oil                  5
Water                   800

## CHAPTER XXI

# TEXTILES

Textile Softener

| | |
|---|---|
| "Aliquat" H226 | 100 |
| Water | 1400 |
| Sodium Acetate | 30-150 |

Antistatic Lubricant
for Wool

| | |
|---|---|
| Mineral Oil | 61.0 |
| "Arquad" 18 | 36.6 |
| "Ethomeen" S/12 | 2.4 |

Synthetic Thread Lubricant
*U. S. Patent 2,870,045*

| | |
|---|---|
| Petrolatum | 12.5 |
| Butyl Stearate | 9.5 |
| Oleic Acid | 8.0 |
| Triethanolamine | 2.5 |
| "Teflon" 30 | 4.0 |
| Water | 63.5 |

Gloss Starch
*German Patent 936,866*

| | | |
|---|---|---|
| *a* | Potato Starch | 45 |
| | Lactic Acid | 45 |

| | | |
|---|---|---|
| *b* | Potato Starch | 8.5 |
| | Paraffin Wax | 1.5 |

Mix *a* together then add *b*.

Dry Bleach
*U. S. Patent 2,897,154*

| | |
|---|---|
| Silver Phosphate | 1 |
| Trichlorocyanoric Acid | 8 |
| Sod. Dodecyl Sulfate | 3 |
| Salt Cake | 88 |

Noncorrosive Nylon Bleach
*U. S. Patent 2,898,179*

| | |
|---|---|
| Sodium Chlorite | 0.20 |
| Citric Acid | 1.30 |
| Sodium Nitrate | 1.60 |
| Sodium Acid Phos- | |
| phate | 0.75 |
| Water | 100.00 |

Bleaching Yellowed Nylon
*U. S. Patent 2,830,868*

| | |
|---|---|
| Sodium Hypochlorite | 1 oz. |
| Acetic Acid | 1/6 oz. |

Water                    1 gal.

Soak fabric for 15 to 40 minutes at 125°F. Remove and rinse.

---

Mercerizing Penetrant

| Butyl "Cellosolve" | 25-50 |
|---|---|
| Pinonic Acid | 50-75 |

---

Heat Stable Enzyme Desizer

| Amylase | 300 |
|---|---|
| Crude Gypsum | 200 |
| Sodium Chloride | 500 |

---

Flameproofing for Porous
Fabrics
*French Patent 1,109,296*

| Vinyl Chloride-Vinyl Acetate Copolymer | 10 g. |
|---|---|
| Antimony Oxide | 10 g. |
| Methylethylketone | 100 cc. |
| Zinc Stearate | 1 g. |

Mix until uniform and apply by impregnation.

---

Fireproofing for Jute
*U. S. Patent 2,852,414*

| Antimony Phosphate | 5-18 |
|---|---|
| Polyvinyl Chloride Dispersion (55%) | 50-76 |

---

Ratproof Cordage
*U. S. Patent 2,822,295-6*

Impregnate cordage with solution of dodecyl alcohol or its acetate.

---

Textile Printing Paste Extender
*U. S. Patent 2,844,547*

| "Guar" Gum | 15 |
|---|---|
| "Triton" X–100 | 14 |
| Ferrous Sulfate | ½ |
| Sodium Hexametaphosphate | ½ |
| Water | 470 |
| Mineral Spirits | 500 |

Add in above order mixing well and then homogenize.

---

Dye Paste Thickener
*German Patent 954,233*

| Carob Bean Flour | 100 |
|---|---|
| Boric Acid | ½ |
| Oxalic Acid | 1 |

---

Accelerator for Color Developer
*U. S. Patent 2,794,741*

| Thiourea | 2.25-3.25 |
|---|---|
| Sodium Carbonate $H_2O$ | 10-20 |
| Benzotriazole | 2-2.5 |
| Water | 500 |

---

Azoic Printing Emulsion

| Carboxymethyl Cellulose HV 120 or "Keltex" S, 4% solution (stabilizer) | 18 gal. |
|---|---|
| "Deceresol" Wetting Agent No. 18 | 3 pt. |
| Caustic Soda, 25% solution | 9 pt. |

Mineral Spirits 20 gal.
Oleic Acid 9 pt.

---

Aniline Black Emulsion

Water 24 gal.
Sodium Chlorate 12½ lb.
Yellow Prussiate of
Soda 25 lb.
Octyl Alcohol 1 pt.
Modified Locust
Bean Gum 3 lb.
"Marasperse" N 7 lb.
Mineral Spirits 15 gal.

---

Flash Aging Emulsion

Modified Locust
Bean Gum 3 lb.
Starch Ether Gum 3 lb.
Sodium Lignin
Sulfonate ¾ lb.
Water 16 gal.
"Aerotex" Resin
P116 12 oz.
Xylol 1 gal.
Mineral Spirits 21 gal.

---

Bleaching Nylon Fabrics

"Textone" ½ lb.
Acetic Acid ½ lb.
Water 100 gal.
Bleach at boil for 30 minutes
in this solution.
Soap 4 lb.
"Duponol" D Paste 4 lb.
Tetrasodium
Pyrophosphate 4 lb.

Water 100 gal.
Scour at boil for 30 minutes
in this solution.

---

Lubricant for Cellulose
Acetate Yarn
U. S. Patent 2,407,105

White Mineral Oil 77.5
Sulfated Olive Oil 5.0
Sperm Oil 2.5
Trihydroxyethylamine
Oleate 15.0

This is easily removed from
the yarn by hot water.

---

Whitening Nylon

Formula No. 1
British Patent 675,156

Pad the nylon with a 0.15%
aqueous solution of 1-naphthyl-
amine-5-sulfonic acid.

No. 2
U. S. Patent 2,619,470

3,7-bis (Anisoyl-
amino) Dibenzo-
thiophene Sulfonic
Acid 0.1-6.0
Alcohol 20.0
Glycerin 1.50-4.25
Water To make 100.0

---

Spray Printing for Woolen
Fabrics

Dyestuff 5-20
Intensifier 150

Water 330
Tragacanth Thickening 300
40% Acetic Acid 50
"Solvadine" BL Conc.
(10% Solution) 50
Glycerin 100

After drying, the prints are steamed for 30 to 50 minutes at ¼ atmosphere pressure. Steaming is essential in order to obtain satisfactory rubbing fastness. "Neolan" dyes are said to be especially suitable for spray printing of woolen upholstery material.

---

Screen-Printing Emulsion for
Cellulose Acetate
*British Patent 647,098*

Kerosene 64.00
Water 26.00
Sodium Lignin
Sulfonate 5.00
Glycerol 2.50
Glacial Acetic Acid 2.50
2,4-Dinitro-3'-Hy-
droxy Diphenylamine 0.01

---

Dyeing Synthetic Fibers

"Orlon," "Acrilan," and "Dynel" are addition polymers based on acrylonitrile, with "Orlon" being based essentially on this monomer only, "Acrilan" containing a small amount of another monomer, and "Dynel" being a copolymer of about 40% acrylonitrile and 60% vinyl chloride. "Dacron" is a condensation product of terephthalic acid with ethylene glycol. All four fibers are alike in showing marked hydrophobic properties and in their comparative lack of affinity for the dyestuffs generally used for such fibers, as wool, silk, cotton, and rayon. However, by the use of modified or new dyeing processes with dyestuffs selected for their suitability, these synthetic fibers can be dyed in a good range of shades with adequate fastness. General recommendations and specific suggestions for these fibers follow.

---

"Orlon"

The acrylic fiber, "Orlon," is available in two forms, "Orlon" 41, which is in staple form before spinning into yarn, and "Orlon" 81, which is a stretched monofilament. "Orlon" 41 is easier to dye than "Orlon" 81 and is better adapted for the production of medium to full shades.

The "Nacelan" dyes may be used on "Orlon" for light to medium shades. Dyeing temperatures near the boil are necessary even for light shades and for heavier dyeings, pressure dyeing up to 15 lb. and about 250°F. will be necessary. The addition of carriers or swelling agents, such as $m$-cresol, aniline, phenol, or $p$-phenyl phenol in amounts ranging from 3 to 10% on the weight of material results in greater absorption of dye and heavier shades, but introduces the additional problem of removal after the dyeing operation, for which a vigorous soaping is necessary.

The following "Nacelan" dyes are suitable for "Orlon" and can be blended with each other as desired for compound shades:

"Nacelan" Fast Yellow CG
"Nacelan" Brilliant Orange 3R
"Nacelan" Scarlet CSB
"Nacelan" Rubine RB
"Nacelan" Violet 5RL
National Blue KGB

For the most complete range of shades, "Orlon" is best dyed with selected acid colors by the cuprous-ion method. After scouring the material, 2.5 to 12.5% copper sulfate is added to the bath at 140°F., followed by two fifths as much hydroxyl ammonium sulfate, neutralized with caustic soda to a $p$H of 5 to 6.

The dissolved dye is added and the temperature raised slowly to the boil in 15 to 20 minutes. Dyeing is continued for 1 hour at the boil, with any subsequent necessary additions of dye being made after cooling to 180°F., followed by ½ hour at the boil. The dyed material is finally thoroughly rinsed. In this method also, the use of elevated pressure and temperature, where possible, will result in higher dye absorption.

The following acid dyes are representative of those suitable for this application:

Chinoline Yellow Extra Conc.
Wool Orange A Conc.
Fast Red S Conc.
Alizarin Violet NRR
Alizarin Fast Blue RB
Wool Fast Blue GL
Alizarin Cyanone Green GN Extra

---

"Dynel"

"Dynel" can also be dyed with the two groups of dyes given for "Orlon," following similar techniques. For the cuprous-ion method with acid dyes, it may be desirable, for medium to full shades, to use a carrier, such as $p$-phenyl phenol, and for "Dynel," there appears to be an advantage in using zinc formaldehyde sulfoxylate rather than hydroxyl ammonium sulfate. A typical formula for the dye bath is as follows:

Add in succession, in solution form:

| | |
|---|---|
| "Nacconol" NR | 1.0 |
| p-Phenyl Phenol (in 1.0% Solution with 0.5% Caustic Soda) | 1.5 |
| 28% Acetic Acid | 7.0 |
| Cupric Acetate (in 4% Solution) | 1.0 |
| "Hydro" AWC (Zinc Formaldehyde Sulfoxylate) | 0.6 |
| Dyestuff | 2.0 |

Immerse the undyed material, raise the temperature to 210°F., and run for 1½ hours, followed by thorough rinsing. If the "Dynel" becomes delustered, the luster can be restored by aftertreatment in a 6% Glauber's salt solution at 200 to 212°F. or by subjecting to dry heat at 250°F. for 20 minutes, in either case followed by soaping at 150°F. and rinsing. The following acid dyes are suitable for htis method.

Milling Yellow 3G
Alizarin Light Brown BL
Cloth Red B

Alizarin Violet NRR
Alizarin Fast Blue RB
Wool Fast Blue GL
Alizarin Cyanone Green GN Extra
Alizarin Light Gray 2BLW

---

## "Acrilan"

"Acrilan" shows more affinity and dyeing properties closer to those of wool than the other acrylics. It can be dyed with selected acid dyes, but, in most cases, more acid is necessary than for wool. For a 3% dyeing, the use of 6 to 8% sulfuric acid is suggested, starting with 2% and adding the balance at ½ hour intervals to the bath cooled at 170 to 180°F. Dyeing is effected near the boil for 1½ to 2 hours. Scouring before dyeing with 1% "Nacconol" NR at 160°F. is recommended.

The following acid dyes are representative of those suitable for application to "Acrilan" by this method:

Milling Yellow 3G
Milling Orange GR
Cloth Red G
Alizarin Violet NRR
Alizarin Fast Blue RB
Alizarin Sapphire 2GL
Alizarin Cyanone Green GN Extra

The acetate dyes can also be applied to "Acrilan" by preparing the dye bath with:

"Nacconol" NR     1
p-Phenyl Phenol (Dissolved with 1.5%
    Caustic Soda)     3
Monosodium or Monoammonium Phosphate 6-8
"Nacelan" Dye (Dispersed with 1%
    "Nacconol" NR)     2

Place the material into the bath, raise the temperature to the boil, and continue boiling for 1 to 1½ hours. Rinse and, in order to remove residual p-phenyl phenol, treat at 180°F. for 10 minutes with 9% caustic soda, rinse, soap at 160°F., and finally rinse.

The following acetate dyes are best adapted for this method of application to "Acrilan":

"Nacelan" Pink B
"Nacelan" Fast Yellow CG
"Nacelan" Printing Blue 2B
    Extra Conc.
"Nacelan" Brown NR

---

### "Dacron"

At ordinary dyeing temperatures and pressure, "Dacron" shows little or no affinity for water-soluble dyes and the dispersed acetate dyes are practically the only group suitable. For heavy shades, these require the use of a carrier, and p-phenyl phenol, monochlorobenzene, benzoic acid, salicylic acid, and others have been used. At elevated pressure and temperature, up to 250°F., the necessity for carriers is eliminated and full shades can be obtained from dyes selected from the acetate, vat dyes, and insoluble azo colors. A dry heat treatment for several seconds at 360 to 420°F. has also been found to yield good dyeings on material impregnated with some dispersed water-soluble dyes.

For dyeing with acetate dyes at normal temperature and pressure, the dye bath may be prepared with:

"Nacconol" NR     1
p-Phenyl Phenol (Dissolved with 1.5%
    Caustic Soda)     3
Monosodium Phosphate     6
"Nacelan" Dye (Dispersed in Water with
    1% "Nacconol" NR)     2-3

Immerse the material at 110°F., raise the temperature to the boil, boil 1 to 1½ hours, and rinse. Remove the carrier by treating for 10 minutes at 180°F. with 9% caustic soda.

Rinse, soap at 160°F. for 10 minutes, and finally rinse.

The following dyes are well adapted for this application:

"Nacelan" Fast Yellow CG
"Nacelan" Yellow 5GK
"Nacelan" Orange 3R
"Nacelan" Pink B
"Nacelan" Scarlet CSB
"Nacelan" Violet 5RL
"Nacelan" Printing Blue 2B Extra Conc.
"Nacelan" Blue KGB

---

### Dyeing of "Orlon" with Vat Dyes

In a package, 100 g. of "Orlon" type 41 raw stock is wet out with 1 g. "Igepon" T gel for 15 minutes at 160 to 180°F. The bath is dropped and the machine is charged with 25 liters of water at 120°F., containing 60 g. sodium bicarbonate and 60 g. "Rongalite."

Just before the addition of the leuco vat solution, 15 g. sodium hydrosulfite is added. The leuco vat solution is prepared in the usual manner. The temperature is quickly raised to 260°F. and maintained with circulation for 40 minutes. The bath is then dropped and the material oxidized and scoured in a fresh bath.

### Dyeing "Acrilan" with Chrome Dyes

"Acrilan" may be dyed in dark shades with good fastness by using a few selected chrome colors, such as:

"Chromogene" Black ETOO Supra CF
Alizarin Blue Black BA Extra
Chrome Yellow A Extra
Anthracene Chromate Brown EBA Extra Conc.

Dyeing is carried out at the boil using 4% of 66°Bé. sulfuric acid on the weight of the fiber, followed by chroming in a fresh bath at the boil with 2% bichromate and 2% of 66°Bé. sulfuric acid.

---

### Dyeing "Dynel" with "Celliton" Dyes

"Dynel" may be dyed satisfactorily with selected "Celliton" colors, such as:

"Celliton" Fast Yellow GA–CF
"Celliton" Fast Brown 3RA Ex CF
"Celliton" Fast Pink RFG
"Celliton" Fast Red Violet RNA–CF
"Celliton" Fast Violet BA–CF

"Celliton" Fast Blue FFRN Extra Conc.

"Celliton" Blue BGF Extra

Dyeings must be carried out at 205°F.

---

Dyeing "Dacron" with "Algosols"

The "Algosol" dye is made up with sodium nitrite, padded on the "Dacron" goods, the goods dried, heat-set at 350°F. for about 1 minute, and then passed through a 2% sulfuric acid solution at 140 to 160°F.

---

Rayon-Tire-Cord Dip

("R F L")

| | |
|---|---|
| Resorcinol-Formaldehyde Master Batch* | 35.6 |
| GR–S Latex Type 2 | 33.4 |
| Water | 29.2 |
| 28% Ammonia | 1.8 |

Add the water, followed by the ammonia, to the GR–S Latex

---

\* Resorcinol - Formaldehyde Master Batch

| | |
|---|---|
| Resorcinol | 4.77 |
| 1% Caustic Soda Solution | 6.40 |
| Formaldeyde | 10.50 |
| Water | 78.33 |

Add the resorcinol and caustic soda to cold (16 to 18°C.) water. Stir until dissolved. Then add the formaldehyde and stir well. Allow to stand for 6 to 8 hours before use.

and mix. Then add the resorcinol-formaldehyde masterbatch and stir until uniform. This can be used immediately after mixing. Dipped cord should be calendered within 2 or 3 weeks after dipping.

---

Home-Laundry Mothproofing Solution

| | |
|---|---|
| DDT | 25 |
| Aromatic Hydrocarbon Solvent | 65 |
| Polyglycol 400 Monolaurate | 10 |

---

Textile Mildewproofing

Copper 3-phenylsalicylate may be easily applied as a dispersion in water. Application by this method normally involves a conventional type padder with adjustable squeeze rolls and a drying unit. The following formulation for an aqueous dispersion is suggested:

| | |
|---|---|
| Copper 3-Phenylsalicylate | 50 |
| "Methocel" (1500 cp.) | 2 |
| "Triton" X–100 | 2 |
| Cold Water | 112 |

To prepare the dispersion, the "Methocel" should be dissolved by adding it to cold water, containing the "Triton," with rapid stirring. A paste is then formed by wetting out the copper salicylate with the "Methocel"–"Triton" solution and ground to a smooth dispersion with a three-roll mill or ball mill. Passage through a pressure-feed homogenizer or colloid mill will have the same effect. If such equipment is used, it may be necessary to reduce the viscosity of the dispersion somewhat by the addition of more water to the formulation. The grinding procedure is similar to that used in dispersing a pigment into a varnish or paint vehicle.

Copper 3-phenylsalicylate may be applied to fabric by the conventional two-bath process. This involves the passage of the fabric through a water solution of the sodium salt of 3-phenylsalicylic acid as the first bath. The goods are then passed through a second bath consisting of a water solution of a copper salt, such as copper acetate or copper sulfate.

The effectiveness of copper 3-phenylsalicylate can be prolonged by the use of certain resinous binders, or by water-repellent treatments on the fabric. These treatments should not require temperatures above

135°C. (275°F.) for curing or setting the binder, since copper 3 - phenylsalicylate is unstable above this temperature.

## Rotproofing Fishnets

### Formula No. 1
*Japanese Patent 604 (1950)*

| | | |
|---|---|---|
| a | Furfural | 96 |
| | Water | 50 |
| | Caustic Soda | 20 |
| b | Water | 3000 |
| c | Copper Sulfate | 5 |
| | Water | 95 |

Mix and warm *a*; then add b. Immerse the net first in the mixture of *a* and *b*, and then in *c*.

### No. 2
*Japanese Patent 3418 (1951)*

| | |
|---|---|
| Coal-Tar Pitch | 40 |
| Coal Tar | 4 |
| Turpentine | 56 |
| Rouge | 5 |
| Cresol Soap | 15 |

Warm and mix until uniform.

---

## Antiseptic Textile

### Formula No. 1

On the basis of a 50% wet pickup, the following solution is prepared:

| | |
|---|---|
| "G–4" Technical | 3.0 lb. |
| 99% Isopropyl Alcohol | 4.0 gal. |

V.M.&P. Naphtha or
Stoddard Solvent 44.0 gal.

The "G–4" is dissolved in the isopropyl alcohol; this solution is then added to the petroleum solvent with constant stirring. The final mixture is padded on at 50°C. Adequate precautions must be taken against health and fire hazards. Solvent applications are advantageous on heavy or tightly woven fabrics, such as stiff webbings, rope, heavy ducks, etc., as the fibers are quickly penetrated. For water repellency, a suitable solvent-soluble repellent should be dissolved in the petroleum solvent in the proportion normally recommended by the manufacturers. The solvent method is the only suitable procedure for treatment of textiles with a combination of "G–4" and copper naphthenate.

### No. 2

| | |
|---|---|
| "G–4" Technical | 25 lb. |
| Pine Oil | 15 gal. |
| Water (180°F.) | 15 gal. |
| "Nekal" AEMA | 10 lb. |

Dissolve the "G–4" in the pine oil and add slowly, with constant stirring, to the "Nekal"–water solution.

### No. 3

| | |
|---|---|
| "G–4" Technical | 10.0 lb. |
| Isopropyl Alcohol | 10.0 lb. |
| Caustic Soda Flakes | 1.4 lb. |
| "Igepal" CA Extra Conc. | 4.0 lb. |
| Water | 9.0 gal. |

While the use of these emulsions is satisfactory for some purposes, they do not contain water repellents and are not recommended where water leaching is involved.

---

### Removing the Starch Binder from Glass Cloth
*U. S. Patent 2,666,720*

| | |
|---|---|
| Urea | 5-20 |
| "Triton" N–100 | 0.1 |
| Water | To make 100 |

---

### Thin-Boiling (Soluble) Starch
*U. S. Patent 2,276,984*

Thin-boiling starch, which, on cooking with water alone, will yield a starch solution of low viscosity, is produced by treating a starch slurry with hydrogen peroxide in the presence of soda ash (sodium carbonate) and a catalyst at temperatures slightly below the gel point of the starch.

*a* Water 2.00 gal.

Soda Ash 0.33 oz.

Copper Sulfate 0.30 g.

Pearl
Cornstarch 10.00 lb.

*b* 35% Hydrogen
Peroxide 1.40 oz.

Water 0.25 gal.

The ingredients under *a* are placed in a stainless steel vessel, equipped with steam coils, in the order given and with constant stirring. After stirring the slurry for 15 minutes at room temperature, *b* is added.

The temperature is then raised to about 160°F., that is, to a temperature very slightly below the gel point of this particular starch. This temperature is maintained for slightly over 2 hours, when a test with titanium sulfate shows that the hydrogen peroxide has been consumed. The starch is then filtered off and dried at about 160°F.

A 3% slurry of the treated starch, heated for about 15 minutes at 175°F., has the following characteristics:

Iodine Color Blue

*p*H 7.1

Viscosity 40 seconds

Dacron-Yarn Sizing
(Single-End)

Soda Ash 3 lb.

"Elchem" 1273 42 lb.

Water To make 100 gal.

---

Orlon and Acrilan Sizing
(Warp)

Soda Ash 3 lb.

"Elchem" 1273 42 lb.

Softener 14-18 lb.

Water To make 100 gal.

Size Pan
Temperature 130°F.

Quetsch-Roll
Pressure 1,200 lb.

Dry-Can
Temperatures 150-200°F.

Speed 25 yds. per min.

Size Pickup
(Dry Basis) 3.5-5.0%

This size will scour off in the usual hot and preferably alkaline afterscour. It will neither discolor nor become fixed against afterscouring if the fabric is heat-set before scouring.

---

Spun Acrylic Yarn Sizing
(Warp)

Formula No. 1

Soda Ash 5.4 lb.

"Elchem" 1273 75.0 lb.

Softener 25.0 lb.

Water To make 100.0 gal.

Dry-Size Pickup 12.5%

No. 2

| | |
|---|---|
| Soda Ash | 2.7 lb. |
| "Elchem" 1273 | 37.5 lb. |
| Softener | 14.5 lb. |
| Water | To make 100.0 gal. |
| Dry-Size Pickup | 7.3% |

### Antiseptic Cord and Rope
### U. S. Patent 2,468,068

While "G–4" can be impregnated into rope and cordage from solvent solution, it is usually preferred to add the "G–4" to the cordage oil used in the manufacturing process. "G–4" may not be soluble in the cordage oil. In some cases, solutions can be prepared by adding the "G–4" to the oil as a 20% solution in dipropylene glycol or similar solvents; the use of wool grease in the oil increases the solubility of the "G–4." Also, "G–4" may be satisfactorily dispersed in the cordage oil if continuous agitation of the oil can be accomplished.

In Government Specification MIL–R–16070 (SHIPS), a combination of copper naphthenate and "G–4" is prescribed for the treatment of rope; it is required that 0.125 –0.025% copper calculated as metal and 0.125–0.025% "G–4" based on the weight of the rope are deposited in the material. If we assume that the pickup of cordage oil by the rope fibers is 12%, the cordage oil must contain 1.0% "G–4." This is equivalent to adding 5 lb. of the 20% solution of "G–4" in dipropylene glycol to 95 lb. of cordage oil containing the proper amount of copper naphthenate.

## Chapter XXII

# APPENDIX

## TABLES

### Weights and Measures
#### Troy Weight

24 grains = 1 pwt.
20 pwts. = 1 ounce
12 ounces = 1 pound

#### Apothecaries' Weight

20 grains = 1 scruple
3 scruples = 1 dram
8 drams = 1 ounce
12 ounces = 1 pound
The ounce and pound are the same as in Troy Weight.

#### Avoirdupois Weight

$27^{11}/_{32}$ grains = 1 dram
16 drams = 1 ounce
16 ounces = 1 pound
2000 lbs. = 1 short ton
2240 lbs. = 1 long ton

#### Dry Measure

2 pints = 1 quart
8 quarts = 1 peck
4 pecks = 1 bushel
36 bushels = 1 chaldron

#### Liquid Measure

4 gills = 1 pint
2 pints = 1 quart
4 quarts = 1 gallon
$31\frac{1}{2}$ gals. = 1 barrel
2 barrels = 1 hogshead
1 teaspoonful = $\frac{1}{8}$ oz.
1 tablespoonful = $\frac{1}{2}$ oz.
16 fluid oz. = 1 pint

#### Circular Measure

60 seconds = 1 minute
60 minutes = 1 degree
360 degrees = 1 circle

#### Long Measure

12 inches = 1 foot
3 feet = 1 yard
$5\frac{1}{2}$ yards = 1 rod
5280 feet = 1 stat. mile
320 rods = 1 stat. mile

### Square Measure

144 sq. in. = 1 sq. ft.
9 sq. ft. = 1 sq. yard
$30\frac{1}{4}$ sq. yds. = 1 sq. rod
43,560 sq. ft. = 1 acre
40 sq. rods = 1 rood
4 roods = 1 acre
640 acres = 1 sq. mile

### Metric Equivalents
#### Length

1 inch = 2.54 centimeters
1 foot = 0.305 meter
1 yard = 0.914 meter
1 mile = 1.609 kilometers
1 centimeter = 0.394 in.
1 meter = 3.281 ft.
1 meter = 1.094 yd.
1 kilometer = 0.621 mile

#### Capacity

1 U. S. fluid oz. = 29.573 milliliters
1 U. S. Liquid qt. = 0.946 liter
1 U. S. dry qt. = 1.101 liters
1 U. S. gallon = 3.785 liters
1 U. S. bushel = 0.3524 hectoliter
1 cu. in. = 16.4 cu. centimeters
1 milliliter = 0.034 U. S. fluid ounce
1 liter = 1.057 U. S. liquid qt.
1 liter = 0.908 U. S. dry qt.
1 liter = 0.264 U. S. gallon
1 hectoliter = 2.838 U. S. bu.
1 cu. centimeter = .061 cu. in.
1 liter = 1000 milliliters or 100 cu. c.

#### Weight

1 grain = 0.065 gram
1 apoth. scruple = 1.296 grams
1 av. oz. = 28.350 grams
1 troy oz. = 31.103 grams
1 av. lb. = 0.454 kilogram
1 troy lb. = 0.373 kilogram
1 gram = 15.432 grains
1 gram = 0.772 apoth. scruple
1 gram = 0.035 av. oz.
1 gram = 0.032 troy oz.
1 kilogram = 2.205 av. lbs.
1 kilogram = 2.679 troy lbs.

## Approximate pH Values

The following tables give approximate pH values for a number of substances such as acids, bases, foods, biological fluids, etc. All values are rounded off to the nearest tenth and are based on measurements made at 25° C.

### pH Values of Acids

| | |
|---|---|
| Hydrochloric, N | 0.1 |
| Hydrochloric, 0.1N | 1.1 |
| Hydrochloric, 0.01N | 2.0 |
| Sulphuric, N | 0.3 |
| Sulphuric, 0.1N | 1.2 |
| Sulphuric, 0.01N | 2.1 |
| Orthophosphoric, 0.1N | 1.5 |
| Sulphurous, 0.1N | 1.5 |
| Oxalic, 0.1N | 1.6 |
| Tartaric, 0.1N | 2.2 |
| Malic, 0.1N | 2.2 |
| Citric, 0.1N | 2.2 |
| Formic, 0.1N | 2.3 |
| Lactic, 0.1N | 2.4 |
| Acetic, N | 2.4 |
| Acetic, 0.1N | 2.9 |
| Acetic, 0.01N | 3.4 |
| Benzoic, 0.1N | 3.1 |
| Alum, 0.1N | 3.2 |
| Carbonic (saturated) | 3.8 |
| Hydrogen Sulphide, 0.1N | 4.1 |
| Arsenious (saturated) | 5.0 |
| Hydrocyanic, 0.1N | 5 1 |
| Boric, 0.1N | 5.2 |

### pH Values of Bases

| | |
|---|---|
| Sodium Hydroxide, N | 14.0 |
| Sodium Hydroxide, 0.1N | 13.0 |
| Sodium Hydroxide, 0.01N | 12.0 |
| Potassium Hydroxide, N | 14.0 |
| Potassium Hydroxide, 0.1N | 13.0 |
| Potassium Hydroxide, 0.01N | 12.0 |
| Lime (saturated) | 12.4 |
| Sodium Metasilicate, 0.1N | 12.6 |
| Trisodium Phosphate, 0.1N | 12.0 |
| Sodium Carbonate, 0.1N | 11.6 |
| Ammonia, N | 11.6 |
| Ammonia, 0.1N | 11.1 |
| Ammonia, 0.01N | 10.6 |
| Potassium Cyanide, 0.1N | 11.0 |
| Magnesia (saturated) | 10.5 |
| Sodium Sesquicarbonate, 0.1N | 10.1 |
| Ferrous Hydroxide (saturated) | 9.5 |
| Calcium Carbonate (saturated) | 9.4 |
| Borax, 0.1N | 9.2 |
| Sodium Bicarbonate, 0.1N | 8.4 |

### pH Values of Foods

| | |
|---|---|
| Apples | 2.9–3.3 |
| Apricots | 3.6–4.0 |
| Asparagus | 5.4–5.8 |
| Bananas | 4.5–4.7 |
| Beans | 5.0–6.0 |
| Beers | 4.0–5.0 |
| Beets | 4.9–5.5 |
| Blackberries | 3.2–3.6 |
| Bread, white | 5.0–6.0 |
| Butter | 6.1–6.4 |
| Cabbage | 5.2–5.4 |
| Carrots | 4.9–5.3 |
| Cheese | 4.8–6.4 |
| Cherries | 3.2–4.0 |
| Cider | 2.9–3.3 |
| Corn | 6.0–6.5 |
| Crackers | 6.5–8.5 |
| Dates | 6.2–6.4 |
| Eggs, fresh white | 7.6–8.0 |
| Flour, wheat | 5.5–6.5 |
| Gooseberries | 2.8–3.0 |
| Grapefruit | 3.0–3.3 |
| Grapes | 3.5–4.5 |
| Hominy (rye) | 6.8–8.0 |
| Jams, fruit | 3.5–4.0 |
| Jellies, fruit | 2.8–3.4 |
| Lemons | 2.2–2.4 |
| Limes | 1.8–2.0 |
| Maple Syrup | 6.5–7.0 |
| Milk, cows | 6.3–6.6 |
| Olives | 3.6–3.8 |
| Oranges | 3.0–4.0 |
| Oysters | 6.1–6.6 |
| Peaches | 3.4–3.6 |
| Pears | 3.6–4.0 |
| Peas | 5.8–6.4 |
| Pickles, dill | 3.2–3.6 |
| Pickles, sour | 3.0–3.4 |
| Pimento | 4.6–5.2 |
| Plums | 2.8–3.0 |
| Potatoes | 5.6–6.0 |
| Pumpkin | 4.8–5.2 |
| Raspberries | 3.2–3.6 |
| Rhubarb | 3.1–3.2 |
| Salmon | 6.1–6.3 |
| Sauerkraut | 3.4–3.6 |
| Shrimp | 6.8–7.0 |
| Soft Drinks | 2.0–4.0 |
| Spinach | 5.1–5.7 |
| Squash | 5.0–5.4 |
| Strawberries | 3.0–3.5 |
| Sweet Potatoes | 5.3–5.6 |
| Tomatoes | 4.0–4.4 |
| Tuna | 5.9–6.1 |
| Turnips | 5.2–5.6 |
| Vinegar | 2.4–3.4 |
| Water, drinking | 6.5–8.0 |
| Wines | 2.8–3.8 |

### pH Values of Biologic Materials

| | |
|---|---|
| Blood, plasma, human | 7.3–7.5 |
| Spinal Fluid, human | 7.3–7.5 |
| Blood, whole, dog | 6.9–7.2 |
| Saliva, human | 6.5–7.5 |
| Gastric Contents, human | 1.0–3.0 |
| Duodenal Contents, human | 4.8–8.2 |
| Feces, human | 4.6–8.4 |
| Urine, human | 4.8–8.4 |
| Milk, human | 6.6–7.6 |
| Bile, human | 6.8–7.0 |

# Interconversion Tables and Chart
## for Units of Volume and Weight, and Energy

| TO CONVERT FROM | MULTIPLY BY | | | | | | | | | | | | | | |
|---|---|---|---|---|---|---|---|---|---|---|---|---|---|---|---|
| | To Cu. in. | To Cu. Ft. | To Cu. Yd. | To Fl. Oz. | To Pint | To Quart | To Gallon | To Grain | To Oz. Troy | To Oz. Av. | To Lb. Troy | To Lb. Av. | To CC. or G. | To Ltr. or Kg. | To Cu. M. |
| Cu. in. | 1.00000 | .0005787 | .0000214 | .55411 | .034632 | .017316 | .004329 | 252.891 | .526857 | .578037 | .043903 | .036126 | 16.3871 | .016387 | .0000163 |
| Cu. Ft. | 1728.00 | 1.00000 | .037037 | 957.505 | 59.8442 | 29.9221 | 7.48052 | 436996 | 910.408 | 998.847 | 75.8674 | 62.4280 | 28316.9 | 28.3169 | .028317 |
| Cu. Yd. | 46656.0 | 27.0000 | 1.00000 | 25852.6 | 1615.79 | 807.896 | 201.974 | 11799000 | 24581.0 | 26968.9 | 2048.42 | 1685.56 | 764556. | 764.556 | .764556 |
| Fl. Oz. | 1.80469 | .001044 | .00003868 | 1.00000 | .062500 | .031250 | .007813 | 456.390 | .950813 | 1.04316 | .079234 | .065199 | 29.5736 | .029573 | .0000295 |
| Pint | 28.8750 | .016710 | .0006189 | 16.0000 | 1.00000 | .500000 | .125000 | 7302.23 | 15.2130 | 16.6906 | 1.26475 | 1.04318 | 473.177 | .473177 | .0004732 |
| Quart | 57.7500 | .033420 | .001238 | 32.0000 | 2.00000 | 1.00000 | .250000 | 14604.5 | 30.4260 | 33.3816 | 2.53550 | 2.08635 | 946.354 | .946354 | .0009463 |
| Gallon | 231.000 | .133681 | .004951 | 128.000 | 8.00000 | 4.00000 | 1.00000 | 58417.9 | 121.704 | 133.527 | 10.1420 | 8.34541 | 3785.42 | 3.78542 | .003785 |
| Grain | .003954 | .0000023 | .0000000847 | .002191 | .0001369 | .00006850 | .00001712 | 1.00000 | .002083 | .002286 | .0001736 | .0001428 | .064799 | .0000648 | .0000000648 |
| Oz. Troy | 1.89805 | .001098 | .00004068 | 1.05173 | .065733 | .032867 | .008217 | 480.000 | 1.00000 | 1.09714 | .083333 | .068571 | 31.1035 | .031104 | .0000311 |
| Oz. Av. | 1.72999 | .001001 | .00003708 | .958608 | .059913 | .029957 | .007489 | 437.500 | .911457 | 1.00000 | .075955 | .062500 | 28.3495 | .028350 | .0000283 |
| Lb. Troy | 22.7766 | .013181 | .0004882 | 12.6208 | .788800 | .394400 | .098600 | 5760.00 | 12.0000 | 13.1657 | 1.00000 | .822857 | 373.242 | .373242 | .0003732 |
| Lb. Av. | 27.6799 | .016018 | .0005933 | 15.3378 | .958611 | .479306 | .119826 | 7000.00 | 14.5833 | 16.0000 | 1.21526 | 1.00000 | 453.593 | .453593 | .0004536 |
| CC or Gram | .061024 | .00003531 | .000001308 | .033814 | .002113 | .001057 | .0002642 | 15.4323 | .032151 | .035274 | .002679 | .002205 | 1.00000 | .001000 | .000001 |
| Liter or Kcg. | 61.0237 | .035315 | .001308 | 33.8140 | 2.11337 | 1.05669 | .264172 | 15432.3 | 32.1507 | 35.2739 | 2.67923 | 2.20462 | 1000.00 | 1.00000 | .001000 |
| Cu. M. | 61023.7 | 35.3146 | 1.30795 | 33814.0 | 2113.37 | 1056.69 | 264.172 | 15432300 | 32150.7 | 35273.9 | 2679.23 | 2204.62 | 1000000 | 1000.00 | 1.00000 |

Note. The small subnumeral following a zero indicates that the zero is to be taken that number of times; thus, $.0_31428$ is equivalent to $.0001428$.

Values used in constructing table:

1 inch = 2.540001 cm.
1 cu. in. = 16.387083 cc. = 16.387083 g $H_2O$ at 4°C.
4°C. = 39°F.

1 lb. av. = 453.5926 g.
∴ 1 gal. = 8.34541 lb.
∴ 1 lb. av. = 27.679386 cu. in. $H_2O$ at 4°C.

1 lb. av. = 7000 grains.
∴ 1 gallon = 58417.87 grains.
∴ 1 gallon = 3785.4162 g.
231 cu. in. = 1 gallon = 3785.4162 g.

| TO CONVERT FROM | MULTIPLY BY | | | | | | | | | | |
|---|---|---|---|---|---|---|---|---|---|---|---|
| | B.T.U. | P.C.U. | Cal. | Ft. Lbs. | Ft. Tons | Kg. M. | HP Hrs. | KW Hrs. | Joules | Lbs. C | Lbs. $H_2O$ |
| B.T.U. | 1.0000 | .55556 | .251996 | 778.000 | .389001 | 107.563 | $.0_3 3929$ | $.0_3 2931$ | 1055.20 | $.0_4 6876$ | .001031 |
| P.C.U. | 1.80000 | 1.00000 | .453593 | 1400.40 | .700202 | 193.613 | $.0_3 7072$ | $.0_3 5276$ | 1899.36 | $.0_3 1238$ | .001855 |
| Calories | 3.96832 | 2.20462 | 1.00000 | 3091.36 | 1.54368 | 426.844 | .001559 | .001163 | 4187.37 | $.0_3 2729$ | .004089 |
| Ft. Lbs. | $.0_2 1285$ | $.0_3 7141$ | $.0_3 3239$ | 1.00000 | $.0_3 500$ | .138235 | $.0_6 0505$ | $.0_6 3767$ | 1.35625 | $.0_7 8840$ | $.0_5 1325$ |
| Ft. Tons | 2.57069 | 1.42816 | .647804 | 2000.00 | 1.00000 | 276.511 | .001010 | $.0_3 7535$ | 2712.59 | $.0_3 1768$ | $.0_2 2649$ |
| Kg. M. | $.0_2 9297$ | $.0_2 5165$ | $.0_2 2343$ | 7.23301 | $.0_2 3617$ | 1.00000 | $.0_5 3653$ | $.0_5 2725$ | 9.81009 | $.0_6 6394$ | $.0_5 9580$ |
| HP Hrs | 2544.99 | 1413.88 | 641.327 | 1980000 | 990.004 | 273747 | 1.00000 | .746000 | 2685473 | .175044 | 2.62261 |
| KW Hrs. | 3411.57 | 1895.32 | 859.702 | 2654200 | 1327.10 | 366959 | 1.34041 | 1.00000 | 3599889 | .234648 | 3.51562 |
| Joules | $.0_3 9477$ | $.0_3 5265$ | $.0_3 2388$ | .737311 | $.0_3 3687$ | .101937 | $.0_6 3724$ | $.0_6 2778$ | 1.00000 | $.0_7 6518$ | $.0_6 9766$ |
| Lbs. C | 14544.0 | 8080.00 | 3665.03 | 11315000 | 5657.63 | 1564396 | 5.71434 | 4.26285 | 15347000 | 1.00000 | 14.9876 |
| Lbs. $H_2O$ | 970.400 | 539.111 | 244.537 | 754971 | 377.487 | 104379 | .381270 | .284424 | 1023966 | .066744 | 1.00000 |

"P. C. U." refers to the "pound-centigrade unit." The ton used is 2000 pounds. "Lbs. C" refers to pounds of carbon oxidized, 100% efficiency, equivalent to the corresponding number of heat units. "Lbs H₂O" refers to pounds of water evaporated at 100°C, =212°F, at 100% efficiency

$C_1$ ⎮⎮⎮⎮⎮⎮⎮⎮⎮⎮⎮⎮⎮⎮⎮⎮⎮⎮⎮⎮⎮⎮⎮

By the use of the foregoing table[1] about 330 interconversions among twenty-six of the standard engineering units of measure can be directly estimated from the alignment chart to three significant figures or calculated by simple multiplication to six figures. The multiplier factor given in the table is located on the center scale "A" giving the point which when aligned with any number point on "C1" determines the product on "C." Imperfections in the scale due to lack of precision in printing should be checked at intervals along "A" scale by actual division of "C" by "C1," the lines being left out so that the reader can do this. A line scratched on a transparent celluloid triangle gives the best medium for making alignments.

When volume and weight interconversions are given, water is the medium the calculations are based upon. By the introduction of specific gravity factors the medium can be changed, giving the weight of any volume of any material, etc.

Courtesy of Chemical and Metallurgical Engineering.

## CONVERSION OF THERMOMETER READINGS

| F° | C° | F° | C° | F° | C° | F° | C° | F° | C° | F° | C° |
|---|---|---|---|---|---|---|---|---|---|---|---|
| —40 | —40.00 | 30 | —1.11 | 80 | 26.67 | 250 | 121.11 | 500 | 260.00 | 900 | 482.22 |
| —38 | —38.89 | 31 | —0.56 | 81 | 27.22 | 255 | 123.89 | 505 | 262.78 | 910 | 487.78 |
| —36 | —37.78 | 32 | 0.00 | 82 | 27.78 | 260 | 126.67 | 510 | 265.56 | 920 | 493.33 |
| —34 | —36.67 | 33 | 0.56 | 83 | 28.33 | 265 | 129.44 | 515 | 268.33 | 930 | 498.89 |
| —32 | —35.56 | 34 | 1.11 | 84 | 28.89 | 270 | 132.22 | 520 | 271.11 | 940 | 504.44 |
| —30 | —34.44 | 35 | 1.67 | 85 | 29.44 | 275 | 135.00 | 525 | 273.89 | 950 | 510.00 |
| —28 | —33.33 | 36 | 2.22 | 86 | 30.00 | 280 | 137.78 | 530 | 276.67 | 960 | 515.56 |
| —26 | —32.22 | 37 | 2.78 | 87 | 30.56 | 285 | 140.55 | 535 | 279.44 | 970 | 521.11 |
| —24 | —31.11 | 38 | 3.33 | 88 | 31.11 | 290 | 143.33 | 540 | 282.22 | 980 | 526.67 |
| —22 | —30.00 | 39 | 3.89 | 89 | 31.67 | 295 | 146.11 | 545 | 285.00 | 990 | 532.22 |
| —20 | —28.89 | 40 | 4.44 | 90 | 32.22 | 300 | 148.89 | 550 | 287.78 | 1000 | 537.78 |
| —18 | —27.78 | 41 | 5.00 | 91 | 32.78 | 305 | 151.67 | 555 | 290.55 | 1050 | 565.56 |
| —16 | —26.67 | 42 | 5.56 | 92 | 33.33 | 310 | 154.44 | 560 | 293.33 | 1100 | 593.33 |
| —14 | —25.56 | 43 | 6.11 | 93 | 33.89 | 315 | 157.22 | 565 | 296.11 | 1150 | 621.11 |
| —12 | —24.44 | 44 | 6.67 | 94 | 39.44 | 320 | 160.00 | 570 | 298.89 | 1200 | 648.89 |
| —10 | —23.33 | 45 | 7.22 | 95 | 35.00 | 325 | 162.78 | 575 | 301.67 | 1250 | 676.67 |
| — 8 | —22.22 | 46 | 7.78 | 96 | 35.56 | 330 | 165.56 | 580 | 304.44 | 1300 | 704.44 |
| — 6 | —21.11 | 47 | 8.33 | 97 | 36.11 | 335 | 168.33 | 585 | 307.22 | 1350 | 732.22 |
| — 4 | —20.00 | 48 | 8.89 | 98 | 36.67 | 340 | 171.11 | 590 | 310.00 | 1400 | 760.00 |
| — 2 | —18.89 | 49 | 9.44 | 99 | 37.22 | 345 | 173.89 | 595 | 312.78 | 1450 | 787.78 |
| 0 | —17.78 | 50 | 10.00 | 100 | 37.78 | 350 | 176.67 | 600 | 315.56 | 1500 | 815.56 |
| 1 | —17.22 | 51 | 10.56 | 105 | 40.55 | 355 | 179.44 | 610 | 321.11 | 1550 | 843.33 |
| 2 | —16.67 | 52 | 11.11 | 110 | 43.33 | 360 | 182.22 | 620 | 326.67 | 1600 | 871.11 |
| 3 | —16.11 | 53 | 11.67 | 115 | 46.11 | 365 | 185.00 | 630 | 332.22 | 1650 | 898.89 |
| 4 | —15.56 | 54 | 12.22 | 120 | 48.89 | 370 | 187.78 | 640 | 337.78 | 1700 | 926.67 |
| 5 | —15.00 | 55 | 12.78 | 125 | 51.67 | 375 | 190.55 | 650 | 343.33 | 1750 | 954.44 |
| 6 | —14.44 | 56 | 13.33 | 130 | 54.44 | 380 | 193.33 | 660 | 348.89 | 1800 | 982.22 |
| 7 | —13.89 | 57 | 13.89 | 135 | 57.22 | 385 | 196.11 | 670 | 354.44 | 1850 | 1010.00 |
| 8 | —13.33 | 58 | 14.44 | 140 | 60.00 | 390 | 198.89 | 680 | 360.00 | 1900 | 1037.78 |
| 9 | —12.78 | 59 | 15.00 | 145 | 62.78 | 395 | 201.67 | 690 | 365.56 | 1950 | 1065.56 |
| 10 | —12.22 | 60 | 15.56 | 150 | 65.56 | 400 | 204.44 | 700 | 371.11 | 2000 | 1093.33 |
| 11 | —11.67 | 61 | 16.11 | 155 | 68.33 | 405 | 207.22 | 710 | 376.67 | 2050 | 1121.11 |
| 12 | —11.11 | 62 | 16.67 | 160 | 71.11 | 410 | 210.00 | 720 | 382.22 | 2100 | 1148.89 |
| 13 | —10.56 | 63 | 17.22 | 165 | 73.89 | 415 | 212.78 | 730 | 387.78 | 2150 | 1176.67 |
| 14 | —10.00 | 64 | 17.78 | 170 | 76.67 | 420 | 215.56 | 740 | 393.33 | 2200 | 1204.44 |
| 15 | — 9.44 | 65 | 18.33 | 175 | 79.44 | 425 | 218.33 | 750 | 398.89 | 2250 | 1232.22 |
| 16 | — 8.89 | 66 | 18.89 | 180 | 82.22 | 430 | 221.11 | 760 | 404.44 | 2300 | 1260.00 |
| 17 | — 8.33 | 67 | 19.44 | 185 | 85.00 | 435 | 223.89 | 770 | 410.00 | 2350 | 1287.78 |
| 18 | — 7.78 | 68 | 20.00 | 190 | 87.78 | 440 | 226.67 | 780 | 415.56 | 2400 | 1315.56 |
| 19 | — 7.22 | 69 | 20.56 | 195 | 90.55 | 445 | 229.44 | 790 | 421.11 | 2450 | 1343.33 |
| 20 | — 6.67 | 70 | 21.11 | 200 | 93.33 | 450 | 232.22 | 800 | 426.67 | 2500 | 1371.11 |
| 21 | — 6.11 | 71 | 21.67 | 205 | 96.11 | 455 | 235.00 | 810 | 432.22 | 2550 | 1398.89 |
| 22 | — 5.56 | 72 | 22.22 | 210 | 98.89 | 460 | 237.78 | 820 | 437.78 | 2600 | 1426.67 |
| 23 | — 5.00 | 73 | 22.78 | 215 | 101.67 | 465 | 240.55 | 830 | 443.33 | 2650 | 1454.44 |
| 24 | — 4.44 | 74 | 23.33 | 220 | 104.44 | 470 | 243.33 | 840 | 448.89 | 2700 | 1482.22 |
| 25 | — 3.89 | 75 | 23.89 | 225 | 107.22 | 475 | 246.11 | 850 | 454.44 | 2750 | 1510.00 |
| 26 | — 3.33 | 76 | 24.44 | 230 | 110.00 | 480 | 248.89 | 860 | 460.00 | 2800 | 1537.78 |
| 27 | — 2.78 | 77 | 25.00 | 235 | 112.78 | 485 | 251.67 | 870 | 465.56 | 2850 | 1565.56 |
| 28 | — 2.22 | 78 | 25.56 | 240 | 115.56 | 490 | 254.44 | 880 | 471.11 | 2900 | 1593.33 |
| 29 | — 1.67 | 79 | 26.11 | 245 | 118.33 | 495 | 257.22 | 890 | 476.67 | 2950 | 1621.11 |

## ALCOHOL PROOF AND PERCENTAGE TABLE

| U. S. Proof at 60° F. | Per cent Alcohol by Volume at 60° F. | Per cent Alcohol by Weight | U. S. Proof at 60° F. | Per cent Alcohol by Volume at 60° F. | Per cent Alcohol by Weight |
|---|---|---|---|---|---|
| 0 | 0.0 | 0.00 | 57 | 28.5 | — |
| 1 | 0.5 | — | 58 | 29.0 | 23.82 |
| 2 | 1.0 | 0.80 | 59 | 29.5 | — |
| 8 | 1.5 | — | 60 | 30.0 | 24.67 |
| 4 | 2.0 | 1.59 | 61 | 30.5 | — |
| 5 | 2.5 | — | 62 | 31.0 | 25.52 |
| 6 | 3.0 | 2.39 | 63 | 31.5 | — |
| 7 | 3.5 | — | 64 | 32.0 | 26.38 |
| 8 | 4.0 | 3.19 | 65 | 32.5 | — |
| 9 | 4.5 | — | 66 | 33.0 | 27.24 |
| 10 | 5.0 | 4.00 | 67 | 33.5 | — |
| 11 | 5.5 | — | 68 | 34.0 | 28.10 |
| 12 | 6.0 | 4.80 | 69 | 34.5 | — |
| 13 | 6.5 | — | 70 | 35.0 | 28.97 |
| 14 | 7.0 | 5.61 | 71 | 35.5 | — |
| 15 | 7.5 | — | 72 | 36.0 | 29.84 |
| 16 | 8.0 | 6.42 | 73 | 36.5 | — |
| 17 | 8.5 | — | 74 | 37.0 | 30.72 |
| 18 | 9.0 | 7.23 | 75 | 37.5 | — |
| 19 | 9.5 | — | 76 | 38.0 | 31.60 |
| 20 | 10.0 | 8.05 | 77 | 38.5 | — |
| 21 | 10.5 | — | 78 | 39.0 | 32.48 |
| 22 | 11.0 | 8.86 | 79 | 39.5 | — |
| 23 | 11.5 | — | 80 | 40.0 | 33.36 |
| 24 | 12.0 | 9.63 | 81 | 40.5 | — |
| 25 | 12.5 | — | 82 | 41.0 | 34.25 |
| 26 | 13.0 | 10.50 | 83 | 41.5 | — |
| 27 | 13.5 | — | 84 | 42.0 | 35.15 |
| 28 | 14.0 | 11.32 | 85 | 42.5 | — |
| 29 | 14.5 | — | 86 | 43.0 | 36.05 |
| 30 | 15.0 | 12.14 | 87 | 43.5 | — |
| 31 | 15.5 | — | 88 | 44.0 | 36.96 |
| 32 | 16.0 | 12.96 | 89 | 44.5 | — |
| 33 | 16.5 | — | 90 | 45.0 | 37.86 |
| 34 | 17.0 | 13.79 | 91 | 45.5 | — |
| 35 | 17.5 | — | 92 | 46.0 | 38.78 |
| 36 | 18.0 | 14.61 | 93 | 46.5 | — |
| 37 | 18.5 | — | 94 | 47.0 | 39.70 |
| 38 | 19.0 | 15.44 | 95 | 47.5 | — |
| 39 | 19.5 | — | 96 | 48.0 | 40.62 |
| 40 | 20.0 | 16.27 | 97 | 48.5 | — |
| 41 | 20.5 | — | 98 | 49.0 | 41.55 |
| 42 | 21.0 | 17.10 | 99 | 49.5 | — |
| 43 | 21.5 | — | 100 | 50.0 | 42.49 |
| 44 | 22.0 | 17.93 | 101 | 50.5 | — |
| 45 | 22.5 | — | 102 | 51.0 | 43.43 |
| 46 | 23.0 | 18.77 | 103 | 51.5 | — |
| 47 | 23.5 | — | 104 | 52.0 | 44.37 |
| 48 | 24.0 | 19.60 | 105 | 52.5 | — |
| 49 | 24.5 | — | 106 | 53.0 | 45.33 |
| 50 | 25.0 | 20.44 | 107 | 53.5 | — |
| 51 | 25.5 | — | 108 | 54.0 | 46.23 |
| 52 | 26.0 | 21.28 | 109 | 54.5 | — |
| 53 | 26.5 | — | 110 | 55.0 | 47.24 |
| 54 | 27.0 | 22.13 | 111 | 55.5 | — |
| 55 | 27.5 | — | 112 | 56.0 | 48.21 |
| 56 | 28.0 | 22.97 | 113 | 56.5 | — |

| U.S. Proof at 60° F. | Per cent Alcohol by Volume at 60° F. | Per cent Alcohol by Weight | U.S. Proof at 60° F. | Per cent Alcohol by Volume at 60° F. | Per cent Alcohol by Weight |
|---|---|---|---|---|---|
| 114 | 57.0 | 49.19 | 158 | 79.0 | 72.38 |
| 115 | 57.5 | — | 159 | 79.5 | — |
| 116 | 58.0 | 50.17 | 160 | 80.0 | 73.53 |
| 117 | 58.5 | — | 161 | 80.5 | — |
| 118 | 59.0 | 51.15 | 162 | 81.0 | 74.69 |
| 119 | 59.5 | — | 163 | 81.5 | — |
| 120 | 60.0 | 52.15 | 164 | 82.0 | 75.86· |
| 121 | 60.5 | — | 165 | 82.5 | — |
| 122 | 61.0 | 53.15 | 166 | 83.0 | 77.04 |
| 123 | 61.5 | — | 167 | 83.5 | — |
| 124 | 62.0 | 54.15 | 168 | 84.0 | 78.23 |
| 125 | 62.5 | — | 169 | 84.5 | — |
| 126 | 63.0 | 55.16 | 170 | 85.0 | 79.44 |
| 127 | 63.5 | — | 171 | 85.5 | — |
| 128 | 64.0 | 56.18 | 172 | 86.0 | 80.62 |
| 129 | 64.5 | — | 173 | 86.5 | — |
| 130 | 65.0 | 57.21 | 174 | 87.0 | 81.90 |
| 131 | 65.5 | — | 175 | 87.5 | — |
| 132 | 66.0 | 58.24 | 176 | 88.0 | 83.14 |
| 133 | 66.5 | — | 177 | 88.5 | — |
| 134 | 67.0 | 59.28 | 178 | 89.0 | 84.41 |
| 135 | 67.5 | — | 179 | 89.5 | — |
| 136 | 68.0 | 60.32 | 180 | 90.0 | 85.69 |
| 137 | 68.5 | — | 181 | 90.5 | — |
| 138 | 69.0 | 61.38 | 182 | 91.0 | 86.99 |
| 139 | 69.5 | — | 183 | 91.5 | — |
| 140 | 70.0 | 62.44 | 184 | 92.0 | 88.31 |
| 141 | 70.5 | — | 185 | 92.5 | — |
| 142 | 71.0 | 63.51 | 186 | 93.0 | 89.65 |
| 143 | 71.5 | — | 187 | 93.5 | — |
| 144 | 72.0 | 64.59 | 188 | 94.0 | 91.02 |
| 145 | 72.5 | — | 189 | 94.5 | — |
| 146 | 73.0 | 65.67 | 190 | 95.0 | 92.42 |
| 147 | 73.5 | — | 191 | 95.5 | — |
| 148 | 74.0 | 66.77 | 192 | 96.0 | 93.85 |
| 149 | 74.5 | — | 193 | 96.5 | — |
| 150 | 75.0 | 67.87 | 194 | 97.0 | 95.32 |
| 151 | 75.5 | — | 195 | 97.5 | — |
| 152 | 76.0 | 68.92 | 196 | 98.0 | 96.82 |
| 153 | 76.5 | — | 197 | 98.5 | |
| 154 | 77.0 | 70.10 | 198 | 99.0 | 98.38 |
| 155 | 77.5 | — | 199 | 99.5 | — |
| 156 | 78.0 | 71.23 | 200 | 100.0 | 100.00 |
| 157 | 78.5 | — | | | |

## Buffer Systems

The following table gives some common buffer systems and the approximate pH of maximum buffer capacity. The zone of effective buffer action will vary with concentration but the general average will be ± 1.0 pH from the value given, for concentrations approximately 0.1 molar.

| | |
|---|---|
| Glycocoll - Sodium Chloride - Hydrochloric Acid | 2.0 |
| Potassium Acid Phthalate-Hydrochloric Acid | 2.8 |
| Primary Potassium Citrate | 3.7 |
| Acetic Acid-Sodium Acetate | 4.6 |
| Potassium Acid Phthalate-Sodium Hydroxide | 5.0 |
| Secondary Sodium Citrate | 5.0 |
| Carbonic Acid-Bicarbonate | 6.5 |
| Primary Phosphate-Secondary Phosphate | 6.8 |
| Primary Phosphate-Sodium Hydroxide | 6.8 |
| Boric Acid-Borax | 8.5 |
| Borax | 9.2 |
| Boric Acid-Sodium Hydroxide | 9.2 |
| Bicarbonate-Carbonate | 10.2 |
| Secondary Phosphate-Sodium Hydroxide | 11.5 |

Courtesy of W. A. Taylor & Company

# International Atomic Weights

| | Symbol | Atomic Number | Atomic Weight[1] | | Symbol | Atomic Number | Atomic Weight[1] |
|---|---|---|---|---|---|---|---|
| Actinium | Ac | 89 | 227 | Molybdenum | Mo | 42 | 95.95 |
| Aluminum | Al | 13 | 26.98 | Neodymium | Nd | 60 | 144.27 |
| Americium | Am | 95 | [243] | Neptunium | Np | 93 | [237] |
| Antimony | Sb | 51 | 121.76 | Neon | Ne | 10 | 20.183 |
| Argon | A | 18 | 39.944 | Nickel | Ni | 28 | 58.69 |
| Arsenic | As | 33 | 74.91 | Niobium | | | |
| Astatine | At | 85 | [210] | (Colubium) | Nb | 41 | 92.91 |
| Barium | Ba | 56 | 137.36 | Nitrogen | N | 7 | 14.008 |
| Berkelium | Bk | 97 | [245] | Nobelium | No | 102 | [...] |
| Beryllium | Be | 4 | 9.013 | Osmium | Os | 76 | 190.2 |
| Bismuth | Bi | 83 | 209.00 | Oxygen | O | 8 | 16 |
| Boron | B | 5 | 10.82 | Palladium | Pd | 46 | 106.7 |
| Bromine | Br | 35 | 79.916 | Phosphorus | P | 15 | 30.975 |
| Cadmium | Cd | 48 | 112.41 | Platinum | Pt | 78 | 195.23 |
| Calcium | Ca | 20 | 40.08 | Plutonium | Pu | 94 | [242] |
| Californium | Cf | 98 | [246] | Polonium | Po | 84 | 210 |
| Carbon | C | 6 | 12.010 | Potassium | K | 19 | 39.100 |
| Cerium | Ce | 58 | 140.13 | Praseodymium | Pr | 59 | 140.92 |
| Cesium | Cs | 55 | 132.91 | Promethium | Pm | 61 | [145] |
| Chlorine | Cl | 17 | 35.457 | Protactinium | Pa | 91 | 231 |
| Chromium | Cr | 24 | 52.01 | Radium | Ra | 88 | 226.05 |
| Cobalt | Co | 27 | 58.94 | Radon | Rn | 86 | 222 |
| Copper | Cu | 29 | 63.54 | Rhenium | Re | 75 | 186.31 |
| Curium | Cm | 96 | [243] | Rhodium | Rh | 45 | 102.91 |
| Dysprosium | Dy | 66 | 162.46 | Rubidium | Rb | 37 | 85.48 |
| Einsteinium | Es | 99 | [254] | Ruthenium | Ru | 44 | 101.7 |
| Erbium | Er | 68 | 167.2 | Samarium | Sm | 62 | 150.43 |
| Europium | Eu | 63 | 152.0 | Scandium | Sc | 21 | 44.96 |
| Fermium | Fm | 100 | [253] | Selenium | Se | 34 | 78.96 |
| Fluorine | F | 9 | 19.00 | Silicon | Si | 14 | 28.09 |
| Francium | Fr | 87 | [223] | Silver | Ag | 47 | 107.880 |
| Gadolinium | Gd | 64 | 156.9 | Sodium | Na | 11 | 22.997 |
| Gallium | Ga | 31 | 69.72 | Strontium | Sr | 38 | 87.63 |
| Germanium | Ge | 32 | 72.60 | Sulfur | S | 16 | 32.066[2] |
| Gold | Au | 79 | 197.2 | Tantalum | Ta | 73 | 180.88 |
| Hafnium | Hf | 72 | 178.6 | Technetium | Tc | 43 | [99] |
| Helium | He | 2 | 4.003 | Tellurium | Te | 52 | 127.61 |
| Holmium | Ho | 67 | 164.94 | Terbium | Tb | 65 | 159.2 |
| Hydrogen | H | 1 | 1.0080 | Thallium | Tl | 81 | 204.39 |
| Indium | In | 49 | 114.76 | Thorium | Th | 90 | 232.12 |
| Iodine | I | 53 | 126.91 | Thulium | Tm | 69 | 169.4 |
| Iridium | Ir | 77 | 193.1 | Tin | Sn | 50 | 118.70 |
| Iron | Fe | 26 | 55.85 | Titanium | Ti | 22 | 47.90 |
| Krypton | Kr | 36 | 83.80 | Tungsten | W | 74 | 183.92 |
| Lanthanum | La | 57 | 138.92 | Uranium | U | 92 | 238.07 |
| Lead | Pb | 82 | 207.21 | Vanadium | V | 23 | 50.95 |
| Lithium | Li | 3 | 6.940 | Xenon | Xe | 54 | 131.3 |
| Lutetium | Lu | 71 | 174.99 | Ytterbium | Yb | 79 | 173.04 |
| Magnesium | Mg | 12 | 24.32 | Yttrium | Y | 39 | 88.92 |
| Manganese | Mn | 25 | 54.93 | Zinc | Zn | 30 | 65.38 |
| Mercury | Hg | 80 | 200.61 | Zirconium | Zr | 40 | 91.22 |

[1] A value given in brackets denotes the mass number of the isotope of longest known half-life.

[2] Because of natural variations in the relative abundances of the isotopes of sulfur the atomic weight of this element has a range of $\pm 0.003$.

# REFERENCES AND ACKNOWLEDGMENTS

Agricultural Chemicals
Alchemist
Alchimist
American Fruit Grower
American Journal of Medical Sciences
American Journal of Pharmacy
American Perfumer
Australasian Engineer

Bakers' Helper
Bakers' Weekly
Bolletino chimico-farmaceutico
British Abstracts
British Plastics
Bulletin of the Institute of Physical and
  Chemical Research (Japan)

Calco Technical Bulletin
Canadian Dairy and Ice Cream Journal
Chemical Abstracts
Chemical Industries
Chemist Analyst
Chemist and Druggist
Copper and Brass Research Association
Cornell Veterinarian

Dioptic Review
Drug and Cosmetic Industry
Durez Plastics and Chemical Co.
Dyestuffs

Electronic Engineering
Esso Oilways

Florida Engineering and Industrial Experimental Station
Fonderie

Hutnik

Ice Cream Field
Ice Cream Trade Journal
Industry and Power

Journal of the American Ceramic Society
Journal of the American Concrete In-
  stitute
Journal of the American Leather Chem-
  ists' Association
Journal of the American Pharmaceutical
  Association
Journal of the Electrochemical Society

Journal of Investigative Dermatology
Journal of Tropical Medicine

The Laboratory
Leather and Shoes

Manufacturing Chemist
Materials and Methods
Metal Finishing
Metal Progress
Modern Lithography
Modern Plastics

N.A.R.D. Journal
National Bureau of Standards Technical
  News Bulletin
National Paint Bulletin
Nature
New Jersey Journal of Pharmacy

Organic Finishing

Paint and Varnish Production Manager
Patent Office Gazette
Perfumery and Essential Oil Record
Pharmaceutica Acta Helvetica
Pharmacy International
Plant Engineering
Power Transmission
Practitioner
Products Finishing
Promyshlennost Energetika
PSA Journal

Rayon and Synthetic Textiles
Resinous Reporter
Revue générale caoutchouc
Rubber Developments
Rubber World

Schimmel Briefs
Schundler and Co., F. E. Bulletins
Science
Science and Mechanics
Soap and Sanitary Chemicals
Soap, Perfumery and Cosmetics

Tin Research Institute

U. S. Dept. of Commerce Bulletin

Water and Sewage Works

# Trade-Mark Chemicals

The practice of marketing raw materials under names which in themselves are chemically not descriptive of the products they represent has become very widespread.  No modern book of formulae could justify its claims either to completeness or modernity without including numerous formulae containing trade-mark chemicals.

Without wishing to enter into a discussion of the justification of trade-marks, the Editor recognizes the tremendous service rendered to commercial chemistry by manufacturers of trade-mark products, both in the physical data supplied and the formulations suggested.

Deprived of the protection afforded their products by this system of nomenclature, these manufacturers would have been forced to stand by helplessly while the fruits of their labor were being filched from them by competitors who, unhampered by expenses of research, experimentation and promotion, would be able to produce something "just as good" at prices far below those of the original producers.

That these competitive products were "just as good" solely in the minds of the imitators would only be evidenced in costly experimental work on the part of the purchaser and, in the meantime, irreparable damage would have been done to the truly ethical product.  It is obvious, of course, that under these circumstances, there would be no incentive for manufacturers to develop new materials.

Because of this, and also because the CHEMICAL FORMULARY is primarily concerned with the physical results of compounding rather than with the chemistry involved, the Editor felt that the inclusion of formulae containing various trade-mark chemicals would be of definite value to the manufacturer of chemical products.  If they had been left out, many ideas and processes would have been automatically eliminated.  Trade-marks are in quotes throughout the book for easy distinction.

As a further service, the trade-mark products used in this volume are included in the list of chemicals and supplies.

## Chemicals and Supplies: Where to Buy Them * †

Numbers on right refer to list of suppliers on pages directly following this list. Thus to find out who supplies borax look in left-hand column, alongside borax, on page 368. The number there is 67. Now turn to page 391 and find number 67. Alongside is the supplier, American Potash & Chemical Corp., New York, N. Y.

| Product | No. |
|---|---|
| **A** | |
| A. A. P. Naphthols | 29 |
| A-Syrup | 855 |
| Abalyn | 557 |
| Abietic Acid | 557 |
| Abopon | 509 |
| Absorption Base | 605 |
| Accelerator 808 | 393 |
| Accelerator 833 | 393 |
| Accelerators, Vulcanization | 779 |
| Accroides, Gum | 941 |
| Acetamide | 123 |
| Acetanilide | 792 |
| Acetic Acid | 793 |
| Acetic Anhydride | 215 |
| Acetoin | 679 |
| Acetone | 978 |
| Acetphenetidine | 731 |
| Acetyl Cellulose...See Cellulose Acetate | |
| Acetyl Salicylic Acid | 565 |
| Acidolene | 707 |
| Acids, Fatty | 382 |
| Acimul | 509 |
| Acrawax | 509 |
| Acrawax B | 509 |
| Acrawax C | 509 |
| Acriflavine | 1 |
| Acrylic Resins | 895 |
| Acryloid | 895 |
| Activated Charcoal | 341 |
| Adeps Lanae | See Lanolin |
| Adheso Wax | 509 |
| Adipic Acid | 393 |
| A. D. M. No. 100 Oil | 87 |
| Advawet 33 | 11 |
| Aerogel | 749 |
| Aerosol | 45 |
| Agar | 721 |
| Agene | 807 |
| AgeRite Alba | 1141 |
| AgeRite Powder | 1141 |
| Akcocene | 45 |
| Aktivin | 15 |
| Albacer | 509 |
| Albalith | 785 |
| Albasol | 773 |
| Albatex | 277 |
| Albolith | 785 |

| Product | No. |
|---|---|
| Albone C | 393 |
| Albron | 21 |
| Albumen | 476 |
| Albusol | 697 |
| Alcohol, Denatured | 301 |
| Alcohol, Pure | 301 |
| Aldehol | 633 |
| Aldehyde $C_{14}$ | 1139 |
| Aldol | 793 |
| Alframine | 743 |
| Alginic Acid | 16 |
| Alizarin | 769 |
| Alkalies | 795 |
| Alkaloids | 602 |
| Alkanet | 577 |
| Alkanol | 393 |
| Alkyd Resins | 489 |
| Alloxan | 411 |
| Almond Oil | 693 |
| Aloes | 870 |
| Aloin | 585 |
| Aloxite | 221 |
| Alperox | 679 |
| Alpha Naphthol | 479 |
| Alphanaphthylthiourea | See Antu |
| Alphasol | 45 |
| Altax | 1141 |
| Alumina | 587 |
| Aluminum | 887 |
| Aluminum Acetate | 793 |
| Aluminum Bronze Powder | 48 |
| Aluminum Chloride | 798 |
| Aluminum Hydrate | 79 |
| Aluminum Hydroxide | |
| See Aluminum Hydrate | |
| Aluminum Nitrate | 545 |
| Aluminum Oleate | 979 |
| Aluminum Silicate | 500 |
| Aluminum Stearate | 1193 |
| Aluminum Sulfate | 483 |
| Aluminum Sulfocarbolate | 901 |
| Aluminum Tristearate | 697 |
| Alundum | 803 |
| Alvar | 977 |
| Amaranth | 721 |
| Amberette | 1065 |
| Amberlac | 895 |
| Amberol | 895 |

* See additional list on p. 385.

† Chemicals not listed here may be located by communicating with Chemical Week, 330 West 42 St., New York 36, N. Y., or consulting their Annual Buyer's Guide issue, or Oil, Paint and Drug Reporter, 30 Church St., New York 7, N. Y.

Dyestuffs included in the various formulae can be obtained from General Dyestuff Corporation, National Aniline Division of Allied Chemical & Dye Corp., or Calco Chemical Division of American Cyanamid Co.

| Product | No. | Product | No. |
|---|---|---|---|
| Cadmolith | 259 | Carbonex | 135 |
| Cajuput Oil | 591 | Carboraffin | 25 |
| Calagum | 509 | Carborundum | 221 |
| Calamine | 946 | Carboseal | 215 |
| Calcene | 299 | Carbowax | 215 |
| Calcium Acetate | 521 | Carboxide | 215 |
| Calcium Arsenate | 173 | Cardamom Seed | 789 |
| Calcium Carbonate | 669 | Carmine | 901 |
| Calcium Carbonate (Precipitated) | 731 | Carnauba Wax | 919 |
| Calcium Chloride | 745 | Carob Bean Flour | 1093 |
| Calcium Chloride (Anhydrous) | 433 | Carragheen | See Irish Moss |
| Calcium Citrate | 853 | Casco | 227 |
| Calcium Cyanamid | 45 | Casein | 227 |
| Calcium Fluoride | 483 | Cassia Oil | 391 |
| Calcium Hydroxide | See Lime Hydrate | Castile Soap | 315 |
| Calcium Lactate | 91a | Castor Oil | 113 |
| Calcium Ligno-Sulfonate | 702 | Castor Oil, Blown | 113 |
| Calcium Myristate | 142 | Castor Oil, Sulfonated | 1197 |
| Calcium Nitrate | 1068 | Castrolite | 923 |
| Calcium Oleate | 773 | Castung | 113 |
| Calcium Oxide | See Lime | Catalin | 33 |
| Calcium Phosphate | 875 | Catalpo | 751 |
| Calcium Polysulfide | 199 | Catechol | 841 |
| Calcium Propionate | 509 | Catylon | 547 |
| Calcium Resinate | 739 | Caustic Soda | 713 |
| Calcium Silicate | 299 | CCH | 713 |
| Calcium Silicofluoride | 28 | Cedar Oil | 383 |
| Calcium Stearate | 1069 | Celascour | 29 |
| Calcium Sulfate | See Plaster of Paris | Celeron | 317 |
| Calcium Sulfide (Luminous) | 57 | Celito | 621 |
| Calcocid | 201 | Cellit | See Cellulose Acetate |
| Calcolac | 201 | Cellosize | 215 |
| Calcoloid | 201 | Cellosolve | 215 |
| Calcozine | 201 | Cellosolve Acetate | 215 |
| Calgon | 191 | Cellosolve Ricinoleate | 509 |
| Calgonite | 203 | Celluloid | 231 |
| Calomel | 1201 | Celluloid Scrap | 973 |
| Calorite | 1029 | Cellulose Acetate | 231 |
| Camphor | 917 | Cellulose Acetate-Butyrate | 1083 |
| Camphor Oil | 693 | Cellulose Ether of Sodium Glycol- | |
| Canaga Oil | 837 | late | 393 |
| Candelilla Wax | 605 | Cellulose Fiber Pulp | 184 |
| Cantharides, Tincture of | 946 | Cellulose Nitrate | 733 |
| Cantharidin | 837 | Cement | 843 |
| Capillary Syrup | See Glucose Syrup | Censteric | 303 |
| Caproic Acid | 215 | Ceraflux | 509 |
| Captax | 1149 | Cercon | 309 |
| Caramel Coloring | 645 | Cerelose | 327 |
| Caraway Oil | 681 | Cereps | 1171 |
| Carbamide | 393 | Ceresalt | 409 |
| Carbitol | 215 | Ceresin Wax | 983 |
| Carbitol Acetate | 215 | Cerium Oxide | 715 |
| Carbofrax | 221 | Cerol | 939 |
| Carbolac | 197 | Cetamin | 509 |
| Carbolic Acid | See Phenol | Cetec | 489 |
| Carbolic Oil | 893 | Cetyl Alcohol | 587 |
| Carbolineum | 219 | Cetyl Trimethyl Ammonium Bro- | |
| Carbon, Activated | 619 | mide | 821 |
| Carbon Bisulfide | 123 | Chalk, Precipitated | 335 |
| Carbon Black | 1115 | Charcoal | 889 |
| Carbon, Decolorizing | 341 | Chemigum | 517 |
| Carbon Tetrachloride | 799 | Chestnut Extract | 894 |

| Product | No. | Product | No. |
|---|---|---|---|
| Nipasol | 937 | Oxyquinoline Sulfate | 149 |
| Nitramon | 393 | Ozokerite Wax | 1057 |
| Nitrated Cotton | See Pyroxylin | | |
| Nitre Cake | 1103 | **P** | |
| Nitric Acid | 749 | Palm Kernel Oil | 117 |
| Nitrobenzol | 201 | Palm Oil | 1193 |
| Nitrocellulose | 395 | Palm Oil Fatty Acids | 1171 |
| Nitrocotton | See Nitrocellulose | Palmitic Acid | 91 |
| Nitroethane | 303 | Pancreas | 91 |
| p-Nitrophenol | 749 | Panoline | 1033 |
| Nitropropane | 301 | Para Aminophenol | 1151 |
| Nonaethylene Glycol | 215 | Parachlormetacresol | 749 |
| Nonaethylene Glycol Laurate | 509 | Parachol | 509 |
| Nonaethylene Glycol Stearate S | 509 | Paracide | 575 |
| Nopco | 773 | Para-dor | 379 |
| Novolak | 111 | Paradura | 827 |
| Nuad | 811 | Paraffin, Chlorinated | 780 |
| Nuba Resin | 781 | Paraffin Oils | 959 |
| Nu-Char | 601 | Paraffin Wax | 815 |
| Nulomoline | 809 | Para-flux | 529 |
| Nuodex | 811 | Paralac | 599 |
| Nuodex Copper, Cobalt | 811 | Paraldehyde | 565 |
| Nusoap | 811 | Paramet | 827 |
| | | Paranitrophenol | 749 |
| | | Paranol | 827 |
| **O** | | Para-Phenylenediamine | 73 |
| | | Parapont | 393 |
| Ocenol | 395 | Parasept | 565 |
| Ochres | 999 | Paratoluene Sulfone Chloride | 749 |
| Octadecane Amide | 91 | Paris Black | 159 |
| Octyl Acetate | 215 | Paris Green | 357 |
| Octyl Alcohol | 215 | Paris White | 1013 |
| Oenanthic Ether | 681 | Parlon | 559 |
| Oil, Citronella | 591 | Paroil | 23 |
| Oil, Mineral | 1029 | Peachol | 509 |
| Oil Root Beer C | 969 | Peanut Oil | 421 |
| Oilate | 811 | Pearl Essence | 725 |
| Oildag | 5 | Pectin | 205 |
| Oilsolate | 895 | Peerless Clay | 1141 |
| Oiticica Oil | 615 | Pegopren | 511 |
| Olate | 873 | Pentacetate | 975 |
| Olein | 241 | Pentachloronitrobenzene | 749 |
| Oleoresins | 969 | Pentachlorphenol | 379 |
| Oleyl Alcohol | 393 | Pentaerythritol | 565 |
| Olive Oil | 659, 1057 | Pentaerythritol Abietate | 559 |
| Olive Oil Substitute | 509 | Pentaerythritol Esters | 565 |
| Olive Oil, Sulfonated | 923 | Pentalyn | 559 |
| Ondulum | 509 | Pentasol | 975 |
| Opal Wax | 393 | Pentawax | 565 |
| Orange Oil | 923 | Penicillin | 301 |
| Oroco | 921 | Pennyroyal Oil | 681 |
| Orvus | 873 | Pentrol | 633 |
| Ortho Dichlorbenzene | 575 | Peppermint Oil | 693 |
| Ortho-Phenylphenate | 379 | Peptone | 369 |
| Orthosil | 849 | Perchloric Acid | 731 |
| Osmo-Kaolin | 457 | Perchloroethylene | 393 |
| Ouricuri Wax | 375 | Perchloron | 849 |
| Oxalic Acid | 759 | Peregal | 479 |
| Oxgall | 781 | Perfume Bases | 595 |
| Oxycholesterol | 42 | Perilla Oil | 625 |
| Oxygen | 265 | Permosalt | 509 |
| Oxynone | 925 | Permosalt A | 509 |

| Product | No. | Product | No. |
|---|---|---|---|
| Propionic Acid | 393 | Resipon | 89 |
| Propyl Alcohol | 1054 | Resoglaz | 11 |
| Propyl p-Hydroxybenzoate | 565 | Resorcin | 841 |
| Propylene Dichloride | 215 | Resorcinol-Formaldehyde Resin | 845 |
| Propylene Glycol | 379 | Revertex | 899 |
| Propylene Glycol Monostearate | 509 | Rezidel | 509 |
| Propylene Glycol Stearate | 509 | Rezinel | 509 |
| Propylene Oxide | 215 | Rezyl | 45 |
| Protease Enzyme | 1165 | Rheolan | 509 |
| Protectoid | 235 | Rhodium | 115 |
| Protoflex | 509 | Rhonite | 915 |
| Protovac | 227 | Rhoplex | 915 |
| Provatol | 413 | Rhotex | 915 |
| Proxate | 673 | Rice Starch | 86 |
| Prussian Blue | 639 | Ricinoleic Acid | 113 |
| Prystal | 33 | Rochelle Salts | 853 |
| Psyllium Seeds | 657 | Rodo | 1141 |
| Puerine | 707 | Rose Water | 681 |
| Pumice | 335 | Rosemary Oil | 823 |
| Pylam Red | 881 | Roseol | 693 |
| Pyrax | 1141 | Rosin | 495 |
| Pyrefume | 837 | Rosin Oil | 777 |
| Pyrethol | 717 | Rosin, Polymerized | 559 |
| Pyrethrum | 837 | Rosin, Polypale | 557 |
| Pyrethrum Extract | 723 | Rosoap A | 509 |
| Pyridin | 647 | Rotenone | 1091 |
| Pyro............See Pyrogallic Acid | | Rouge | 273 |
| Pyrogallic Acid | 1207 | Rubber | 405 |
| Pyrolusite | 381 | Rubber Hydrochloride | 703 |
| Pyrophyllite | 1181 | Rubber Latex | 677 |
| Pyroxylin | 557 | Rubber Resin | 509 |
| Pyroxylin Solutions | 415 | Rubber, Synthetic ........393, 515, 517 | |
| | | Rubidium Salts | 689 |

**Q**

| | | Rum Ether | 469 |
|---|---|---|---|
| Quakersol | 839 | Rutgers 612 | 215 |
| Quartz Sand | 273 | Rutile | 1097 |
| Quassia Extract | 946 | | |
| Quebracho | 47 | **S** | |
| Quince Seed | 577 | "S" Syrup | 855 |
| Quinine Bisulfate | 523 | Saccharine | 565 |
| Quinine Hydrochloride | 731 | Sal Soda | 275 |
| Quinine Salicylate | 792 | Salicylanilide | 479 |
| Quinoline | 135 | Salicylic Acid | 379 |
| | | Salt | 755 |

**R**

| | | Salt Cake | 45 |
|---|---|---|---|
| | | Saltpeter | 331 |
| Raisin Seed Oil | 901 | Santicizers | 749 |
| Rancidex | 509 | Santobane | 749 |
| Rapeseed Oil | 125 | Santocel | 749 |
| Rapidase | 1165 | Santolite | 749 |
| Rauzene | 1127 | Santomask | 749 |
| Rayox | 1141 | Santomerse | 749 |
| Red Oil | 241 | Santowax | 749 |
| Red Oil Soap | 648 | Santox | 749 |
| Red Squill | 577 | Sapamine | 277 |
| Redmanol | 111 | Saponin | 627 |
| Reogen | 1141 | Sardine Oil | 675 |
| Resin DA 1 | 509 | Savolin | 509 |
| Resin R-H-35 | 393 | Schultz Silica | 251 |
| Resinox | 897 | Selenium | 61 |
| Resins, Natural | 45 | Sellatan A | 477 |
| Resins, Synthetic | 891 | Sepia............See Cuttle Fish Bone | |

| Product | No. |
|---|---|
| Stannous Chloride.....See Tin Chloride | |
| Stannous Fluoborate | 483 |
| Starch | 1045 |
| Staybelite | 559 |
| Staybelite Ester | 557 |
| Stearacol | 509 |
| Stearamide | 91 |
| Stearic Acid | 241 |
| Stearin | 1175 |
| Stearin Pitch | 527 |
| Stearite | 1193 |
| Stearol | 863 |
| Stearoricinol | 509 |
| Stearyl Alcohol | 901 |
| Stetsol | 593 |
| Stoddard Solvent | 347 |
| Storax | 693 |
| Stramonium | 837 |
| Stripolite | 923 |
| Stripper, T. S. | 89 |
| Stroba Wax | 509 |
| Strontium Hydrate | 130 |
| Strontium Nitrate | 519 |
| Strontium Sulfate | 130 |
| Strychnine | 853 |
| Styrax.....................See Storax | |
| Styrene | 379 |
| Styrene Dibromide | 379 |
| Styrex | 379 |
| Subcarbonate of Iron | 697 |
| Sublan | 509 |
| Succinic Acid | 769 |
| Sucrose Octoacetate | 793 |
| Sulfadiazine | 201 |
| Sulfanilamide | 201 |
| Sulfanole | 1169 |
| Sulfatate | 509 |
| Sulfated Castor Oil | 547 |
| Sulfathiazole | 201 |
| Sulfite Liquor | 724 |
| Sulfite Lye | 724 |
| Sulfo Turk A | 509 |
| Sulfo Turk B | 509 |
| Sulfo Turk C | 509 |
| Sulfogene | 393 |
| Sulfonamide | 201 |
| Sulfonated Castor Oil | 189 |
| Sulfonated Coconut Oil | 923 |
| Sulfonated Fatty Alcohol | 509 |
| Sulfonated Hydrogenated Castor Oil | 773 |
| Sulfonated Liquid Petrolatum | 1007 |
| Sulfonated Mineral Oil | 509 |
| Sulfonated Olive Oil | 1197 |
| Sulforicinol | 155 |
| Sulfur | 1047 |
| Sulfur Chloride | 1047 |
| Sulfur Dioxide | 1157 |
| Sulfuric Acid | 733 |
| Sulfurized Oils | 1021 |
| Sulfurized Sperm Oil | 773 |
| Sunoco Spirits | 1061 |
| Sunsoy | 1017 |

| Product | No. |
|---|---|
| Superba Black | 159 |
| Superphosphate | 1155 |
| Surfex | 879 |
| Suspendite | 509 |
| Suspensone | 509 |
| Syncrolite | 627 |
| Syntex | 293 |
| Synthane | 1067 |
| Synthenol | 1017 |
| Synvarite Resin | 1070 |

## T

| Product | No. |
|---|---|
| T-Tree Oil | 193 |
| Talc | 335 |
| Talcum.....................See Talc | |
| Tall Oil | 249, 601 |
| Tallow | 1171 |
| Tamal | 915 |
| Tamol | 915 |
| Tanax | 45 |
| Tannic Acid | 1207 |
| Tannin Extract | 47 |
| Tantalum | 437 |
| Tar Acid Oil | 135 |
| Tartar Emetic | 85 |
| Tartaric Acid | 523 |
| Tartrazine | 881 |
| Tea Seed Oil | 683 |
| Tegin | 513 |
| Teglac | 45 |
| Tegofan | 513 |
| Telloy | 1141 |
| Tellurium | 1141 |
| Tellurium Oxide | 69 |
| Tenex | 495 |
| Tenite | 1083 |
| Tepidone | 393 |
| Tergitol | 215 |
| Terpesol | 559 |
| Terpineol | 591 |
| Tetrachlorethane | 379 |
| Tetrachlorethylene | 393 |
| Tetradecanol | 215 |
| Tetralin | 393 |
| Tetrasodium Pyrophosphate | 266 |
| Tetrone | 393 |
| Textac | 559 |
| Textone | 713 |
| Thallium Bromide | 417 |
| Thallium Iodide | 417 |
| Thallium Sulfate | 627 |
| Thanite | 557 |
| Theop | 509 |
| Thermoplex | 895 |
| Thiamin Hydrochloride | 731 |
| Thiocarbamilid | 749 |
| Thioglycollic Acid | 1043 |
| Thionex | 393 |
| Thiourea | 627 |
| Thorium Salts | 1173 |
| Thyme Oil | 823 |
| Thymol | 981 |

| Product | No. |
|---|---|
| Thyroprotein | 1 |
| Ti-Cal | 649 |
| Tidolith | 1119 |
| Timonex | 1087 |
| Tin | 1113 |
| Tin Chloride | 971 |
| Tin Oxide | 719 |
| Tinctures | 829 |
| Tintite | 509 |
| Ti-Sil | 649 |
| Titanium Dioxide | 373 |
| Titanium Tetrachloride | 1047 |
| Titanox | 1097 |
| Ti-Tone | 649 |
| Ti-Tree Oil | 101 |
| Tocopherol | 943 |
| Tollac | 781 |
| Tolu Balsam Oil | 377 |
| Toluene | 135 |
| p-Toluene Sulfone Chloride | 749 |
| o-Toluidine | 639 |
| Toluol | 623 |
| Toluquinone | 731 |
| Toners | 1039 |
| Tonsil | 935 |
| Tornesit | 557 |
| Triacetin | 793 |
| Triamylamine | 975 |
| Tributyl Citrate | 303 |
| Tributyl Phosphite | 301 |
| Tributyrin | 1083 |
| Trichlorethylene | 393 |
| Trichlorobutyl Alcohol | 1139 |
| Triclene | 393 |
| Tricresol | 749 |
| Tricresyl Phosphate | 523 |
| Triethanolamine | 215 |
| Triethanolamine Lactate | 509 |
| Triethanolamine Naphthenate | 509 |
| Triethanolamine Oleate | 509 |
| Triethanolamine Phthalate | 509 |
| Triethanolamine Stearate | 509 |
| Triethylene Glycol Di-2-Ethyl Hexoate | 215 |
| Triethylene Glycol Ester of Hydrogenated Rosin | 557 |
| Trigamine | 509 |
| Trigamine Stearate | 509 |
| Trihydroxyethylamine See Triethanolamine | |
| Trikalin | 509 |
| Trimethylene Glycol | 873 |
| Triphenylguanidine | 393 |
| Triphenylphosphate | 749 |
| Tripoli | 1075 |
| Trisodium Phosphate | 1153 |
| Tritolyl Phosphate | 750 |
| Triton | 915 |
| Triton NE | 915 |
| Troluoil | 77 |
| Truline Binder | 557 |
| Tuads | 1141 |

| Product | No. |
|---|---|
| Tung Oil | See China Wood Oil |
| Tungsten | 437 |
| Tunguran A | 11 |
| Turkelene | 509 |
| Turkerol | 509 |
| Turkey Red Oil | 773 |
| Turmeric | 837 |
| Turpentine | 83 |
| Turpentine Substitute | 77 |
| Turpentine (Venice) | 777 |
| Turtle Oil | 967 |
| Tween | 103 |
| Twitchell Base | 427 |
| Tysenite | 1141 |

## U

| Product | No. |
|---|---|
| Uformite | 895 |
| Ultramarine Blue | 1039 |
| Ultranate | 97 |
| Ultrasene | 97 |
| Ultravon | 277 |
| Umbers | 441 |
| Undecalactone | 1139 |
| Undecylenic Acid | 779 |
| Unilith | 1119 |
| Union Solvent | 1111 |
| Unyte | 861 |
| Uranium Nitrate | 545 |
| Uranyl Nitrate | 697 |
| Urea | 981 |
| Urea-Formaldehyde Resin | 45 |
| Ureka C | 925 |
| Ursulin | 45 |
| Uversol | 545 |

## V

| Product | No. |
|---|---|
| Valex | 193 |
| Vanadium Pentoxide | 1136 |
| Van Dyke Brown | 1181 |
| Vandex | 1141 |
| Vanilla Beans | 1093 |
| Vanillal | 993 |
| Vanillin | 969 |
| Vanzyme | 1141 |
| Varcrex | 211 |
| Varcum | 1147 |
| Varnish | 757 |
| Varnish Gums and Resins | 45 |
| Varnolene | 1033 |
| Varsol | 1033 |
| Vaso | 1157 |
| Vat Colors | 29 |
| Vatsol | 45 |
| Veegum | 1141 |
| Vegetable Colors | 885 |
| Vegetable F Wax | 1055 |
| Velsicol 1068 | 1148 |
| Verdigris | 381 |
| Vermiculite | 567 |
| Vermilion | 441 |
| Victron | 1131 |

| Product | No. |
|---|---|
| Vinapas | 11 |
| Vinsol | 557 |
| Vinyl Acetate | 793 |
| Vinyl Chloride | 215 |
| Vinylite | 215 |
| Virifoam | 509 |
| Viscogum | 509 |
| Viscoloid | 393 |
| Vistac | 11 |
| Vistanex | 11 |
| Vitamins | 731 |
| Vitroil | See Sulfuric Acid |
| V. M. P. Naphtha | 1033 |
| Volclay | 43 |
| Vultex | 1161 |

## W

| Product | No. |
|---|---|
| Water Glass | See Sodium Silicate |
| Wax L33 | 509 |
| Wax, Microcrystalline | 1206 |
| Wax, Synthetic | 509 |
| Wetanol | 509 |
| Wetting Out Agents | 509 |
| Whale Oil | 1171 |
| White Arsenic | 857 |
| White Lead | 771 |
| White Oil | See Mineral Oil |
| White Wax | See Beeswax |
| Whitex Clay | 751 |
| Whiting | 299 |
| Witch Hazel Extract | 365 |
| Witco #1 | 1193 |
| Witco Yellow | 1193 |
| Wood Flour | 675 |
| Wood Oil | See China Wood Oil |
| Woodruff Essence | 595 |
| Wool Fat | See Lanolin |
| Wool Wax | 165 |
| Wool Wax Alcohols | 42 |
| Wyo-Jel | 1203 |

## X

| Product | No. |
|---|---|
| X-13 | 483 |
| Xerol | 471 |

| Product | No. |
|---|---|
| Xylene | See Xylol |
| Xylerol | 509 |
| Xylol | 135 |
| Xynomine | 821 |

## Y

| Product | No. |
|---|---|
| Yeast | 1027 |
| Yeast Extract | 1027 |
| Yelkin | 921 |
| Yellow Wax | See Beeswax |
| Ylang Ylang | 451 |
| Yumidol | 509 |

## Z

| Product | No. |
|---|---|
| Zein | 59 |
| Zelan | 393 |
| Zenite | 393 |
| Zikol | 791 |
| Zimate | 1141 |
| Zinc | 553 |
| Zinc Carbonate | 1193 |
| Zinc Chloride | 1193 |
| Zinc Chromate | 1163 |
| Zinc Fluoborate | 483 |
| Zinc Fluosilicate | 1060 |
| Zinc Lactate | 85 |
| Zinc Oleate | 545 |
| Zinc Oxide | 731 |
| Zinc Peroxide | 731 |
| Zinc Phosphide | 817 |
| Zinc Resinate | 791 |
| Zinc Silicofluoride | 545 |
| Zinc Stearate | 731 |
| Zinc Sulfate | 927 |
| Zinc Sulfide | 785 |
| Zinc Undecylenate | 142 |
| Zinol | 791 |
| Zirconium | 1095 |
| Zirconium Hydrate | 245 |
| Zirconium Oxide | 453 |
| Zirconium Oxychloride | 1095 |
| Zirex | 791 |
| Zopaque | 259 |

# Addenda to Chemicals and Supplies: Where to Buy Them

| Product | No. |
|---|---|
| **A** | |
| Abcovel | 1216 |
| Abitol | 557 |
| A-C Polyethylene | 1210 |
| Accelerator 552 | 393 |
| Accelerator E | 1282 |
| Acetate Flourescent Blue I | 201 |
| Acetulan | 42 |
| Acrysol | 915 |
| Adipol | 813 |
| Admerol 400 RS | 87 |
| Advance | 11 |
| Aerotex | 45 |
| Akroflex C | 393 |
| Alcolan | 1314 |
| Aldo | 509 |
| Aldocet | 509 |
| Aldrin | 1266 |
| Alfrax | 221 |
| Algipin | 1240 |
| Aliquat | 1255 |
| Alkaterge C | 301 |
| Alox | 1211 |
| Alpine Violet | 1277 |
| Alro Amine | 477 |
| Alropon | 1212 |
| Alrosene | 1212 |
| Alrosol | 1212 |
| Alumileaf | 1213 |
| Alumina C-730 | 21 |
| Amberite 115 Cement | 1261 |
| Amberlite | 915 |
| Ambropur | 1241 |
| Amerchol | 42 |
| Amerse | 1333 |
| Amine 220 | 215 |
| Aminophthal Hydrazide | 407 |
| Ammonyx | 821 |
| Anaesthesin | 149 |
| Ano | 487 |
| Anodex | 1279 |
| Antaron | 479 |
| Antarox | 1214 |
| Aqualized Gums | 509 |
| Ardanco | 87 |
| Aridye | 1269 |
| Armac | 91 |
| Armeen | 91 |
| Armid | 91 |

| Product | No. |
|---|---|
| Armowax | 91 |
| Arochem | 1127 |
| Aroclor | 749 |
| Arofene | 1127 |
| Arolene | 529 |
| Aroplas | 1127 |
| Arquad | 91 |
| ASA | 769 |
| Aseptex | 509 |
| ASP | 1287 |
| ASP 400 | 1246 |
| Astrotone BR | 1312 |
| ASU | 87 |
| Atlas G-3300 | 103 |
| Atlox | 103 |
| Atomite | 1325 |
| Attaclay | 1217 |
| Ayr Trap | 1263 |
| **B** | |
| BAC Latex | 393 |
| Badex | 1218 |
| Baker's P-8 | 113 |
| Barca No. 10 | 1239 |
| Bareco Wax | 1219 |
| Barnsdall Wax | 1219 |
| Barolan L | 42 |
| Beckacite | 891 |
| Bentone | 771 |
| Belloid | 477 |
| Benzene Hexachloride | 301 |
| Bismate | 1141 |
| BJF | 749 |
| Bon Ami | 1224 |
| Brij | 103 |
| Burgess Iceberg Pigment | 1229 |
| Burnok 3540 | 1335 |
| Butanol | 301 |
| Butrol | 1228 |
| Butvar | 977 |
| **C** | |
| C-700 | 1219a |
| C-1035 | 1219a |
| Camel-Carb | 1232 |
| Canawax | 1270 |
| Carbo-Lok | 1225 |
| Carbon Black. Oiled | 1115 |

384

| Product | No. | Product | No. |
|---|---|---|---|
| Carbopol | 515 | Diphenyl Guanidine (DPG) | 45 |
| Carbose | 1339 | Dithane | 915 |
| Carbosota | 1220 | Dixie Clay | 1141 |
| Cardis Wax | 1169 | Dodecyl Amine | 91 |
| Carnea Oil | 978 | Dow 276-V2 | 379 |
| Castroleum | 432 | Dow Latex | 379 |
| CDB 60,70,85 | 1252 | Dowanol | 379 |
| Celanese Solvent 901 | 231 | Drapex | 1215 |
| Cellite Superfloss | 621 | Dresinate | 557 |
| Collolyn | 557 | Dryfol | 1003 |
| Cellosolve Ricinoleate | 509 | DS-207 | 771 |
| Celluflex | 231 | Duco | 393 |
| Celogen | 779 | Duomeen | 91 |
| Cerese | 1004 | Durmont | 1242 |
| Cetiol | 1240 | Duroflex | 1226 |
| Cetrimide | 1235 | Duroxon | 1242 |
| Chemice-Lume | 1330 | Dustex | 1075 |
| Chlorasol | 215 | Dutrex | 978 |
| Chlordane | 1148 | Dylex K-34 | 647 |
| Chlorhydrol | 1310 | Dyphos | 771 |
| Chlorowax | 363 | Dypnone | 215 |
| Cinacoa | 1326 | | |
| Clorafin | 557 | **E** | |
| Collatone | 661 | | |
| Comperlan | 1240 | Elchem | 393 |
| Copper Naphthenate | 1193 | El-Sixty | 393 |
| Coray 230 | 432 | Elvacet | 393 |
| Cosol | 781 | Elvanol | 393 |
| CPR Dust Base | 1127 | Emcol M.S.-16 | 429a |
| Crag | 1328 | Emulgade | 1240 |
| Cresantol | 749 | Emulphogene | 487 |
| Crill | 1238 | Emulsifier BCO | 393 |
| Crown Wax 23 | 850 | Emulsifier G-1255 | 103 |
| Cumate | 1141 | Emulsifier S1131 | 509 |
| Cumene Hydroperoxide | 557 | Endothal | 1061 |
| Cunilate | 1317 | Endrin | 978 |
| | | Enjay Butyl 218 | 1248 |
| **D** | | Epolene | 1245 |
| 2,4-D | 646 | Epon | 978 |
| D-65A | 915 | Epotuf | 891 |
| Dantoin | 509 | Erusticator | 849 |
| Darex | 359 | Escalol | 1143 |
| Decroline | 487 | Estynox | 113 |
| Defoamer ED | 1247 | Ethazate | 779 |
| Dehydag Wax | 1240 | Ethofat | 91 |
| Diamine | 487 | Ethomeen | 91 |
| Dichlordimethyl Hydantoin | 509 | Ethyl Ethanol Amine | 975 |
| Dieldrin | 1266 | Eutanol | 1240 |
| Diethyl Amine Ethanol | 975 | Everflex | 359 |
| Dilan | 301 | EVT 50 | 1233 |

| Product | No. |
|---|---|
| London Rosin Oil | 777 |
| Lopar | 432 |
| Lorite | 771 |
| Lorolamine | 393 |
| Lubriplate | 1252 |
| Ludox | 393 |
| Luperco | 679 |
| Lupersol | 679 |
| Lustrex | 749 |
| Lykanut | 1326 |

**M**

| Product | No. |
|---|---|
| Magnesol | 1337 |
| Marasperse | 702 |
| Marbon | 703 |
| Marco | 1282 |
| Marco Resin | 1282 |
| Mark | 1215 |
| Marvinal | 1283 |
| MC Catalyst | 1282 |
| McNamee Clay | 1141 |
| MDI-50 | 393 |
| Mekon Y-20 | 1334 |
| Melmac Resin | 45 |
| Mentor Oil | 1033 |
| Metalead | 1285 |
| Methocel | 379 |
| Methoxychlor | 393 |
| Metrolin | 1227 |
| Michelene D.S. | 743 |
| Micro-Cel | 621 |
| Micromite | 1257 |
| Mifin | 477 |
| Milmer | 749 |
| Mineralite | 1286 |
| Miranol | 1288 |
| Mirapon RK | 1288 |
| Mistron | 1221 |
| Mitin | 477 |
| Modulan | 42 |
| Monamine ACO | 1289 |
| Monoplex S-71 | 915 |
| Monopol Soap | 1197 |
| Moropol | 1290 |
| MPS-500 | 575 |
| MR Resin | 1282 |
| Multicell | 1075 |
| Multiflex | 363 |
| MXP Wetting Agent | 749 |
| Myrj 52 | 103 |

| Product | No. |
|---|---|

**N**

| Product | No. |
|---|---|
| Naarden | 1291 |
| Nacelan | 769 |
| Nalco 211 | 765 |
| National Dyes | 769 |
| Neo-Fat | 91 |
| Nervan CS | 1249 |
| Neutronyx | 821 |
| Nevillac | 781 |
| Nevilloid | 781 |
| Nevsol | 781 |
| Nicotrol | 31 |
| Ninol 737 | 1295 |
| Nitrilo-Triacetic Acid | 487 |
| Non-Fer-Al | 363 |
| Nonic | 849 |
| Nonisol | 1212 |
| Nopcogen 14-11 | 773 |
| Norlin L-70 | 1297 |
| Nostrip | 1280 |
| Novacite | 1329 |
| NPX | 1328 |
| Nuact | 811 |
| Nullapon | 487 |
| Nupercaine | 277 |
| Nylon | 393 |
| Nypene | 781 |
| Nytal | 1141 |

**O**

| Product | No. |
|---|---|
| Octa-Klor | 1266 |
| Octylphenol | 915 |
| Oil (Z$_3$) | 87 |
| Oronite | 1298 |
| Ortholeum | 393 |
| Orvus | 873 |
| Ottasept | 1299 |

**P**

| Product | No. |
|---|---|
| P-33 | 1141 |
| P-4C, 8 | 113 |
| Palatine | 487 |
| Palatone | 379 |
| Pale 16 Oil | 113 |
| Palmeto Wax | 1270 |
| Panaflex | 1300 |
| Panapol | 1300 |
| Paracril | 1248 |
| Paradene | 781 |
| Paraplex G60 | 915 |

| Product | No. | Product | No. |
|---|---|---|---|
| Paratac | 1248 | Preservatol | 1143 |
| Parathion | 779 | Pril | 1223 |
| Parolite | 923 | Primol | 1023 |
| Parzate | 393 | Proflo | 1326 |
| PE100 | 1219a | Promoter No. 3 | 779 |
| Peg 42 | 509 | Protegin | 513 |
| Pelletex | 481 | Publicker Detergent | 877 |
| Pendit | 1309 | Purecal M | 1339 |
| Penetrol | 633 | Puzzolith | 1284 |
| Penford Gum | 1301 | Pyrenone | 1127 |
| Penglo | 1262 | | |
| Perhydrol | 11 | **Q** | |
| Permalite | 1258 | Quadrafos | 1316 |
| Permalux | 393 | Quaternary Ammonium Com- | |
| Permosol Base | 1264 | pounds | 821 |
| Petronauba C | 1219 | Quilon | 393 |
| Philblack | 1303 | | |
| P.H.O. | 781 | **R** | |
| Phygon | 779 | R & R 551 | 921 |
| Picco | 845 | Ratfold T | 1308 |
| Pike's Peak 9T66 | 1256 | RC | 1315 |
| Pip Pip | 749 | Red Arrow | 717 |
| Piperonyl Butoxide | 377 | Renex | 103 |
| Plasticizer C-2 | 215 | Resedalia | 485 |
| Plasticizer Pg-16 | 113 | Resimene | 749 |
| Plasticizer SC | 382 | Resloom | 749 |
| Plasto Dyes | 769 | Resyn 12K51 | 1292 |
| Plastolein | 427 | Retrol | 1260 |
| Pliovic | 517 | Rio Resin | 1141 |
| Plumb-O-Sil | 771 | Robane | 1313 |
| Pluramine | 1273 | Rodicide | 1311 |
| Pluronic | 1339 | Rodine | 37 |
| Plyophen | 891 | Rooticide | 1311 |
| Polar | 1220 | Rotax | 1141 |
| Polawax | 1238 | Rozye | 45 |
| Polyac | 393 | RPA No. 3 | 393 |
| Polycin | 113 | | |
| Polyethylene Glycol Mono- | | **S** | |
| laurate | 509 | Salol | 749 |
| Polyethylene Glycol Mono- | | Santobrite | 749 |
| stearate | 509 | Santocure | 749 |
| Poly Fon | 1338 | Saran | 379 |
| Polyglycol | 509 | Satin HT Clay | 500 |
| Poly Ox | 1328 | SC | 1236 |
| Polyplastex | 1305 | Scalecide | 1307 |
| Poly-Tex | 189 | Scopol | 1323 |
| Polythene | 393 | Seco, Emulsifying Oil | 1061 |
| Pontacyl | 393 | Sesamine | 1305 |
| Pontamine | 393 | Set Sit No. 5 | 1141 |
| P.P.D. | 749 | | |

| Product | No. | Product | No. |
|---|---|---|---|
| Shanco | 1318 | Tergitol Penetrant No. 4 | 215 |
| Sharsol | 975 | Tervan | 1248 |
| Silene | 859 | Tetrine | 509 |
| Silicone Oil L-41 | 1278 | Tetronic | 1339 |
| Sindar | 1319 | Texapon | 1240 |
| Snow Flake | 1325 | Texavon | 45 |
| Snow Floss | 621 | Thermax | 1141 |
| Sodium Tripolyphosphate | 1222 | Theroflex | 393 |
| Solulan | 42 | Thiofide | 749 |
| Sol-U-Pro | 1191 | Thiokol | 1324 |
| Solvadine | 277 | Thiurad | 749 |
| Sonojell | 1007 | Thiuram | 749 |
| Sorbo | 103 | Thixcin | 113 |
| Soybean Oil, Blown | 113 | Ti Bar | 393 |
| Spergon | 1294 | Tolura Oil | 432 |
| Stabelan | 1321 | Toxaphene | 1294 |
| Stabilizer 53,85,10-17 | 1153 | Transphalt | 845 |
| Stabilizer G-18 | 503 | Trex | 1259 |
| Stabilizer Mark I | 1215 | Tribase | 771 |
| Statex | 1275 | Trihydroxyethylamine Stearate | 509 |
| Staybelite | 557 | Trisamine | 301 |
| Staybelite Ester | 557 | Triton X-155, 200 | 915 |
| Stayrite | 1193 | Tylose | 479 |
| Stenol | 1240 | | |
| Sterox | 749 | **U** | |
| Strobane | 515 | | |
| Stuk-Rok Cement | 1261 | Ubatol 2003 | 1327 |
| Sulfocide | 1307 | Ubitol | 1327 |
| Sulfolane | 978 | Ucon | 215 |
| Sulfonated Neatsfoot Oil | | Ultrawet | 97 |
| (Grade CR) | 1101 | Unads | 1141 |
| Sun Emulso N | 1061 | Univul | 1214 |
| Sun Oil | 1061 | Uraform Cement | 1261 |
| Sunarome | 439 | | |
| Superfloss | 621 | **V** | |
| Superjet Lampblack | 1189 | | |
| Superla Wax | 1031 | Vanfre | 1141 |
| Supranol | 487 | Vanstay | 1141 |
| Surfonic | 1254 | Vapoton | 1231 |
| Sweetose | 1019 | Var 70 | 87 |
| Sylkyd | 380 | Vaseline | 255 |
| Synklor | 1294 | Veecote | 1141 |
| Synpron | 1230 | Veegum | 1141 |
| Syntex Resin | 1272 | Vejin | 1332 |
| Synvar | 1070 | Velvex | 1319a |
| | | Versene | 379 |
| **T** | | Vibrin | 779, 1131 |
| | | Victor | 1153 |
| Teflon | 393 | Victowet | 1153 |
| Tegacid | 513 | Vigilan | 1250 |
| Teresso | 1033 | Vinyon | 215 |

# SELLERS OF CHEMICALS AND SUPPLIES*

| No. | Name | Address |
|-----|------|---------|
| 1. | Abbott Laboratories | North Chicago, Ill. |
| 5. | Acheson Colloids Corp | Port Huron, Mich. |
| 7. | Acheson Graphite Corp | Niagara Falls, N. Y. |
| 9. | Acme Oil Corp | Chicago, Ill. |
| 11. | Advance Solvents & Chem. Corp | Jersey City, N. J. |
| 13. | Ajax Metal Co | Philadelphia, Pa. |
| 15. | Aktivin Corp | New York, N. Y. |
| 16. | Algin Corp. of America | New York, N. Y. |
| 17. | Allied Asphalt & Mineral Corp | New York, N. Y. |
| 21. | Aluminum Co. of America | Pittsburgh, Pa. |
| 23. | Amecco Chemicals, Inc | New York, N. Y. |
| 25. | American Active Carbon Co | Columbus, O. |
| 27. | American Agar Co., Inc | San Diego, Calif. |
| 28. | American Agricultural Chemical Co | New York, N. Y. |
| 29. | American Aniline Products, Inc | New York, N. Y. |
| 31. | American-Brit. Chem. Supplies, Inc | New York, N. Y. |
| 33. | American Catalin Corp | New York, N. Y. |
| 35. | American Cellulose Co | Indianapolis, Ind. |
| 37. | American Chemical Paint Co | Ambler, Pa. |
| 41. | American Chlorophyll, Inc | New York, N. Y. |
| 42. | American Cholesterol Products, Inc | Milltown, N. J. |
| 43. | American Colloid Co | Chicago, Ill. |
| 45. | American Cyanamid & Chem. Co | New York, N. Y. |
| 47. | American Dyewood Co | New York, N. Y. |
| 48. | American Firstoline Corp | New York, N. Y. |
| 49. | American Fluoride Corp | New York, N. Y. |
| 51. | American Insulator Corp | New Freedom, Pa. |
| 53. | American Lanolin Corp | Lawrence, Mass. |
| 57. | American Luminous Products Co | Huntington Park, Calif. |
| 59. | American Maize Products Co | New York, N. Y. |
| 61. | American Metal Co. | New York, N. Y. |
| 63. | American Mineral Spirit Co | New York, N. Y. |
| 64. | American Molasses Co | New York, N. Y. |
| 65. | American Plastics Corp | New York, N. Y. |
| 67. | American Potash & Chem. Corp | New York, N. Y. |
| 69. | American Smelting & Refining Co | New York, N. Y. |
| 71. | American Zinc Co | New York, N. Y. |
| 73. | Amido Products Co | New York, N. Y. |
| 77. | Anderson Prichard Oil Corp | Oklahoma City, Okla. |
| 79. | Ansbacher-Siegle Corp | Rosebank, N. Y. |
| 81. | Ansul Chemical Co | Marinette, Wis. |
| 83. | Antwerp Naval Stores Co., Inc | Boston, Mass. |
| 85. | Apex Chemical Co | New York, N. Y. |
| 86. | Arabol Mfg. Co | New York, N. Y. |
| 87. | Archer-Daniels-Midland Co | Minneapolis, Minn. |
| 89. | Arkansas Co. | Newark, N. J. |
| 91. | Armour & Co | Chicago, Ill. |
| 91a. | Armstrong Inc., C. M. | New York, N. Y. |
| 92. | Aromatic Products, Inc | New York, N. Y. |
| 93. | Asbury Graphite Mills | Asbury Park, N. J. |

* Just when this list was ready to go to press, it was carefully checked and it was found that some of the firms listed went out of business. Their names have been cancelled, and this accounts for the discontinuity in numbering.

| No. | Name | Address |
|---|---|---|
| 95. | Atlantic Gelatine Co. | Woburn, Mass. |
| 97. | Atlantic Refining Co. | Philadelphia, Pa. |
| 99. | Atlantic Research Associates | Newtonville, Mass. |
| 103. | Atlas Powder Co. | Wilmington, Del. |
| 109. | Badcock, Robert & Co. | New York, N. Y. |
| 111. | Bakelite Corp. | New York, N. Y. |
| 113. | Baker Castor Oil Corp. | Jersey City, N. J. |
| 115. | Baker & Co., Inc. | Newark, N. J. |
| 117. | Baker, Franklin Co. | Hoboken, N. J. |
| 119. | Baker, H. J. & Bro. | New York, N. Y. |
| 121. | Baker, J. E., Co. | York, Pa. |
| 123. | Baker, J. T. Chem. Co. | Philipsburg, N. J. |
| 125. | Balfour, Guthrie & Co., Ltd. | New York, N. Y. |
| 127. | Barada & Page, Inc. | Kansas City, Mo. |
| 129. | Barber Asphalt Co. | Philadelphia, Pa. |
| 130. | Barium Chemicals, Inc. | Willoughby, O. |
| 131. | Barium Reduction Corp. | Charleston, W. Va. |
| 133. | Barnsdall Tripoli Corp. | Seneca, Mo. |
| 135. | Barrett Co. | New York, N. Y. |
| 139. | Battelle & Renwick | New York, N. Y. |
| 141. | Battleboro Oil Co. | Battleboro, N. C. |
| 142. | Beacon Co. | Boston, Mass. |
| 143. | Beck, Koller & Co. | Detroit, Mich. |
| 145. | Belmont Smelting & Refining Wks. | Brooklyn, N. Y. |
| 149. | Benzol Products Co. | Newark, N. J. |
| 149a. | Berk & Co. Inc., F. W. | Wood Ridge, N. J. |
| 150. | Bernard Color & Chem. Co. | New York, N. Y. |
| 151. | Bersworth Labs., F. C. | Framingham, Mass. |
| 153. | Beryllium Corp. of America | New York, N. Y. |
| 155. | Bick & Co., Inc. | Reading, Pa. |
| 157. | Bilhuber-Knoll Corp. | New York, N. Y. |
| 159. | Binney & Smith | New York, N. Y. |
| 161. | Bisbee Linseed Co. | Philadelphia, Pa. |
| 165. | Bopf-Whittam Corp. | Linden, N. J. |
| 167. | Borax Union, Inc. | San Francisco, Calif. |
| 169. | Borne Scrymser Co. | New York, N. Y. |
| 171. | Bowdlear Co., W. H. | Syracuse, N. Y. |
| 173. | Bowker Chemical Corp. | New York, N. Y. |
| 175. | Bradley & Baker | New York, N. Y. |
| 177. | Brazil Oiticica, Inc. | New York, N. Y. |
| 179. | British Drug Houses, Ltd. | London, England |
| 181. | British Xylonite Co. | London, England |
| 183. | Brooke, Fred L. Co. | Chicago, Ill. |
| 184. | Brown Co. | Portland, Me. |
| 185. | Brush Beryllium Co. | Cleveland, O. |
| 187. | Buffalo Electro Chem. Co. | Buffalo, N. Y. |
| 189. | Burkard-Schier Chem. Co. | Chattanooga, Tenn. |
| 191. | Buromin Corp. | Pittsburgh, Pa. |
| 193. | Bush, W. J. & Co., Inc. | New York, N. Y. |
| 194. | Byerlyte Corp. | Cleveland, O. |
| 197. | Cabot, Godfrey L., Inc. | Boston, Mass. |
| 199. | Calcium Sulfide Corp. | Damascus, Va. |
| 201. | Calco Chemical Co. | Bound Brook, N. J. |
| 203. | Calgon, Inc. | Pittsburgh, Pa. |
| 205. | Calif. Fruit Growers Exchange | Ontario, Calif. |
| 206. | Calif. Milk Products Co. | Philadelphia, Pa. |
| 207. | Campbell, C. W. Co., Inc. | New York, N. Y. |
| 209. | Campbell, John & Co. | New York, N. Y. |
| 211. | Campbell Rex & Co. | London, England |
| 213. | Carbic Color & Chemical Co. | New York, N. Y. |
| 215. | Carbide & Carbon Chem. Corp. | New York, N. Y. |
| 219. | Carbolineum Wood Preserving Co. | Milwaukee, Wis. |

| No. | Name | Address |
|---|---|---|
| 221. | Carborundum Co. | Niagara Falls, N. Y. |
| 223. | Carey, Philip Co. | Lockland, Ohio |
| 225. | Carus Chem. Co., Inc. | La Salle, Ill. |
| 229. | The Casein Mfg. Co. of Amer., Inc. | New York, N. Y. |
| 231. | Celanese Corp. | New York, N. Y. |
| 239. | Central Scientific Co. | Chicago, Ill. |
| 241. | Century Stearic Acid & Candle Wks. | New York, N. Y. |
| 245. | Ceramic Color & Chem. Mfg. Co. | New Brighton, Pa. |
| 247. | Cerro. de Pasco Copper Corp. | New York, N. Y. |
| 249. | Champion Paper & Fiber Co. | Canton, N. C. |
| 251. | Chaplin-Bibbo | New York, N. Y. |
| 253. | Chazy Marble Lime Co., Inc. | Chazy, N. Y. |
| 255. | Chesebrough Mfg. Co. | New York, N. Y. |
| 257. | Chemical & Pigment Co. | Baltimore, Md. |
| 263. | Chemical Solvents, Inc. | New York, N. Y. |
| 265. | Cheney Chem. Co. | Cleveland, Ohio |
| 266. | Chew, John A., Ine. | New York, N. Y. |
| 267. | Chicago Apparatus Co. | Chicago, Ill. |
| 269. | Chicago Copper & Chem. Co. | Blue Island, Ill. |
| 271. | Chipman Chem. Co., Inc. | Bound Brook, N. J. |
| 272. | Chiris, Antoine Co. | New York, N. Y. |
| 273. | Chrystal, Charles B. Co., Inc. | New York, N. Y. |
| 275. | Church & Dwight Co., Inc. | New York, N. Y. |
| 277. | Ciba Co., Inc. | New York, N. Y. |
| 279. | Cinelin Co. | Indianapolis, Ind. |
| 281. | Clarke, John & Co. | New York, N. Y. |
| 283. | The Cleveland-Cliffs Iron Co. | Cleveland, Ohio |
| 285. | Climax Molybdenum Co. | New York, N. Y. |
| 287. | Clinton Co. | Clinton, Ia. |
| 289. | Coleman & Bell Co. | Norwood, Ohio |
| 293. | Colgate-Palmolive-Peet Co. | Jersey City, N. J. |
| 295. | Colledge, E. W., Inc. | Cleveland, Ohio |
| 297. | Colonial Beacon Oil Co. | Everett, Mass. |
| 299. | Columbia Alkali Corp. | New York, N. Y. |
| 301. | Commercial Solvents Corp. | New York, N. Y. |
| 305. | Commonwealth Color & Chem. Co. | Brooklyn, N. Y. |
| 307. | Compagnie Duval | New York, N. Y. |
| 309. | Conewango Refining Co. | Warren, Pa. |
| 311. | Consolidated Chem. Sales Corp. | Newark, N. J. |
| 313. | Consolidated Feldspar Corp. | Trenton, N. J. |
| 315. | Conti Products Corp. | New York, N. Y. |
| 317. | Continental Diamond Fiber Co. | Bridgeport, Pa. |
| 319. | Continental Oil Co. | Ponca City, Okla. |
| 321. | Cook Swan Co., Inc. | New York, N. Y. |
| 325. | Coopers Creek Chem. Co. | W. Conshohocken, Pa. |
| 327. | Corn Products Refining Co. | New York, N. Y. |
| 329. | Cowles Detergent Co. | Cleveland, Ohio |
| 330. | C. P. Chemical Solvents, Inc. | New York, N. Y. |
| 331. | Croton Chem. Corp. | Brooklyn, N. Y. |
| 333. | Crowley Tar Products Co. | New York, N. Y. |
| 335. | Crystal, Charles B. Co., Inc. | New York, N. Y. |
| 336. | Crystal Soap & Chem. Co. | Philadelphia, Pa. |
| 337. | Cudahy Packing Co. | Chicago, Ill. |
| 339. | Danco, Gerard J. | New York, N. Y. |
| 341. | Darco Corp. | New York, N. Y. |
| 343. | Darling & Co. | Chicago, Ill. |
| 345. | Davison Chem. Corp. | Baltimore, Md. |
| 347. | Deep Rock Oil Corp. | Chicago, Ill. |
| 349. | De Lore, C. P. Co. | St. Louis, Mo. |
| 351. | Delta Chem. Mfg. Co. | Baltimore, Md. |
| 353. | Delta Chem. & Iron Co. | Wells, Mich. |
| 355. | Denver Fire Clay Co. | Denver, Colo. |

| No. | Name | Address |
|-----|------|---------|
| 357. | Devoe & Reynolds Co. | New York, N. Y. |
| 359. | Dewey & Almy Chem. Co. | Boston, Mass. |
| 363. | Diamond Alkali Co. | Pittsburgh, Pa. |
| 364. | Dicalite Co. | New York, N. Y. |
| 365. | Dickinson, E. E. Co. | Essex, Conn. |
| 367. | Dickinson, J. Q. & Co. | Malden, W. Va. |
| 369. | Difco Laboratories, Inc. | Detroit, Mich. |
| 371. | Digestive Ferments Co. | Detroit, Mich. |
| 373. | Marshall Dill. | San Francisco, Calif. |
| 375. | Distributing & Trading Co. | New York, N. Y. |
| 377. | Dodge & Olcott Co. | New York, N. Y. |
| 379. | Dow Chemical Co. | Midland, Mich. |
| 380. | Dow Corning Corp. | Midland, Mich. |
| 381. | Drakenfeld, B. F. & Co. | New York, N. Y. |
| 382. | Drew, E. F. Co. | New York, N. Y. |
| 383. | Dreyer, P. R. Co. | New York, N. Y. |
| 385. | Dreyfus Co., L. A. | Rosebank, N. Y. |
| 387. | Drury, A. C. & Co., Inc. | Chicago, Ill. |
| 389. | Ducas, B. P. Co. | New York, N. Y. |
| 391. | Duche, T. M. & Sons. | New York, N. Y. |
| 393. | DuPont, E. I., de Nemours & Co., Inc. | Wilmington, Del. |
| 397. | Durite Plastics. | Philadelphia, Pa. |
| 401. | Eagle-Picher Lead Co. | Cincinnati, Ohio |
| 403. | Eakins, J. S. & W. R., Inc. | Brooklyn, N. Y. |
| 405. | Earle Bros. | New York, N. Y. |
| 407. | Eastman Kodak Co. | Rochester, N. Y. |
| 409. | Economic Materials Co. | Chicago, Ill. |
| 411. | Edwal Labs. | Chicago, Ill. |
| 413. | Eff Laboratories Inc. | Cleveland, Ohio |
| 415. | Egyptian Lacquer Co. | Kearney, N. J. |
| 417. | Eimer & Amend. | New York, N. Y. |
| 421. | Elbert & Co. | New York, N. Y. |
| 423. | Electro Bleaching Gas Co. | New York, N. Y. |
| 425. | Electro-Metallurgical Co. | New York, N. Y. |
| 427. | Emery Industries, Inc. | Cincinnati, Ohio |
| 429. | Empire Distilling Corp. | New York, N. Y. |
| 429a. | Emulsol Corp. | Chicago, Ill. |
| 430. | Enco Chemical Corp. | New York, N. Y. |
| 431. | Enterprise Animal Oil Co. | Philadelphia, Pa. |
| 431a. | Essential Aromatics Corp. | New York, N. Y. |
| 432. | Esso Marketers | New York, N. Y. |
| 432a. | Fairmount Chemical Co. | Newark, N. J. |
| 433. | Fales Chem. Co., Inc. | Cornwall Landing, N. Y. |
| 435. | Falk & Co. | Pittsburgh, Pa. |
| 437. | Fansteel Metallurgical Corp. | No. Chicago, Ill. |
| 439. | Felton Chemical Co. | Brooklyn, N. Y. |
| 441. | Fezandie & Sperrle, Inc. | New York, N. Y. |
| 443. | Fiberloid Corp. | Indian Orchard, Mass. |
| 445. | Filtrol Co. | Los Angeles, Calif. |
| 445a. | Fine Organics, Inc. | New York, N. Y. |
| 446. | Firestone Tire & Rubber Co. | Akron, Ohio |
| 447. | Fishbeck, Chas. Co. | New York, N. Y. |
| 449. | Fisher Scientific Co. | Pittsburgh, Pa. |
| 451. | Florasynth Laboratories | New York, N. Y. |
| 453. | Foote Mineral Co. | Philadelphia, Pa. |
| 455. | Formica Insulation Co. | Cincinnati, Ohio |
| 457. | Fougera, E. & Co. | New York, N. Y. |
| 459. | France, Campbell & Darling. | Kenilworth, N. J. |
| 461. | Franco-American Chemical Wks. | Carlstadt, N. J. |
| 463. | Frank-Vliet Co. | New York, N. Y. |
| 465. | Franks Chem. Products Co., Inc. | Brooklyn, N. Y. |
| 467. | French Potash Co. | New York, N. Y. |

| No. | Name | Address |
|-----|------|---------|
| 469. | Fries, Alex. & Bro | Cincinnati, Ohio |
| 471. | Fries Bros | New York, N. Y. |
| 473. | Fritzchie Bros | New York, N. Y. |
| 475. | Garrigues, Stewart & Davies, Inc | New York, N. Y. |
| 476. | Gartenberg, H. & Co., Inc | Chicago, Ill. |
| 477. | Geigy Co., Inc | New York, N. Y. |
| 479. | General Aniline Works, Inc | New York, N. Y. |
| 481. | General Atlas Carbon Co | New York, N. Y. |
| 483. | General Chemical Co | New York, N. Y. |
| 485. | General Drug Co | New York, N. Y. |
| 487. | General Dyestuffs Corp | New York, N. Y. |
| 489. | General Electric Co | Pittsfield, Mass. |
| 491. | General Electric Co | Schenectady, N. Y. |
| 493. | General Magnesite & Magnesia Co | Philadelphia, Pa. |
| 495. | General Naval Stores Co | New York, N. Y. |
| 497. | General Plastics Corp | London, England |
| 499. | General Plastics, Inc | No. Tonawanda, N. Y. |
| 499a. | General Refractories Co | Philadelphia, Pa. |
| 500. | Georgia Kaolin Co | Elizabeth, N. J. |
| 501. | Girdler Corp | Louisville, Ky. |
| 503. | Givaudan-Delawanna, Inc | New York, N. Y. |
| 505. | Glidden Co | Cleveland, Ohio |
| 507. | Globe Chem. Co | Cincinnati, Ohio |
| 509. | Glyco Products Co., Inc | Brooklyn, N. Y. |
| 513. | Goldschmidt Corp | New York, N. Y. |
| 514. | Goldsmith Bros. Smelt. & Refining Co | New York, N. Y. |
| 515. | Goodrich, B. F. Co | Akron, Ohio |
| 517. | Goodyear Tire & Rubber Co | Akron, Ohio |
| 519. | Grasselli Chemical Co | Cleveland, Ohio |
| 521. | Gray, W. S. Co | New York, N. Y. |
| 523. | Greeff, R. W. & Co | New York, N. Y. |
| 525. | Griffith Laboratories | Chicago, Ill. |
| 527. | Gross, A. & Co | New York, N. Y. |
| 529. | Hall, C. P. & Co | Akron, Ohio |
| 531. | Halowax Corp | New York, N. Y. |
| 533. | Hammill & Gillespie, Inc | New York, N. Y. |
| 535. | Hamilton, A. K | New York, N. Y. |
| 537. | Hammond Drierite Co | Yellow Springs, Ohio |
| 539. | Handy & Harman | New York, N. Y. |
| 540. | Hardesty Chemical Co | New York, N. Y. |
| 541. | Hardy, Charles, Inc | New York, N. Y. |
| 543. | Harrison Mfg. Co | Rahway, N. J. |
| 545. | Harshaw Chemical Co | Cleveland, Ohio |
| 547. | Hart Products Corp | New York, N. Y. |
| 549. | Haskelite Mfg. Corp | Chicago, Ill. |
| 551. | Haveg Corp | Newark, Del. |
| 553. | Hegeler Zinc Co | Danville, Ill. |
| 555. | Heine & Co | New York, N. Y. |
| 557. | Hercules Powder Co | New York, N. Y. |
| 561. | Heveatex Corp | Melrose, Mass. |
| 563. | Hewitt, C. B. & Bro | New York, N. Y. |
| 565. | Heyden Chemical Corp | New York, N. Y. |
| 567. | Hill Bros. Chemical Co | Los Angeles, Calif. |
| 569. | Hillside Fluor Spar Mines | Chicago, Ill. |
| 570. | Hilo Varnish Co | Brooklyn, N. Y. |
| 571. | Holland Aniline Dye Co | Holland, Mich. |
| 573. | Hommel, O. Co | Pittsburgh, Pa. |
| 575. | Hooker Electro-Chemical Co | New York, N. Y. |
| 577. | Hopkins, J. L. & Co | New York, N. Y. |
| 579. | Hord Color Products | Sandusky, Ohio |
| 581. | Horn Jefferys & Co | Burbank, Calif. |
| 583. | Horner, James B., Inc | New York, N. Y. |

| No. | Name | Address |
|---|---|---|
| 585. | Huisking, Chas. L. & Co., Inc. | New York, N. Y. |
| 587. | Hummel Chemical Co., Inc. | New York, N. Y. |
| 589. | Hurst, Adolph & Co., Inc. | New York, N. Y. |
| 591. | Hutchinson, D. W. & Co., Inc. | New York, N. Y. |
| 593. | Hycar Corp. | Akron, Ohio |
| 595. | Hymes, Lewis Associates. | New York, N. Y. |
| 599. | Imperial Chem. Industries. | London, England |
| 601. | Industrial Chem. Sales Co. | New York, N. Y. |
| 602. | Inland Alkaloid Co. | Tipton, Ind. |
| 603. | Innes, O. G. Corp. | New York, N. Y. |
| 605. | Innis Speiden Co. | New York, N. Y. |
| 607. | International Pulp Corp. | New York, N. Y. |
| 609. | International Selling Corp. | New York, N. Y. |
| 611. | Interstate Color Co., Inc. | New York, N. Y. |
| 613. | Iowa Soda Products Co. | Council Bluffs, Ia. |
| 615. | Jackson, L. N. & Co. | New York, N. Y. |
| 617. | Jacobson, C. A., W. Va. Univ. | Morgantown, W. Va. |
| 619. | Jennison-Wright Co. | Toledo, Ohio |
| 621. | Johns-Manville Corp. | New York, N. Y. |
| 623. | Jones & Laughlin Steel Corp. | Pittsburgh, Pa. |
| 625. | Jones, S. L. & Co. | San Francisco, Calif. |
| 629. | Kali Mfg. Co. | Philadelphia, Pa. |
| 633. | Kay Fries Chem., Inc. | New York, N. Y. |
| 635. | Kelco Co. | San Diego, Calif. |
| 637. | Kentucky Clay Mining Co. | Mayfield, Ky. |
| 639. | Kentucky Color & Chem. Co. | Louisville, Ky. |
| 641. | Kessler Chem. Corp. | Philadelphia, Pa. |
| 643. | Kinetic Chem., Inc. | Wilmington, Del. |
| 645. | Kohnstamm, H. & Co. | New York, N. Y. |
| 646. | Kolker Chem. Works. | Newark, N. J. |
| 647. | Koppers Products Co. | Pittsburgh, Pa. |
| 648. | Kranich Soap Co. | Brooklyn, N. Y. |
| 649. | Krebs Pigment & Color Corp. | Newark, N. J. |
| 651. | Kuhlman, Establs. | Paris, France |
| 655. | Lattimer-Goodwin Chem. Co. | Grand Junction, Ohio |
| 657. | Laxseed Co. | New York, N. Y. |
| 659. | Leghorn Trading Co., Inc. | New York, N. Y. |
| 661. | Lehn & Fink Corp. | New York, N. Y. |
| 663. | Leonhard Wax, Theo. Co., Inc. | Haledon, Paterson, N. J. |
| 665. | Lewis, C. H. & Co. | New York, N. Y. |
| 667. | Lewis, John D., Inc. | Providence, R. I. |
| 669. | Limestone Products Corp. of America | Newton, N. J. |
| 671. | Lincks, Geo. H. | New York, N. Y. |
| 672. | Lindsay Light & Chem. Co. | Chicago, Ill. |
| 673. | Liquid Carbonic Corp. | Chicago, Ill. |
| 675. | Litter, D. H., Co. | New York, N. Y. |
| 677. | Littlejohn & Co., Inc. | New York, N. Y. |
| 679. | Lucidol Corp. | Buffalo, N. Y. |
| 681. | Lueders, Geo. & Co. | New York, N. Y. |
| 683. | Lundt & Co. | New York, N. Y. |
| 685. | Maas & Waldstein. | Newark, N. J. |
| 687. | MacAndrews & Forbes Co. | New York, N. Y. |
| 689. | Mackay, A. D. | New York, N. Y. |
| 691. | Magnetic Pigment Co. | New York, N. Y. |
| 693. | Magnus, Mabee & Reynard, Inc. | New York, N. Y. |
| 695. | Makalot Corp. | Boston, Mass. |
| 697. | Mallinckrodt Chemical Works. | St. Louis, Mo. |
| 699. | Malt Diatase Co. | Brooklyn, N. Y. |
| 701. | Manchester Oxide Co. | Manchester, England |
| 702. | Marathon Corp. | Rotschild, Wis. |
| 703. | Marbon Corp. | Gary, Ind. |
| 705. | Marine Magnesium Prod. Corp. | S. San Francisco, Calif. |

| No. | Name | Address |
|---|---|---|
| 707. | Martin Dennis Co | Newark, N. J. |
| 709. | Martin, L. Co | New York, N. Y. |
| 711. | Martin Laboratories | New York, N. Y. |
| 713. | Mathieson Alkali Co | New York, N. Y. |
| 715. | Maywood Chemical Works | Maywood, N. J. |
| 717. | McCormick & Co | Baltimore, Md. |
| 719. | McGean Chem. Co | Cleveland, Ohio |
| 721. | McKesson & Robbins, Inc | New York, N. Y. |
| 723. | McLaughlin, Gormley, King & Co | Minneapolis, Minn. |
| 724. | Mead Corp | Chillicothe, Ohio |
| 725. | Mearl Corp | New York, N. Y. |
| 727. | Mechling Bros. Chem Co | Camden, N. J. |
| 731. | Merck & Co | Rahway, N. J. |
| 734. | Metal & Thermit Corp | New York, N. Y. |
| 735. | Metasap Chem. Co | Harrison, N. J. |
| 737. | Metro-Nite Co | Milwaukee, Wis. |
| 739. | Meyer & Sons, J | Philadelphia, Pa. |
| 741. | Mica Insulator Co | New York, N. Y. |
| 743. | Michel Export Co | New York, N. Y. |
| 745. | Michigan Alkali Co | New York, N. Y. |
| 747. | Miller, Carl F. Co | Seattle, Wash. |
| 749. | Monsanto Chem. Works | St. Louis, Mo. |
| 750. | Montrose Chemical Corp | Newark, N. J. |
| 751. | Moore-Munger, Inc | New York, N. Y. |
| 753. | Morningstar, Nicol, Inc | New York, N. Y. |
| 755. | Morton Salt Co | Chicago, Ill. |
| 757. | Murphy Varnish Co | Newark, N. J. |
| 759. | Mutual Chem. Co. of America | New York, N. Y. |
| 763. | Mutual Citrus Products Co | Anaheim, Calif. |
| 765. | National Aluminate Corp | Chicago, Ill. |
| 767. | National Ammonia Co., Inc. | Philadelphia, Pa. |
| 769. | National Aniline & Chem. Wks | New York, N. Y. |
| 771. | National Lead Co | New York, N. Y. |
| 773. | National Oil Products Co | Harrison, N. J. |
| 775. | National Pigments & Chem. Co | St. Louis, Mo. |
| 777. | National Rosin Oil & Size Co | New York, N. Y. |
| 779. | Naugatuck Chem. Co | Naugatuck, Conn. |
| 780. | Naylee Chemical Co | Philadelphia, Pa. |
| 781. | Neville Co | Pittsburgh, Pa. |
| 783. | N. J. Laboratory Supply Co | Newark, N. J. |
| 785. | N. J. Zinc Co | New York, N. Y. |
| 789. | Newmann-Buslee & Wolfe, Inc | Chicago, Ill. |
| 791. | Newport Industries, Inc | New York, N. Y. |
| 792. | New York Quinine & Chem. Wks., Inc | Brooklyn, N. Y. |
| 793. | Niacet Chem. Co | Niagara Falls, N. Y. |
| 795. | Niagara Alkali Co | New York, N. Y. |
| 797. | Niagara Chemicals Corp | Niagara Falls, N. Y. |
| 798. | Niagara Chlorine Products Co | Lockport, N. Y. |
| 799. | Niagara Smelting Corp | Niagara Falls, N. Y. |
| 800. | Norda Essential Oil & Chem. Co | New York, N. Y. |
| 801. | Northwestern Chem. Co | Wauwatosa, Wis. |
| 803. | Norton Co | Worcester, Mass. |
| 805. | Norwich Pharmacal Co | Norwich, N. Y. |
| 807. | Novadel-Agene Corp | Newark, N. J. |
| 809. | Nulomoline Co | New York, N. Y. |
| 811. | Nuodex Products, Inc | Elizabeth, N. J. |
| 813. | Ohio-Apex, Inc | Nitro, W. Va. |
| 815. | Oil States Petroleum Co | New York, N. Y. |
| 817. | Oldbury Electro-Chem. Co | New York, N. Y. |
| 819. | Olive Branch Minerals Co | Cairo, Ill. |
| 821. | Onyx Oil & Chem. Co | Jersey City, N. J. |
| 823. | Orbis Products Corp | New York, N. Y. |

| No. | Name | Address |
|-----|------|---------|
| 824. | Osborn, C. J. Co. | New York, N. Y. |
| 824a. | Owens-Corning Fiberglas Corp. | New York, N. Y. |
| 825. | Papermakers' Chem. Corp. | Wilmington, Del. |
| 827. | Paramet Chem. Corp. | Long Island City, N. Y. |
| 829. | Parke, Davis & Co. | Detroit, Mich. |
| 831. | Parker Rust Proof Co. | Detroit, Mich. |
| 833. | Patent Chemicals, Inc. | New York, N. Y. |
| 835. | Peek & Velsor, Inc. | New York, N. Y. |
| 837. | Penick, S. B. & Co. | New York, N. Y. |
| 839. | Penn. Alcohol Corp. | Philadelphia, Pa. |
| 841. | Penn. Coal Products Co. | Petrolia, Pa. |
| 843. | Penn.-Dixie Cement Corp. | New York, N. Y. |
| 845. | Penn. Industrial Chem. Corp. | Clairton, Pa. |
| 847. | Penn. Refining Co. | Butler, Pa. |
| 849. | Penn. Salt Mfg. Co. | Philadelphia, Pa. |
| 850. | Petrolite Corp. | New York, N. Y. |
| 851. | Pfaltz-Bauer, Inc. | New York, N. Y. |
| 853. | Pfizer, Chas. & Co., Inc. | New York, N. Y. |
| 855. | Phila. Quartz Co. | Philadelphia, Pa. |
| 857. | Philipp Bros. | New York, N. Y. |
| 859. | Pittsburgh Plate Glass Co. | Pittsburgh, Pa. |
| 861. | Plaskon Corp. | Toledo, Ohio |
| 863. | Plymouth Organic Labs. | New York, N. Y. |
| 867. | Powhatan Mining Corp. | Woodlawn, Baltimore, Md. |
| 869. | Pray, W. P. | New York, N. Y. |
| 870. | Prentiss, R. J. Co. | New York, N. Y. |
| 871. | Prior Chem. Corp. | New York, N. Y. |
| 873. | Procter & Gamble Co. | Cincinnati, Ohio |
| 875. | Provident Chem. Wks. | St. Louis, Mo. |
| 877. | Publicker, Inc. | Philadelphia, Pa. |
| 879. | Pure Calcium Products Co. | Painesville, Ohio |
| 881. | Pylam Products Co. | New York, N. Y. |
| 883. | Quaker Oats Co. | Chicago, Ill. |
| 885. | Ransom, L. E. Co. | New York, N. Y. |
| 885a. | Rare Metal Products Co. | Belleville, N. J. |
| 886. | Republic Chemical Corp. | New York, N. Y. |
| 887. | Reynolds Metals Co., Inc. | New York, N. Y. |
| 889. | Read, Chas. L. & Co., Inc. | New York, N. Y. |
| 891. | Reichhold Chemicals, Inc. | Detroit, Mich. |
| 893. | Reilly Tar & Chem. Corp. | Indianapolis, Ind. |
| 894. | Republic Chem. Corp. | New York, N. Y. |
| 895. | Resinous Prod. & Chem. Co. | Philadelphia, Pa. |
| 897. | Resinox Corp. | New York, N. Y. |
| 899. | Revertex Corp. | Brooklyn, N. Y. |
| 901. | Revson, R. F. Co. | New York, N. Y. |
| 903. | Rhone-Poulenc, Inc. | Paris, France |
| 905. | Richards Chem. Works. | Jersey City, N. J. |
| 907. | Riverside Chem. Co. | No. Tonawanda, N. Y. |
| 909. | Robeson Process Co. | New York, N. Y. |
| 911. | Rochester Gas & Elec. Corp. | Rochester, N. Y. |
| 913. | Rogers & McClellan. | Boston, Mass. |
| 915. | Rohm & Haas. | Philadelphia, Pa. |
| 917. | Rosenthal-Bercow Co. | New York, N. Y. |
| 919. | Ross, Frank B. Co., Inc. | New York, N. Y. |
| 921. | Ross-Rowe, Inc. | New York, N. Y. |
| 923. | Royce Chem. Co. | Carlton Hill, N. J. |
| 925. | Rubber Service Labs. Co. | Akron, Ohio |
| 927. | Russell, W. R. & Co. | New York, N. Y. |
| 929. | Russia Cement Co. | Gloucester, Mass. |
| 931. | Ryland, H. C., Inc. | New York, N. Y. |
| 933. | Saginaw Salt Products Co. | Saginaw, Mich. |
| 935. | Salomon, L. A. & Bro. | New York, N. Y. |

| No. | Name | Address |
|-----|------|---------|
| 939. | Sandoz Chem. Works | New York, N. Y. |
| 941. | Scheel, Wm. H. | New York, N. Y. |
| 943. | Schering Corp. | Bloomfield, N. J. |
| 946. | Schieffelin & Co. | New York, N. Y. |
| 947. | Schimmel & Co. | New York, N. Y. |
| 951. | Schofield-Daniel Co. | New York, N. Y. |
| 953. | Scholler Bros., Inc. | Philadelphia, Pa. |
| 955. | Schundler, F. E. & Co. | Joliet, Ill. |
| 957. | Schuylkill Chem. Co. | Philadelphia, Pa. |
| 959. | Schwabacher, S. & Co., Inc. | New York, N. Y. |
| 961. | Scientific Glass Apparatus Co. | Bloomfield, N. J. |
| 965. | Seacoast Laboratories | New York, N. Y. |
| 967. | Edwin Seebach Co. | New York, N. Y. |
| 969. | Seeley & Co., Inc. | New York, N. Y. |
| 971. | Seldner & Enequist, Inc. | Brooklyn, N. Y. |
| 973. | Serinsky, Moses Co. | Indianapolis, Ind. |
| 975. | Sharples Chemicals, Inc. | Philadelphia, Pa. |
| 977. | Shawinigan, Ltd. | New York, N. Y. |
| 978. | Shell Chem. Co. | New York, N. Y. |
| 979. | Shepherd Chem. Co. | Norwood, Cincinnati, O. |
| 981. | Sherka Chem. Co., Inc. | Bloomfield, N. J. |
| 982. | Sherwin-Williams Co. | Cleveland, Ohio |
| 983. | Sherwood Petroleum Co. | Englewood, N. J. |
| 985. | Shields, Thomas J. Co. | New York, N. Y. |
| 987. | Siemon Colors, Inc. | Newark, N. J. |
| 989. | Siemon & Co. | Bridgeport, Conn. |
| 991. | Silica Products Co. | Kansas City, Mo. |
| 993. | Silver, Geo., Import Co. | New York, N. Y. |
| 994. | Simons, Harold L. Co. | Long Island City, N. Y. |
| 995. | Sinclair Refining Co. | Olmstead, Ill. |
| 997. | Skelly Oil Co. | Chicago, Ill. |
| 999. | Smith Chem. & Color Co. | Brooklyn, N. Y. |
| 1001. | Smith & Nichols, Inc. | New York, N. Y. |
| 1003. | Smith, Werner G. Co. | Cleveland, Ohio |
| 1004. | Socony-Vacuum Co. | New York, N. Y. |
| 1005. | Solvay Sales Corp. | New York, N. Y. |
| 1007. | Sonneborn, L., Sons. | New York, N. Y. |
| 1009. | Southern Mica Co. | Franklin, N. C. |
| 1011. | Southern Pine Chem. Co. | Jacksonville, Fla. |
| 1013. | Southwark Mfg. Co. | Camden, N. J. |
| 1015. | Sparhawk Co. | Sparkhill, N. Y. |
| 1017. | Spencer Kellogg & Sons Sales Corp. | Buffalo, N. Y. |
| 1019. | Staley, A. E. Mfg. Co. | Decatur, Ill. |
| 1021. | Stamford Rubber Supply Co. | Stamford, Conn. |
| 1023. | Stanco Distributors. | New York, N. Y. |
| 1025. | Standard Alcohol Co. | New York, N. Y. |
| 1027. | Standard Brands, Inc. | New York, N. Y. |
| 1027. | Standard Oil Co. of Calif. | San Francisco, Cal. |
| 1031. | Standard Oil Co. of Indiana | Chicago, Ill. |
| 1033. | Standard Oil Co. of N. J. | New York, N. Y. |
| 1035. | Standard Oil Co. of N. Y. | New York, N. Y. |
| 1037. | Standard Silicate Co. | Pittsburgh, Pa. |
| 1039. | Standard Ultramarine Co. | Huntington, W. Va. |
| 1041. | Stanley, John T. Co. | New York, N. Y. |
| 1043. | Stanton Lab. | Wyncote, Pa. |
| 1045. | Starch Products Co. | New York, N. Y. |
| 1047. | Stauffer Chem. Co. | New York, N. Y. |
| 1049. | Stauffer Chem. Co. of Texas. | Freeport, Texas |
| 1051. | Stein, Hall & Co. | New York, N. Y. |
| 1053. | Stokes & Smith Co. | Philadelphia, Pa. |
| 1054. | Stoney-Mueller, Inc. | Lyndhurst, N. J. |
| 1055. | Strahl & Pitsch. | New York, N. Y. |

| No. | Name | Address |
|-----|------|---------|
| 1059. | Strohmeyer & Arpe Co. | New York, N. Y. |
| 1059. | Stroock & Wittenberg Corp. | New York, N. Y. |
| 1060. | Sundheimer, Henry, Inc. | New York, N. Y. |
| 1061. | Sun Oil Co. | Philadelphia, Pa. |
| 1064. | Swift & Co. | Chicago, Ill. |
| 1065. | Synfleur Scientific Labs. | Monticello, N. Y. |
| 1067. | Synthane Corp. | Oaks, Pa. |
| 1068. | Synthetic Nitrogen Products Co. | New York, N. Y. |
| 1069. | Synthetic Products Co. | Cleveland, Ohio |
| 1070. | Synvar Corp. | Wilmington, Del. |
| 1071. | Tainton Trading Co. | New York, N. Y. |
| 1073. | Takamine Laboratory, Inc. | Clifton, N. J. |
| 1075. | Tamms Silica Co. | Chicago, Ill. |
| 1077. | Tanners Supply Co. | Grand Rapids, Mich. |
| 1079. | Tannin Corp. | New York, N. Y. |
| 1081. | Tennant & Sons, C. Co. of N. Y. | New York, N. Y. |
| 1083. | Tenn. Eastman Corp. | Kingsport, Tenn. |
| 1085. | Texas Chem. Co. | Houston, Texas |
| 1087. | Texas Mining & Smelting Co. | Laredo, Texas |
| 1088. | Theobold Industries, Inc. | Kearney, N. J. |
| 1089. | Thomas, Arthur H. Co. | Philadelphia, Pa. |
| 1091. | Thorocide, Inc. | St. Louis, Mo. |
| 1093. | Thurston & Braidich. | New York, N. Y. |
| 1095. | Titanium Alloy Mfg. Co. | Niagara Falls, N. Y. |
| 1097. | Titanium Pigment Corp. | New York, N. Y. |
| 1099. | Tobacco By-Products & Chem. Corp. | Louisville, Ky. |
| 1101. | Trask, Arthur C. Co. | Chicago, Ill. |
| 1103. | Trojan Powder Co. | Allentown, Pa. |
| 1105. | Turner, Joseph & Co. | Ridgefield, N. J. |
| 1107. | Uhe, George Co. | New York, N. Y. |
| 1109. | Uhlich, Paul Co. | New York, N. Y. |
| 1111. | Union Oil Co. | Los Angeles, Calif. |
| 1113. | Union Smelting & Refining Co., Inc. | Newark, N. J. |
| 1115. | United Carbon Co. | Charleston, W. Va. |
| 1117. | United Clay Mines Corp. | Trenton, N. J. |
| 1119. | United Color & Pigment Co. | Newark, N. J. |
| 1121. | U. S. Bronze Powder Works, Inc. | New York, N. Y. |
| 1123. | U. S. Gypsum Co. | Chicago, Ill. |
| 1125. | U. S. Industrial Alcohol Co. | New York, N. Y. |
| 1127. | U. S. Industrial Chem. Co. | New York, N. Y. |
| 1129. | U. S. Phosphoric Prod. Corp. | New York, N. Y. |
| 1131. | U. S. Rubber Products, Inc. | New York, N. Y. |
| 1133. | U. S. Smelting, Refining & Mining Co. | New York, N. Y. |
| 1135. | Utah Gilsonite Co. | St. Louis, Mo. |
| 1136. | Vanadium Corp. of America. | New York, N. Y. |
| 1137. | Van Allen, L. R. & Co. | Chicago, Ill. |
| 1139. | Van-Ameringen Haebler, Inc. | New York, N. Y. |
| 1141. | Vanderbilt, R. T. Co. | New York, N. Y. |
| 1143. | Van Dyk & Co., Inc. | Jersey City, N. J. |
| 1145. | Van Schaack Bros. Chem. Co. | Chicago, Ill. |
| 1147. | Varcum Chem. Corp. | Niagara Falls, N. Y. |
| 1148. | Velsicol Corp. | Chicago, Ill. |
| 1149. | Verley, Albert & Co. | Chicago, Ill. |
| 1151. | Verona Chem. Co. | Newark, N. J. |
| 1153. | Victor Chem. Works. | Chicago, Ill. |
| 1155. | Virginia-Carolina Chem. Corp. | Richmond, Va. |
| 1157. | Virginia Smelting Works. | W. Norfolk, Va. |
| 1159. | Vitro Mfg. Co. | Pittsburgh, Pa. |
| 1161. | Vultex Chem. Co. | Cambridge, Mass. |
| 1162. | Wah-Chang Trading Corp. | New York, N. Y. |
| 1163. | Waldo, E. M. & F., Inc. | Muirkirk, Md. |
| 1165. | Wallerstein Co., Inc. | New York, N. Y. |

| No. | Name | Address |
|-----|------|---------|
| 1167. | The Warner Chem. Co. | New York, N. Y. |
| 1169. | Warwick Chem. Co. | West Warwick, R. I. |
| 1170. | Washine National Sands, Inc. | Long Island City, N. Y. |
| 1171. | Welch, Holme & Clark Co. | New York, N. Y. |
| 1173. | Welsbach & Co. | Gloucester, N. J. |
| 1175. | Werk, M. Co. | Cincinnati, Ohio |
| 1177. | Western Charcoal Co. | Chicago, Ill. |
| 1179. | Westinghouse Elec. & Mfg. Co. | E. Pittsburgh, Pa. |
| 1181. | Whittaker, Clark & Daniels | New York, N. Y. |
| 1183. | Wiffen & Co., Sons, Ltd. | London, England |
| 1185. | Wilckes-Martin-Wilckes Co. | New York, N. Y. |
| 1187. | Will & Baumer Candle Co. | New York, N. Y. |
| 1189. | Williams, C. K. & Co. | Easton, Pa. |
| 1191. | Wilson Laboratories | Chicago, Ill. |
| 1192. | Winthrop Chemical Corp. | New York, N. Y. |
| 1193. | Witco Chem. Co. | New York, N. Y. |
| 1195. | Woburn Chem. Corp. | Harrison, N. J. |
| 1197. | Wolf, Jacques & Co. | Passaic, N. J. |
| 1199. | Wood Flour, Inc. | Manchester, N. H. |
| 1201. | Wood Ridge Mfg. Co. | Wood Ridge, N. J. |
| 1203. | Wyodak Chem. Co. | Cleveland, Ohio |
| 1205. | Young, J. S. & Co. | Hanover, Pa. |
| 1206. | Ziegler, G. S. & Co. | New York, N. Y. |
| 1207. | Zinsser, Wm. & Co. | New York, N. Y. |
| 1209. | Zophar Mills, Inc. | Brooklyn, N. Y. |

## Addenda to Sellers of Chemicals and Supplies

| No. | Name | Address |
|-----|------|---------|
| 1210 | Allied Chemical Corp. | New York, N. Y. |
| 1211 | Alox Corp. | New York, N. Y. |
| 1212 | Alrose Chemical Co. | Cranston, R. I. |
| 1213 | Alumileaf Co. | Los Angeles, Calif. |
| 1214 | Antara Corp. | New York, N. Y. |
| 1215 | Argus Chemical Laboratory | Brooklyn, N. Y. |
| 1216 | Arnold Hoffman Co. | Providence, R. I. |
| 1217 | Attapulgus Clay Co. | Philadelphia, Pa. |
| 1218 | Baker Chemical Co., J. T. | North Phillipsburg, N. J. |
| 1219 | Bareco Oil Co. | Tulsa, Okla. |
| 1219a | Bareco Wax Co. Div., Petrolite Corp. | Kilgore, Tex. |
| 1220 | Barrett Division, Allied Chemical & Dye Corp. | New York, N. Y. |
| 1221 | Berkshire Color & Chem. Corp. | Springfield, Mass. |
| 1222 | Blockson Chemical Co. | Joliet, Ill. |
| 1223 | Böhme Fettchemie Ges. | Hamburg, Germany |
| 1224 | Bon Ami Co. | New York, N. Y. |
| 1225 | Brookdale Laboratories | Bloomfield, N. J. |
| 1226 | Brown Co., Andrew | Los Angeles, Calif. |
| 1227 | Brown Oil & Chemical Corp. | Port Richmond, N. Y. |
| 1228 | Buckman Labs. Inc. | Memphis, Tenn. |
| 1229 | Burgess Pigment Co. | Paterson, N. J. |
| 1230 | Caldwell Co. | Akron, Ohio |
| 1231 | California Spray Chemical Corp. | Richmond, Calif. |
| 1232 | Campbell Sons Corp. Harry T. | Towson, Md. |
| 1233 | Cargill, Inc. | Minneapolis, Minn. |
| 1234 | Carroll Inc. R. E. | Trenton, N. J. |
| 1235 | Chemo Puro Mfg. Corp. | Newark, N. J. |
| 1236 | Circo Products | Cleveland, Ohio |
| 1237 | Columbia-Southern Chemical Corp. Subs. of Pittsburgh Plate Glass Co. | Pittsburgh, Pa. |
| 1238 | Croda, Inc. | New York, N. Y. |
| 1239 | Deecy Products Co. | Cambridge, Mass. |
| 1240 | Dehydag c/o Fallek Products | New York, N. Y. |
| 1241 | Dragoco | Holyminden, Germany |
| 1242 | Dura Commodities Corp. | New York, N. Y. |
| 1243 | Durez Chemicals Inc. | No. Tonawanda, N. Y. |
| 1244 | Dye Specialties Inc. | Jersey City, N. J. |
| 1245 | Eastman Chemical Products, Inc. | Kingsport, Tenn. |
| 1246 | Edgar Bros. | Metuchen, N. J. |
| 1247 | El Dorado Defoamer Div. Foremost Food & Chemical Co. | Oakland, Calif. |
| 1248 | Enjay Co., Inc. | New York, N. Y. |
| 1249 | Evans & Rais, Ltd. | Norman, Manchester, England |
| 1250 | Fanning Chemical Corp. | Newark, N. J. |
| 1251 | Ferro Chemical Corp. | Bedford, Ohio |
| 1252 | Fiske Bros. Refining Co. | Newark, N. J. |
| 1253 | Food Machy. & Chemical Corp. | San Jose, Calif. |
| 1254 | General Tire & Rubber Co. | Akron, Ohio |

| No. | Name | Address |
|-----|------|---------|
| 1255 | General Mills, Inc. | Minneapolis, Minn. |
| 1256 | General Reduction Co. | Chicago, Ill. |
| 1257 | Georgia Marble Co. | Tate, Ga. |
| 1258 | Great Lakes Carbon Corp. | Los Angeles, Calif. |
| 1259 | Griffin Chemical Co. | San Francisco, Calif. |
| 1260 | Gustin-Bacon Mfg. Co. | Kansas City, Mo. |
| 1261 | Hanson-Van Winkle-Munning Co. | Matawan, N. J. |
| 1262 | Heyden Newport Chemical Corp. | New York, N. Y. |
| 1263 | Horn Co., A. C. | Long Island City, N. Y. |
| 1264 | E. F. Houghton & Co. | Philadelphia, Pa. |
| 1265 | Huron Milling Co. | New York, N. Y. |
| 1266 | Hyman & Co., Julius | Denver, Colo. |
| 1267 | Indoil Chemical Co. | Chicago, Ill. |
| 1268 | Industrial Raw Materials Corp. | New York, N. Y. |
| 1269 | Interchemical Corp. | New York, N. Y. |
| 1270 | International Wax Refining Co. | Brooklyn, N. Y. |
| 1271 | Jefferson Chemical Co., Inc. | Houston, Tex. |
| 1272 | Jones Dabney Co. | Louisville, Ky. |
| 1273 | Kearny Mfg. Co. | Kearny, N. J. |
| 1274 | Kotal Co. | New York, N. Y. |
| 1275 | Kraft Chemical Co. | Chicago, Ill. |
| 1276 | Lanaetex Products, Inc. | New York, N. Y. |
| 1277 | B. Laporte, Ltd. | Luton, England |
| 1278 | Linde Air Products Co. | New York, N. Y. |
| 1279 | MacDermid Inc. | Waterbury, Conn. |
| 1280 | Maguire Industries | New York, N. Y. |
| 1281 | Malmstrom Chemical Corp. | Newark, N. J. |
| 1282 | Marco Chemicals Inc. | Sewaren, N. J. |
| 1283 | Martin Co., Glenn L. | Baltimore, Md. |
| 1284 | Master Builders Corp. | Cleveland, Ohio |
| 1285 | Metalead Products Corp. | San Francisco, Calif. |
| 1286 | Mineralite Sales Corp. | New York, N. Y. |
| 1287 | Minerals & Chemicals Corp. of America | Menlo Park, N. J. |
| 1288 | Miranol Chemical Co. | Irvington, N. J. |
| 1289 | Mona Industries Inc. | Paterson, N. J. |
| 1290 | Moretex Chemical Prod. Inc. | Spartanburg, S. C. |
| 1291 | Naarden N.V. Chem. Fabr. | Naarden, Holland |
| 1292 | National Adhesives Div. of National Starch Products, Inc. | New York, N. Y. |
| 1293 | National Southern Products Corp. | New York, N. Y. |
| 1294 | Naugatuck Chemical Div. U. S. Rubber Co. | New York, N. Y. |
| 1295 | Ninol Labs. | Chicago, Ill. |
| 1297 | The Northwest Linseed Co. | Minneapolis, Minn. |
| 1298 | Oronite Chemical Co. | San Francisco, Calif. |
| 1299 | Ottawa Chemical Co. | Toledo, Ohio |
| 1300 | Pan American Refining Co. | New York, N. Y. |
| 1301 | Penick & Ford Ltd. Inc. | New York, N. Y. |
| 1302 | Perfumery Associates Inc. | New York, N. Y. |
| 1303 | Phillips Chemical Co. | Bartlesville, Okla. |
| 1304 | Pittsburgh Agricultural Chemical Co. | Pittsburgh, Pa. |

| No. | Name | Address |
|-----|------|---------|
| 1305 | Polyplastex United | Elmhurst, N. Y. |
| 1306 | Powell & Co., John | New York, N. Y. |
| 1307 | Pratt Co., B. G. | Hackensack, N. J. |
| 1308 | Raffold Corp. | Andover, Mass. |
| 1309 | Rayette, Inc. | St. Paul, Minn. |
| 1310 | Reheis Co., Inc. | Berkeley Heights, N. J. |
| 1311 | Rhodes Chemical Corp. | Plainfield, N. J. |
| 1312 | Rhodia Inc. | New York, N. Y. |
| 1313 | Robeco Chemicals Inc. | New York, N. Y. |
| 1314 | Robinson Wagner Co. | New York, N. Y. |
| 1315 | Rubber Corp. of America | Hicksville, N. Y. |
| 1316 | Rumford Chemical Works | Rumford, R. I. |
| 1317 | Scientific Oil Compounding Co. | Chicago, Ill. |
| 1318 | Shanco Plastics & Chemicals Inc. | Tonawanda, N. Y. |
| 1319 | Sindar Corp. | New York, N. Y. |
| 1319a | Southeastern Clay Co. | Aiken, S. C. |
| 1320 | Squibb & Son, E. R. | New York, N. Y. |
| 1321 | Stabelan Chemical Co. | Toledo, Ohio |
| 1322 | Standard Oil Co. of Illinois | Chicago, Ill. |
| 1323 | Styrene Co.-Polymers, Ltd. | Manchester, England |
| 1324 | Thiokol Corp. | Trenton, N. J. |
| 1325 | Thompson-Weinman & Co. | Cartersville, Ga. |
| 1326 | Traders Oil Co. | Ft. Worth, Tex. |
| 1327 | U B S Chemical Corp. | Cambridge, Mass. |
| 1328 | Union Carbide Chemical Co. | New York, N. Y. |
| 1329 | Vapor Blast Co. | Milwaukee, Wisc. |
| 1330 | Varnition Co. | Burbank, Calif. |
| 1331 | Vegetable Oil Products Co. Inc. | Wilmington, Calif. |
| 1332 | Vejin Inc. | Cincinnati, Ohio |
| 1333 | Vestal Laboratories, Inc. | St. Louis, Mo. |
| 1334 | Warwick Wax Co. Inc. | Long Island City, N.Y. |
| 1335 | T. F. Washburn Co. | Chicago, Ill. |
| 1336 | Waylite Co. | Bethlehem, Pa. |
| 1337 | Westvaco Chemical Corp. | New York, N. Y. |
| 1338 | West Virginia Pulp & Paper Co. | New York, N. Y. |
| 1339 | Wyandotte Chemical Corp. | Wyandotte, Mich. |